From the Valley to the Summit

The Inseparable Connection Between the Sacramento Shops and the Transcontinental Railroad

by Stephen E. Nemeth

Nemeth, Stephen E.

From the Valley to the Summit
The Inseparable Connection Between the Sacramento Shops
and the Transcontinental Railroad.
Includes bibliographical references and index
ISBN-13: 978-0-578-41485-0

1. United States—History Railroad 2. United States—West (California) State History 3. Transportation—Railroads History 4. Technology and Engineering—Civil Transportation 5. Technology and Engineering—History

All bulletin articles from the Union Pacific Railroad are in the public domain.

Photos are courtesy of the author, or used with permission of the
Sacramento State Railroad Museum, 2017

Cover design by Karen Phillips

Interior design by Lisa Ham for Two Songbirds Press 2018

Table of Contents

Introduction

Historians and those fascinated by the development about California could discuss many topics, including indigenous cultures, immigration, agriculture, and maritime commerce.

However, a conversation of the development of industry in California cannot be undertaken without talking about the railroad and the role that the Sacramento Locomotive Works (the "Shops") played as the first major industrial complex in the West.

In 2019, the Shops are long gone, its former location a place showing little evidence of the innovation and growth that occurred there. This will be too easily erased as plans for a new development move ahead.

Fortunately, for many years, the Southern Pacific Company published a bulletin, monthly in most cases, which described the news, events, and history of the Central Pacific and Southern Pacific railroads, and often featured the Shops in its pages.

The research for this book was undertaken to uncover and make sense of the Central Pacific and Southern Pacific railroads' operations around Sacramento with an emphasis on the wide reach of the accomplishments occurring at the Shops, and uses the railroad bulletin articles as the launch points for the discussion.

The book is intended to help preserve the history of the Shops as well as touch on the far-reaching impact of what was achieved there, and re-introduce these stories in a way accessible to a contemporary audience.

Chapter One
From Those That Were There

Theodore Judah, the dreamer and planner of the line over the Sierra, campaigned for financing from the wealthy in San Francisco. Some of them had financed the Sacramento Valley Railroad and lost money. So, another railroad was not of interest to them. Judah's pursuit of funding led to a meeting in Sacramento's St. Charles Hotel in November of 1860. About thirty local businessmen attended the meeting. A quote by Leland Stanford found in the book *High Road to Promontory* by George Kraus (1969) describes the gathering and Stanford's first meeting with Judah:

> At first, I had no personal acquaintance with him. The first time that my attention was called to the question of the construction of the railroad was by a gentleman by the name of James Bailey (a Sacramento jeweler), who was afterward the secretary of the Central Pacific. Mr. Bailey came down to see me and told me that Judah had discovered in the mountains a pass over which a railroad could be built, and desired that I should see Mr. Judah. I told him I would be glad to see Mr. Judah, but did not know that we would care to do anything in this matter. I told him, however, that I would talk with Mr. Huntington and others about it. We fell into conversation upon the subject, and the result was that we agreed to have a meeting the same evening at my house. We then fully considered the subject and discussed the matter one way and another. We afterward met again, when Mr. Hopkins joined us, and the result was that we concluded that we would

make the acquaintance of Mr. Judah. Neither of us was acquainted with him, and we invited him to meet us and make our acquaintance. He came down with Mr. Bailey and we had a conversation.

❧ ❧ ❧

In summer of 1861, Leland Stanford, Collis Huntington, and Charles Crocker, three of the "Associates" funding the railroad, saw firsthand the proposed Donner Pass route. These three men, with Mark Hopkins, became known as the "Big Four." Numerous people, some who supported the railroad and many who were critical, realized the ominous task of building a railroad over a summit higher than seven thousand feet. After the summer trip, Stanford provided his impressions, which were from the perspective of a civil engineer and entrepreneur:

> I remember that while we were making our explorations, we came to the summit, and at Donner Pass we looked down on Donner Lake, 1,200 feet below us, and then looked up at the drifts above us, 2,000 feet, and I must confess that it looked very formidable. We there and then discussed the question of the paying qualities of the enterprise and we came to this conclusion: That if there was a way by which a vessel could start from San Francisco or from New York and sail around Cape Horn in behind those mountains, we could not afford to compete. If this could not be done, however, and if we had only the ox and mule teams to compete with, we saw that we could obtain such a rate for carrying freight

and passengers that we could afford to build the road with the prospect of further developments in Nevada. At that time the business of Nevada was very promising and we had an idea, like everybody else on this side, that most of the mountains in Nevada were filled with mineral wealth. (Kraus, 34)

During May of 1861, Judah wrote to his wife, Anna, with a comment that reflected the challenge ahead: "I am running a line through the most difficult country ever conceived of for a railroad." (Bain, 97)

❦ ❦ ❦

Friction between Judah and the Big Four was common. There were disagreements before the line had even left the city limits of Sacramento. Huntington spoke later about the problems laying track in Sacramento:

It had been understood that the Central Pacific Railroad was to run up I Street in Sacramento and I had given order that it was to go up that Street to Fifth and thence to B Street and out to the levee. Work had commenced, however, in running it by a point on the American River, where water over flowed every year, and would require more of the expensive rip-rap to be brought down from Folsom to protect the embankment. Work had been going on for several days when I walked out to see how affairs were progressing. Mr. Judah came up and said that the board of directors had given orders to have the work done that way but, I replied, it will cost $200,000 more at least to put the road here; it must go up I Street. I then sent for Mr. Cody and ordered him to move the road to I Street. (Kraus, 55)

In this case, Judah's wishes prevailed and the line was built along the American River.

❦ ❦ ❦

Judah continued to have disagreements with those holding the cash. In May of 1863, Judah wrote to his friend, Dr. Daniel W. Strong:

I had a blow out about two weeks ago and freed my mind, so much so that I looked for instant decapitation. I called things by their right name and invited war; but counsels of peace prevailed and my head is still on; my hands are tied however. We have no meetings of the board nowadays, except the regular monthly meetings, which, however, was not had this month; but here have been any quantity of private conferences to which I have not been invited. I try to think it is all for the best, and devote myself with additional energy to my legitimate portion of the enterprise. (Kraus, 62)

Frustrated to his limit, Judah left for the east coast in September of 1863 in an attempt to get other backing for the project. According to Judah's wife, Anna, one of the people he planned to visit was Cornelius Vanderbilt who already had control of railroad lines between Chicago and New York. On October 9, Judah wrote again to Dr. Strong:

I have a feeling of relief in being away from the scenes of contention and strife which it has been my lot to experience for the past year, and to know that the responsibilities of events, so far as regards Pacific Railroad, do not rest on my shoulders. (Bain, 142)

Judah's story ended on the East Coast. He contracted yellow fever crossing Panama and died November 2 of that same year.

❦ ❦ ❦

The early years of construction were not smooth in regard to finding adequate funding. It is easy to think the opposite given the outcome and tremendous wealth that fell on the Big Four. Charles Crocker recalled the

4

difficult times:

> We could not borrow a dollar on the faith of the company. Mr. Stanford, Mr. Huntington and myself had to give our personal obligations for the money necessary to carry us from month to month. There was not a bank that would loan the company a cent. They had no faith in it. We bought the first fifty miles of iron on our own personal obligations. We procured from D.O. Mills, who was personally known to all of us, a paper testifying to our responsibility and our honor as men and merchants, and that whatever we agreed to do, he believed we would faithfully adhere to. Mr. Huntington bought the iron and gave our personal obligations for it, and put up the bonds of the company besides as security; and we entered into an agreement that we would be personally responsible as individuals for ten years for the payment of the interest on those bonds. Those were the responsibilities we took, and if we had not done so there would have been no railroad. I would have been glad, when we had thirty miles of road built, to have got a clean shirt and absolution from my debts. I owed everybody that would trust me, and would have been glad to have them forgive my debts and taken everything I had, even the furniture of my family, and to have gone into the world and started anew. (Kraus, 48)

❦ ❦ ❦

Though there was progress being made laying track towards the summit, everything was not going smoothly. A.P. Partridge, a veteran Central Pacific construction worker, recalled difficulties near Cisco, which was ninety-two miles from Sacramento at an elevation of 5,941 feet:

> Two miles west of Cisco, we put up a high trestle bridge, one hundred feet high. With a heavy

storm and a big fall of rain there was a lake on the hill above the bridge and when it filled it broke loose. Down everything came and swept away four bents from the center of our bridge. That bridge had to be replaced, with the road blocked to Blue Canyon. Well, we went to the woods and hewed the timber, hauled it to the track by main force, then got some ox teams and hauled it to the bridge and repaired the break. (Bain, 305)

❦ ❦ ❦

Albert Deane Richardson, who traveled the West for the *New York Tribune*, recalled the construction work in his book *Beyond the Mississippi*: "Chinamen were brought in, and in the spring of 1865 they began to swarm up the Sierra like flies upon a honeycomb."

❦ ❦ ❦

In early May of 1868, the line from Reno to Truckee was completed and there was still about seven miles of work to do closer to the summit. In April, Huntington responded to a note from Hopkins with his thoughts about the public's attitude when the line was finally complete:

> I notice that you write that everybody is in favor of a railroad until they get it built and then everyone is against it unless the railroad company will carry them and theirs for nothing. (Kraus, 196)

❦ ❦ ❦

An early traveler on the Central Pacific, W.L. Humason of Hartford, Connecticut made a trip to the West Coast. The date of his account is unknown, but, it seems to have been made in late spring or early summer; his words describe the descent from the summit:

> As we descended the mountains, the snow storm turned into a rain storm and we reached Sacramento in the midst of it; took the steamer and

sailed down the Sacramento River, looking with wonder at the immense piles of salmon that lined the shores at every landing. We took on board great loads of them, which had just been caught in the river. We were also surprised at the advanced stage of the crops and vegetables along the shore. Strawberries had long been plentiful and wheat was almost ready for harvest.

❦ ❦ ❦

When Charles Crocker was ten years old, he earned money delivering newspapers to help his father buy a farm in Indiana where the family moved in 1836. After two years of helping work the soil, he worked in a sawmill and, later, on a forge. So, his early years were filled with physical work. He settled in Sacramento in 1850, developed a dry goods business and was elected to the state legislature in 1860. After one term, developing the Pacific Railroad became his passion. Great wealth followed. During his time with the railroad, he was the leader of the working force that carried the ties and pounded the spikes. With this exposure to the laborers, he saw the physical work being done in all seasons. He knew that workers were being lowered down a vertical shaft into Summit Tunnel to light fuses and be hoisted out before the explosion. He saw people with ideas and innovation and knew the pay each received. Crocker's business philosophy, which was clearly drawn from his work as a youth, his observations during the growth of the railroad, his astounding wealth as an adult, and his realization that good fortune came his way too, led him to comment: "One man works hard all his life and ends up a pauper. Another man, no smarter, makes twenty million dollars. Luck has a hell of a lot to do with it." (Kraus)

Chapter Two
Floods, Ground Breaking, and Early Central Pacific Construction

For Americans living in 1845 east of the Mississippi River, a trip to California was at least a two thousand mile wagon ride over the Rocky Mountains, across the Wasatch Range, across the desert of Nevada and over the 7,000-feet-high Sierra Range.

George Kraus, in *High Road to Promontory*, quotes Daniel Webster's address to the United States Senate,

> What do we want with this region of savages and wild beasts, of deserts of shifting sands and whirlwinds of dust, of cactus and prairie dogs? To what use could we ever put those endless mountain ranges, impenetrable and covered to their bases with eternal snow? What could we do with the western coast line three thousand miles away, rock-bound, cheerless and uninviting?

He was entitled to his opinion.

In 1848, when gold was discovered about twenty miles east of Sacramento, profit-minded people flocked west, both to mine their own gold and to sell supplies to the miners. Agricultural and maritime interests grew, and soon entrepreneurs of the 1850s began considering the profit potential of a railroad that would connect Sacramento with the more established businesses and the larger populations of eastern states. Slowly, the image of the wagon ride over rough terrain began to be replaced by the idea of an easier, shorter trip west, in a railroad car.

The first essay of this chapter describes Mother Nature's threat to the planned railroad terminus in Sacramento.

The second essay describes the initial breaking of ground for the transcontinental railroad in Sacramento and the enthusiasm, with some skepticism, that prevailed.

Also included in the chapter are four articles from Southern Pacific's *The Bulletin* that describe major construction accomplishments and the monument dedicated to the surveyor who helped determine the route over the Sierra into Nevada, Theodore Judah.

These four articles are:

"When Building of the Central Pacific Started," May 1922

"How the CP Built Ten Miles in One Day," June 1919

"A Railroad Record That Defies Defeat," May 1928

"Judah Monument Dedicated at Sacramento," May 1930

Floods Before the Railroad

It is a puzzle as to why investors, in 1861, incorporated a railroad whose maintenance facilities, yard, locomotives, cars, and ticket offices were at the junction of two rivers that had a recent history of flooding the city. Tributaries to the Sacramento River, the Pit and Mc-

Cloud rivers are about 180 miles north of downtown Sacramento. Water from the Feather and Yuba Rivers joins these flows to comprise the Sacramento River. In addition, the American River adds to the flow at Sacramento. With no regulation to the flow, floods resulted.

During the twelve years before the Central Pacific was formed, there were noteworthy floods in 1850, 1852, 1853, and 1861. The one in 1861 occurred just three months before the official incorporation of the railroad.

Dr. John F. Morse, who might have been the only Sacramento historian at the time, gave his account of the 1850 flood:

> The Sacramento River and the American River were rising rapidly, and the back country seemed to be filling up and cutting off all communication with the highlands; but still, everyone was inclined to believe the ridiculous and false assurances of safety, which could scarcely be extinguished when the city was actually under water. Hence, when the water began to rush in and overwhelm the place, there was no adequate means of escape for life and property; and consequently some were drowned in their beds, and many died in consequence of the terrible exposures to which they were subjected. (Morse, 58)

After the flood of 1850, the sentiment that the city needed to be protected grew. One idea was to add height to the river banks. Levees—embankments along the river banks that raise the level of the channel to constrain the flow—became the solution, and have been part of the flood prevention effort ever since. They have not been impervious to high water, however, as subsequent floods in the area have proven.

Another inundation of the city occurred in 1861 when the river's elevation reached the highest level since 1853. The water from Sutter's Lake overtopped its usual bounds and cut a channel through First Street to the American River. Swift's Bridge and Lisle's Bridge over the American River were destroyed. Norris's Bridge became unusable and ferries had to be used to cross the American River. (Sutter's Lake eventually was filled to provide more acreage and became part of the Central Pacific's yard.)

A remarkable effort, in addition to the levee construction, was to raise the street level thereby creating a situation where rooms at the new street level had been the second floor. Mark Twain, in 1866, jokingly wrote about the concept for the *Territorial Enterprise*: "It makes the first floor shady, very shady." He also brought up the fact that residents that were on the second floor now lived at street level. In his words, "It benefits the plebeian second floor boarders at the expense of the bloated aristocracy of the first—that is to say, it brings the plebeians down to the first floor and degrades the aristocrats to the cellar."

Not joking, he added:

> The energy the Sacramentans have shown in making this expensive grade improvement and raising their houses up to its level is in every way creditable to them, and is a sufficient refutation of the slander so often leveled at them that they are discouraged by the floods, lack confidence in their ability to make their town a success, and are without energy. A lazy and hopeless population would hardly enter upon such costly experiments as these when there is so much high ground in the State which they could fly to if they chose.

The same comments could apply to those building the railroad. They had heard about, and likely witnessed, past floods that inundated the land on which the railroad was to begin and have its center of operations, but, they did build and expand on that land to create one of the largest employers in the West.

❦ ❦ ❦

The Skies Smiled – 690 Miles to Go

When the first dirt was moved in Sacramento as

part of the new grade heading toward the Sierra, there must have been the widest possible range of optimism and pessimism. From the earliest rumors, there were skeptics of the plan who doubted the possibility of success, and there were invigorated supporters who envisioned economic wonders for California. One of the individuals leery of the whole idea was Collis Huntington, one of the famous Big Four. He did not attend the ground-breaking event and explained the attitude later by saying,

> If you want to jubilate in driving the first spike here, go ahead and do it. I don't. These mountains look too ugly and I see too much work ahead. We may fail and I want to have as few people know it as we can. Any little nobody can drive the first spike but there are months of hard labor and unrest between the first and the last spike. The last spike is the one we'll celebrate. (Bain, 145)

Filling ravines, boring tunnels, and building bridges were only some of the obstacles. Money was scarce. Darius O. Mills, a financial leader of the 1860s, spoke before the Pacific Railway Commission in 1887 and, with a great retrospective, spoke of how the business world of the early 1860s viewed the venture of the Big Four:

> The difficulties were very great and rendered their credit very poor. It was a constant struggle, and the sense of community as well as my own was against their being able to carry out the enterprise. They borrowed in every way, on their own credit, on the credit of the road, and any credit they could use. They were generally understood to be borrowers to any amount they could get while constructing the road. They had difficulty always in borrowing on the first mortgage bonds when the road was unfinished because it was never classed as a first-class security. Loans made to the company were more on the individual credit of the parties than upon anything else.

Prior to the ground breaking, it was known that all the iron, rolling stock, and railroad material would have to be manufactured in the Atlantic states and shipped. Transportation time around Cape Horn was generally eight to ten months (later in the construction, some material was shipped across the Isthmus of Panama).

As ground was being broken on January 8, 1863 at the corner of Front and K Streets in Sacramento, hope and optimism, at least in the minds of some, prevailed. But no one at the commemorative event could have known that the future would hold these things:

The first thirty feet of Summit Tunnel would progress at the rate of about one foot per day.

The work of the Central Pacific would extend 690 miles from Sacramento to Promontory, Utah.

On April 28, 1869, a crew comprised of about four thousand men would lay over ten miles of track (25,800 ties) west of Promontory, Utah.

The rails would eventually join with the Union Pacific after Theodore Judah, possibly the project's greatest advocate, said in a note to his wife written while on a survey expedition, "I am running a line through the most difficult country ever conceived of for a railroad line."

In 1921, there would be close to thirty-two miles of snow sheds between Blue Canyon and Truckee.

One hundred and forty years after the railroad was completed over the Sierra, trains nearly one mile long would pass over the same right of way carrying cell phones, televisions, cars, computers, and airplane parts.

On January 9, 1863, the *Sacramento Union* described the activities of the previous day:

> The skies smiled yesterday upon a ceremony of vast significance in Sacramento, California and the Union. With rites appropriate to the occasion, and in presence of the dignitaries of

the State, representatives of every portion of the commonwealth, and a great gathering of citizens, ground was formally broken at noon for the commencement of the Central Pacific Railroad – the California link of the continental chain that is to untie American communities now divided by thousands of miles of trackless wilderness. Among the assemblage were pioneers who had assisted in laying the foundations of the Golden State, who had dreamed, toiled and schemed for years in behalf of this grand enterprise, and cling with steady faith, through many depressing defeats, to the belief that they would live to witness the consummation of their hopes; men who had more recently determined to devote their energies and their means to the execution of the project; representatives of the various sections of the State who appreciated the importance of the work to the whole Pacific Coast, no matter where the line should be located; high officials whose presence and earnest approval enhanced the dignity of the occasion; divines to invoke blessings on the work; and last, but not least, directors and contractors, who gave substantial assurance that the brain, the muscle, the gold and the iron were ready to make the railroad a reality.

The choice of scene for the ceremony was not favorable to the presence of the gentler sex, but the balconies opposite – on Front Street, above K – were adorned with a fair delegation. The great preponderance of pantaloons was a disagreeable necessity of the 'situation'. A stand was erected near the levee, a short distance above K Street, and the ends were adorned with the national flag. A general distribution of bundles of hay gave a comparatively dry footing to the crowd in the immediate vicinity. The Sacramento Union's Brass Band was stationed on the balcony of the American Exchange Hotel, and

between the addresses enlivened the proceedings by playing the national airs and peculiarly 'Wait for the Wagon'. Two wagons adorned with flags, drawn by horses that were also decorated with the national colors, were stationed near the rostrum, with earth ready to be shoveled out for the railroad embankment. On one of these wagons was a large banner bearing a representation of hands clasped across the continent from the Atlantic to the Pacific, with the prayer of every loyal heart, 'May the Bond Be Eternal'. Shortly after twelve o'clock p.m., Governor Stanford appeared upon the stand, and the ceremonies of the occasion were commenced. C. Crocker introduced to the assemblage Leland Stanford, Governor of the state of California. In his speech, Governor Stanford said in part: 'I congratulate you upon the commencement of the great work which, in its results to the state of California and the Pacific Coast, and to the Nation itself, is to be what the Erie Canal was to New York and the Eastern States. This work will go on from this side to completion as rapidly as possible. There will be no delay, no backing, no uncertainty in the continued progress. We may now look forward with confidence to the day, not far distant, when the Pacific Coast will be bound to the Atlantic Coast by iron bonds that shall consolidate and strengthen the ties of nationality, and advance with giant strides the prosperity of the State and the Country.' After a brief prayer by the Reverend J. A. Benton, Stanford was allowed to turn the first spadeful of earth for the project. The two wagons loaded with earth were driven up in front of the rostrum, and Governor Stanford, with a zeal and athletic vigor that showed his heart was in the work and his muscle in the right place, seized the shovel, and amid cheering of the crowd deposited the first earth for the embankment. The enthusiastic Charles Crocker promply called for 'nine cheers' and the crowd, sharing his enthusiasm, cherringly responded. The sun smiled brightly, and everybody felt happy because, after so many years of dreaming, scheming, talking and toiling, they saw with their own eyes the actual commencement of the Pacific Railroad.

During the next six years and four months, the work produced rails to the summit of the Sierra over seven thousand feet above Sacramento and nearly to Ogden, Utah.

Almost all of the work during summer, winter, day, and night, was done by a workforce made up primarily of Chinese people. According to the May 1927 issue of *The Bulletin*: "Governor Stanford held the Chinese workers in such high esteem that he provided in his will for the permanent employment of a large number. Some of these are still living and working on lands now owned by Stanford University."

The same issue of *The Bulletin* referred to the Chinese diet:

> The Chinese diet included dried oysters, dried cuttlefish, sweet rice crackers, dried bamboo sprouts, salted cabbage, sugar, four kinds of dried fruit, five kinds desiccated vegetables, vermicelli, dried seaweed, Chinese bacon, dried abalone, and peanut oil. The white laborers on the grade relieved their thirst drank [sic] water, which was not always the best, and which, at times, in spite of all precautions, was the source of sickness. The Chinese drank lukewarm tea. This tea, beside the grade, was in thirty and forty-gallon whiskey barrels and was always on tap. Several times a day, a Chinese mess attendant would bring fresh tea and pour it into the big barrel. These reinforcements of the beverage were carried in powder kegs suspended across a Celestial shoulder on each end of a bamboo pole.

THE BULLETIN

When Building of C. P. Started

Old Newspaper File Tells of Ceremonies in Sacramento January 16, 1863

May 10, 1869, is a red letter day in the history of the Southern Pacific Company, because on that date the last spike was driven uniting the tracks of the Central Pacific, the Southern Pacific's parent organization, with the tracks from the east, forming the first trans-continental railroad line. But there is another date which should be remembered—January 16, 1863, for it was then that ground was broken for the construction of the Central Pacific line.

An account of the ceremonies incident to the commencement of this work, given by the Sacramento Weekly Union of January 17, 1863, is of interest both to the veterans of the Company who are acquainted with the earlier history of the road and to the younger generation, for the article clearly shows the importance placed by the public in the project then being started.

THE skies smiled yesterday upon a ceremony of vast significance to Sacramento, California, and the Union. With rites appropriate to the occasion, and in the presence of the dignitaries of the state, representatives of every portion of the commonwealth, and a great gathering of citizens, ground was formally broken for the commencement of the Central Pacific Railroad—the California link of the continental chain that is to unite American communities now divided by thousands of miles of trackless wilderness.

"Among the assemblage were pioneers, who had assisted in laying the foundations of the Golden State, who had dreamed, toiled and schemed for years in behalf of this grand enterprise, and clung with steady faith, through many depressing defeats, to the belief that they would live to witness the consummation of their hopes; men who had more recently determined to devote their energies and their means to the execution of the project; representatives of the various sections of the State who appreciated the importance of the work to the whole Pacific Coast, no matter where the line should be located; high officials whose presence and earnest approval enhanced the dignity of the occasion; divines to envoke blessings on the work; and last, but not least, directors and contractors, who gave substantial assurance that the brain, the muscle, the gold and iron, were ready to make the railroad a reality.

Patriotic Phase

"A noticeable feature of the inauguration was the patriotic character it assumed. The orators ascended from the level of material considerations to the contemplation of the work a 'bond of union,' and took occasion to rejoice over the recent declarations that henceforth the Union is to be indissolubly wedded to Liberty. All such allusions, together with incidental references to the dignity of the free labor that was to build the continental railway, were received with enthusiasm and applause by the assembled crowd.

"The choice of scene for the ceremony was not favorable to the presence of the gentler sex, but the balconies opposite—on Front Street, above K—were adorned with a fair delegation. The great preponderance of pantaloons was a disagreeable necessity of the 'situation'.

"A stand was erected near the level, a short distance above K Street, and the ends were adorned with the National flag. A generous distribution of bundles of hay gave a comparatively dry footing to the crowd in the immediate vicinity.

"The Sacramento Union Brass Band was stationed on the balcony of the American Exchange Hotel, and between the addresses enlivened the proceedings by playing National airs and the peculiarly appropriate 'Wait for the Wagon.'

"Two wagons adorned with flags. drawn by horses that were also decorated with the National colors, were stationed near the rostrum with earth ready to be shoveled out for the railroad embankment. On one of these wagons was a large banner bearing a representation of hands clasped across the continent from the Atlantic to the Pacific, with the prayer of every loyal heart.

"May the Bond be Eternal"

"Shortly after 12 o'clock Governor Stanford appeared upon the stand and the ceremonies of the occasion commenced. C. Crocker introduced to the assemblage Leland Stanford, Governor of the State of California.

"Governor Stanford then made a speech. After Stanford's speech Rev. Dr. J. A. Benton offered prayer. At the close of the prayer C. Crocker announced that 'the Governor of the State of California will now shovel the first earth for the Great Pacific Railroad.' (Cheers)

First Earth Turned

"The two wagons loaded with earth were driven up in front of the rostrum and Governor Stanford, with a zeal and athletic vigor that showed his heart was in the work and his muscle in the right place, seized the shovel, and amid the lusty cheering of the crowd, deposited the first earth for the embankment.

"The enthusiastic Charles Crocker promptly called for 'nine cheers' and the crowd, sharing his enthusiasm, cheeringly responded. The sun smiled brightly, and everybody felt happy because, after so many years of dreaming, scheming, talking and toiling, they saw with their own eyes the actual commencement of a Pacific Railroad.

"This ceremony ended, the Governor returned to the platform. A. M. Crane of Alameda, President Pro Tem, of the Senate was introduced and addressed the assemblage. Then followed speeches by:—

"J. H. Warwick, of Sacramento, J. A. Banks, of San Francisco, member of the Assembly; Rev. Dr. J. T. Peck, W. H. Spears. Assemblyman from Nevada; Newton Booth, Senator from Sacmento; Dr. J. F. Morse and Charles Crocker."

G. W. REAR APPOINTED NEW ENGINEER OF BRIDGES

G. W. Rear, formerly General Bridge Inspector, has been apointed Engineer of Bridges with jurisdiction over all bridges, trestles, wharves, tunnels, and ferry slips on the Pacific System of the Southern Pacific. Mr. Rear received much prominence in engineering circles in 1912 when he supervised the erection of the Sacramento River bridge which is 834 feet in length; weighs 8,200,000 lbs. and cost $730,000.

Mistakes

When a plumber makes a mistake, he charges twice for it.

When a lawyer makes a mistake, it's just what he wanted, because he has a chance to try the case all over again.

When the doctor makes a mistake, he buries it.

When a tailor makes a mistake, some one has a bad fit.

When a watchmaker makes a mistake, he has the time to fix it.

When a preacher makes a mistake nobody knows the difference.

When an electrician makes a mistake, he blames it on induction. Nobody knows what that is.

BUT—When an engineer makes a mistake,—GOOD NIGHT!—Ex.

Two Waiters Have Been On Owl Diner Eight Years

AN interesting record of service with the Southern Pacific Company is that of Charles Alexander, a waiter, who has been with the company 23 years and has just finished his eighth consecutive year as waiter on the Owl dining car running out of Oakland. William Patton, another waiter in the same car has also had the same period of service on the Owl diner, but has not been with the Company as long as has Alexander.

Both men are familiar figures with the traveling public.

HOW THE C. P. BUILT 10 MILES IN ONE DAY

In connection with the celebration last month of the fiftieth anniversary of the driving of the last spike, the following account of the celebrated feat of the Central Pacific in laying ten miles of track in a single day is reprinted from a guide book, known as "Crofutt's Overland Tours," published in the early eighties:

"Proceeding westward from Promontory four miles can be seen close to the road, on the south side, a signboard which reads:

"'Ten Miles of track in one day.'

"Again on the same side, ten miles farther west, another with the same inscription will appear. These boards mark the track which was laid by track layers of the Central Pacific Company in one day, under the immediate charge of J. H. Strowbridge, superintendent of construction; H. H. Minkler, track layer, and James Campbell, superintendent of division. This undoubtedly is the most extraordinary feat of the kind ever accomplished in this or any other country.

Why It Was Done

"During the building of the road a great rivalry existed between the two companies as to which could lay the most track in one day. This rivalry commenced early in the year 1868, the Union laying six miles; soon after the Central laid seven miles, and then again the Union seven and a half miles. The Central men, not to be outdone, announced that they could lay ten miles in a day. Mr. Durant, vice-president of the Union, offered to bet $10,000 that it couldn't be done, and the Central resolved that it should be done.

"Consequently, on the 29th day of April, 1869, when only fourteen miles of track remained to be laid to meet the Union at Promontory Point, and in the presence of Governor Stanford and many prominent men from the East and West, and a committee from the Union Pacific Company to note the progress, the work commenced.

How It Was Done.

"When the car loaded with rails came to the end of the track, the two outer rails on either side were seized with iron nippers, hauled forward off the car and laid on the ties by four men who attended exclusively to this. Over these rails the car was pushed forward, and the process repeated. Behind these men came a gang of men who half drove the spikes and screwed on the fishplates. A short interval behind these came a gang of Chinamen, who drove home the spikes already inserted and added the rest. Behind these came a second squad of Chinamen, two deep on each side of the track. The inner men had shovels, the outer ones picks. Together they ballasted the track.

"The average rate of speed at which all these processes were carried on was one minute and 47½ seconds to every 240 feet of track laid down.

"Those unacquainted with the enormous amount of material required to build ten miles of railroad can learn something from the following figures:

"It requires 25,800 cross ties, 3500 iron rails, 55,000 spikes, 7040 fishplates, and 14,080 bolts, the whole weighing 4,362,000 pounds. This material is required for a single track, exclusive of turnouts.

"To bring this material forward and place it in position, over 4000 men and hundreds of cars and wagons were employed. The discipline acquired in the four years since the commencement of the road enabled the force to begin at the usual time in the morning, calm and unexcited, and march steadily on to 'Victory'—as the place where they rested at 1:30 p. m. was called—having laid eight miles of track in six hours.

"Here this great Central army must be fed, but Campbell was equal to the requirements. The camp and water train was brought up at the proper moment, and the whole force took dinner, including many distinguished guests. After the hour nooning, the army was again on the march, and at precisely 7 p. m. ten miles and 200 feet of track had been completed.

"When this was done the Union committee expressed their satisfaction and returned to their camp and Campbell sprang upon the engine and ran it over the ten miles of track in 40 minutes, thus demonstrating the work was well done."

---•••---

OPPORTUNITY MONOGRAPHS.

The Federal Board for Vocational Education is issuing a number of opportunity monographs for disabled soldiers, sailors and marines. As an aid in selecting a vocation the little booklets are excellent, and the attention of all who appear handicapped and to whom they may be an aid should be directed. The booklets may be obtained by writing to the Federal Board for Vocational Education, Washington, D. C.

(Continued from Page 3)
rendered an excellent program at their tabernacle. At the Weber Club, various railroad officials were the guests at a banquet and delivered speeches. They were introduced by Toastmaster W. H. Wattis, and were: Mayor Browning of Ogden, William Hood, Chief Engineer, Southern Pacific Company; E. E. Calvin, Federal Manager, Union Pacific; President Heber J. Grant of the Mormon Church, Vice-President Shoup of the Southern Pacific Company; H. V. Platt, Vice-President and General Manager, Oregon Short Line; H. Bissell, a pioneer construction engineer; J. A. Monroe, General Traffic Manager, Union Pacific Railroad.

VETERANS AT LUNCHEON IN SAN FRANCISCO.

Coincident with the celebration at Ogden, the veterans of the Southern Pacific on Saturday, May 10th, gathered at noon at the Palace Hotel, San Francisco, as the luncheon guests of the Company and the United States Railroad Administration. Close to 600 people attended the luncheon, applauded the toasts to the memory of the "Big Four," listened intently to reminiscences connected with the story of the Last Spike, and laughed heartily at the antics of "Fatty" Arbuckle in the comedy film which formed part of the program.

As usual the "old boys" entered enthusiastically into the spirit of the occasion, and their wives and daughters enjoyed the reunion no less. Resolutions of appreciation were passed by the veterans and a rising vote of thanks extended to the company, the Administration and the committee of arrangements.

The fast-dwindling ranks of those who actually saw the historic "wedding" of the Union Pacific and Central Pacific at Promontory, Utah, May 10, 1869, were represented at the luncheon by A. B. Cooper, who was a boy of 9 at the time; R. A. Murphy, fireman on the old engine "Jupiter"; R. W. Baxter, foreman of the Union Pacific construction crew; and A. L. Bowsher, in charge of telegraph construction, who witnessed the ceremony from the top of a telegraph pole.

There were undoubtedly others present who could recall the stirring days of old. Several did, in fact, among them R. L. Fulton, former land agent, who described life in Wyoming when the West was really "wild and woolly"; H. V. Blasdell, representing the Union Pacific pensioners; S. W. Bones, R. C. Martin, and others.

It was a matter of general regret that A. H. Macdonald, for years chairman of the Veterans' Banquet Committee; E. Black Ryan, secretary to Senator Stanford; J. H. Strowbridge, superintendent of construction; R. H. Pratt, superintendent—all men whose personalities are linked with the olden days—were unable to be present. Their messages were read to the assembled guests.

Telegrams of greeting and affectionate interest were received from President Julius Kruttschnitt, District Director Wm. Sproule, Federal Manager W. R. Scott, Vice-President Paul Shoup, General Manager J. H. Dyer, W. H. Crocker, C. J. Millis, Col. John P. Irish, and others whose duties prevented their attendance but who were there in spirit. R. J. Clancy, assistant to the General Manager, and scheduled for an address, was caught in an avalanche of eleventh-hour subscriptions to the Victory Loan, and as Chairman of the General Office Committe was compelled to stay on the job while the big building went "over the top."

Among the guests of honor were
(Continued on Page 6)

A Railroad Record That Defies Defeat

How Central Pacific laid ten miles of track in one day back in 1869

By Erle Heath, *Associate Editor*

FIFTY-NINE years ago a squad of eight Irishmen and a small army of Chinese coolies made a record in track laying that has never been equalled. In one day, on April 28, 1869, these men, fired with the enthusiasm of the greatest railroad construction race in the history of the world, laid ten miles and fifty-six feet of track in a little less than twelve hours to bring the railhead of the Central Pacific three and one-half miles from Promontory, Utah, where connection was made a few days later with the Union Pacific to form the first transcontinental railroad.

The names of the Irish rail handlers have been passed down through the years. Their super-human achievement will be remembered as long as there is railroad history.

ing 14,000 men, had pierced the snow-covered, granite walls of the high Sierra and extended their trail of iron over the barren plains 675 miles eastward. For more than five years the Union Pacific had been building westward. Officers of both roads were awake to the future advantages that would accrue to the company having the longest mileage. This rivalry ex-

This is a page copied from the time book kept by George Coley, foreman of the Central Pacific crew that laid ten miles and fifty-six feet of railroad track in one day on April 28, 1869, setting a record that has never been equalled. The foreman's notations give the mileage stations between which the track was laid, and the names of the men who did the job, also the fact that the men received four days' pay for the day's work. The last two men on the list handled the track gauges for the eight rail handlers. The original book has been preserved by Coley's daughter, Mrs. Jennie Yeates, of Oakland, Cal., who also has the tape measure used by her father. The picture of Coley (left) was taken in 1911, not long before he died, while on the pension rolls of Southern Pacific. J. H. Strobridge (above) was construction superintendent in charge of the Central Pacific forces during the building of the first transcontinental railroad.

close together. The story is told that Vice President Durant of the Union Pacific bet $10,000 that it could not be done, and that his money was "covered."

For several days Crocker and his construction superintendent, J. H. Strobridge, marshalled their forces and laid their plans. Ties were hauled ahead by two-horse teams and distributed along the right-of-way. For some distance the ties were spaced on the already graded road bed. Rails and track materials were moved up from the rear and held in trains ready to advance. More than 4000 men and hundreds of horses and wagons were on the spot. Every man knew his particular job, taught by many months of track work. No one would get in the other fellow's way. In the Central Pacific camp, the patient and methodical Orientals were stirred to a pitch of excitement never shown before, and shared with the few hundred whites the anxiety to "get at the job."

So, too, will that day's work of "John Chinaman" be recalled as the most stirring event in the building of the railroad.

During six years the builders of the Central Pacific, at times number-

tended through the ranks from the presidents to the track laborers. There followed some marvelous feats of track laying.

The Challenge

One day the Union Pacific broke all records by laying six miles of track. Charles Crocker and his Chinese "pets" were invited to match that. They beat it by a mile. Then the Union Pacific came back with seven and a half miles, working from three in the morning until almost midnight. But the Central Pacific was not to be beaten.

Crocker declared his men would lay ten miles of rail in a day. Such a ridiculous thing was scoffed at in the rival camps, which were now drawing

Mishap First Day

April 27, 1869, was the day selected, but an engine off the track early in the proceedings compelled a postponement until the next day. This mishap brought many laughs from the Union Pacific side, but only served town stretched out. In the valley below, continuous trains of wagons and mounted work shops moved along in parallel lines. It could only be compared to the advance of an army.

When a halt was called for the midday meal, six miles of track had been laid and the men were confident they would reach their goal. A number of Union Pacific officers had lunch with Stanford, Crocker, and others of the Central Pacific. They were ready to extend congratulations. "Victory" was the name given the spot where lunch was taken. The station is now called Rozel.

Grades and Curves

After lunch the work went on, but not so rapidly. The ascending grade on the west slope of Promontory Mountain was more difficult than the section covered during the morning and there were many curves. Considerable time was lost in bending rails, which was done by placing the rail on two blocks and forcing it into the desired curve by blows of a heavy hammer.

When the forward march was halted at 7 o'clock, ten miles and 56 feet of new track had been added to the Central Pacific. Jim Campbell, boarding boss and later superintendent of the division, jumped into a locomotive and ran it back over the new line at a clip of 40 miles an hour just to prove that the job had been well done.

If the roadway had been perfectly level and straight, these men could have laid fifteen miles of track. The task had involved bringing up and putting into position 25,800 ties, 3520 rails averaging 560 pounds each, 55,000 spikes, 14,080 bolts, and other material making a total of 4,462,000 pounds.

Workers Acclaimed

Each of the rail handlers lifted 125 tons of iron during the day, in addition to carrying the weight of their heavy rail tongs. They walked many feet more than the ten miles forward. Their's was a wonderful exhibition of skill and strength, and they richly deserved the acclaim showered on them when they proudly rode in a wagon as a feature of Sacramento's railroad celebration a few days later. When the parade was over, their wagon was filled with flowers thrown to them by men and women, boys and girls.

Ten miles of railroad track laying in one day! It is a record that will probably never be challenged. It is not likely there will ever again be such a spirited race for railroad supremacy as the one that inspired the Central Pacific and Union Pacific to such marvelous feats in those early days. Never will there be assembled such an army of railroad workers.

With the eight sons of Erin and the sons of "John Chinaman" rest the palms of a great track-laying victory.

Watsonville Veterans Claim a Record

IF anyone would doubt the satisfaction of employes to remain on the Coast Division as the best on Pacific

These four men have worked continuously at the Watsonville station for a combined total of 106 years. Left to right: R. H. Davis, agent, 21 years; Otto G. Albrecht, cashier, 22 years; Sherman French, chief clerk, 39 years; and Fred J. Lobdell, ticket clerk, 24 years.

Lines, their attention is directed to the station force at Watsonville Junction.

Four men have worked at that station continuously for a total of 106 years, and are still going strong. Barring accidents, they are good for 24 years more, because the oldest of the quartet has six years more before old man pension gets him.

Sherman French, chief clerk, has been at the station 39 years. Next in point of service is F. J. Lobdell, ticket clerk, with 24 years. Otto G. Albrecht, cashier, has 22 years' service, and R. H. Davis, agent, has been there 21 years. For good measure, we might add that there are two others, Fred J. Burdick, bill clerk, and G. F. Rice, warehouse foreman, who have each been at Watsonville station a measly ten years.

So far as we have been able to learn, this is a record on Southern Pacific Lines.

Three of the employes—French, Lobdell, and Albrecht—have never worked at any other station, each having started at the bottom and been promoted gradually to their present positions.

Watsonville is the third largest station on the Coast Division and the largest apple shipping point on the system, which indicates that these men deserve more than passing mention for not bidding out into easier jobs.

Oregon Trainmaster Honored For His Military Service

Colonel John L. May, trainmaster, Portland Division, a veteran of the Oregon National Guard and now an officer of the National Guard Reserve, was awarded a medal for 25 years' faithful service at the quarterly muster and inspection of National Guard troops at the Multnomah County Armory March 31.

Colonel May joined the National Guard 25 years ago and worked his way up to the rank he now holds. He commanded the 162nd Infantry regiment (3rd Oregon) during the World War and he was a lieutenant-colonel with Oregon troops on the border. Previous to that he commanded troops during the Spanish-American War and the Phillipine Insurrection. Following the World War, Colonel May was adjutant-general of Oregon for a brief period. The service medal was presented to Colonel May by Brigadier-General George A. White, Commander of the Guard of the State.

Walter Hanna Handles Crowds at San Francisco Ferry

Walter J. Hanna, who has been station master at the San Francisco Ferry since the retirement of W. H. Voiles last December, has a wide circle of railroad friends in the San Francisco Bay District who are well pleased over Walter's advancement. Hanna entered the service as apron tender at the Ferry Station in March, 1913. Later, he was wharfinger, then passenger director and in 1918 was appointed assistant station master, which position he held until his recent promotion.

Every Nine Days Company Buys One New Locomotive

By purchasing a new locomotive every nine days, a new passenger car every six days, and a new freight car every two and one-half hours during the last five years, Southern Pacific has more than kept pace with the rapid growth of the Pacific Coast.

In the years 1923-27 inclusive the Pacific Lines of the Southern Pacific put into service 194 new locomotives of all types, at a cost of $15,197,969. During the same period 318 new passenger cars costing $9,519,844 were required, and 16,699 freight cars purchased at a cost of $36,388,174. In all $61,105,715 was spent in five years for locomotives, freight and passenger cars.

This total does not include 13,647 new refrigerator cars purchased at a cost of $44,380,933 by the Pacific Fruit Express, in which Southern Pacific owns half interest.

Lacking adequate new motive power, cars and other facilities, railroads could not long continue to serve the public efficiently and well. It must be remembered that railroads, to make these large and necessary expenditures for equipment, must be able to earn sufficient to render railway securities safe and attractive to the average man or woman with a dollar to invest.

Worth a Try

Fugitive—Quick! The police are after me. Where can I find a place to hide?

Office Employe—Jump into the filing cabinet; no one can ever find anything there.—Ex.

Judah Monument Dedicated at Sacramento

American Society of Civil Engineers Join Railroaders in Honoring Pioneer Promoter of First Transcontinental Rail Line

DISTINGUISHED civil engineers of the nation, prominent citizens of Sacramento, and veteran railroaders joined Southern Pacific men and women on April 25 in formally dedicating a monument being erected at Sacramento in memory of Theodore Dehone Judah, pioneer engineer who mapped the route for the western link of the country's first transcontinental railroad and whose efforts resulted in the founding of the Central Pacific Railroad Company, parent organization of the present Southern Pacific.

The ceremony was held under the auspices of the American Society of Civil Engineers during the annual convention of that body in Sacramento, it being the first meeting of the seventy-eight-year-old national society ever held on the Pacific Coast. Judah became a member of the society in 1854 and it was particularly fitting that the organization participate in the ceremony honoring one of its most illustrious members. Thomas E. Stanton, Jr., president of the Sacramento chapter of the society, presided, and John F. Coleman, national president, was on the speakers' platform.

An address by W. H. Kirkbride, engineer of maintenance of way and structures, extolling the accomplishments of the pioneer railroad builder, was the high light of the dedicatory

Jos. M. Graham, right, is the sole survivor of the construction engineers who built the first transcontinental railroad east from Sacramento. He is now 88 years old and resides in Berkeley, California. Theodore Judah, left, of San Francisco, took part in the ceremony at Sacramento when the memorial monument to his distinguished great granduncle was officially dedicated. The model shows the monument as it will appear when completed.

program. In behalf of the officers and employes, whose contributions furnished the funds for erecting the memorial, Mr. Kirkbride presented the monument to the city of Sacramento. The acceptance speech was made by Mayor C. H. S. Bidwell.

Theodore Judah, of San Francisco, great grandnephew of the distinguished early-day engineer, concluded the ceremony by placing into a niche in the monument a copper box containing historic documents and photographs pertaining to Judah's activities between the years 1854 and 1863. Music during the ceremony was furnished by the Firemen's Band of Sacramento.

When completed, the monument will be an imposing tribute to the man who was first chief engineer of the Central Pacific Company. It is located in the municipal park facing the Sacramento passenger station and is built up from huge granite rocks taken from the Sierra Nevada mountains over which Judah tramped in the early '60's searching for a suitable pass for the railroad. The concrete base is fourteen feet square. The main rock, on which will be carvings symbolic of the rugged country over which the railroad was built, is 11 feet high, 12 feet wide, and weighs about 40 tons. At the base of this central rock will be a bronze bust of Judah and a plaque on which will be inscribed a brief record of his accomplishments.

Among the several hundred people

Speakers' platform at the ceremony on April 25 when the Theodore D. Judah memorial monument was formally dedicated at Sacramento. The small model in front of the huge granite stone shows the monument as it will appear when finished ready for unveiling. The ceremony honoring the pioneer railroad engineer was held under the auspices of the American Society of Civil Engineers. Thos. E. Stanton, Jr., president of the Sacramento chapter of the society (the man seated near the center of the front row, wearing the light suit) presided. On

Mr. Stanton's right is John F. Coleman, national president of the society; and next is W. H. Kirkbride, engineer of maintenance of way and structures, who delivered the principal address and presented the monument to the city of Sacramento in behalf of the men and women of Southern Pacific. The acceptance speech was made by Mayor C. H. S. Bidwell, who is shown at Mr. Stanton's left. C. W. Anderson, city councilman and pioneer resident of Sacramento, is to the left of the mayor. Descendants of the Judah family were also present.

who witnessed the dedicatory ceremony were a number of veteran railroaders who helped to build the railroad over the Sierra and across the plains into Utah more than sixty years ago. Prominent among these pioneers were Joseph M. Graham, of Berkeley; J. O. Wilder and Amos. L. Bowsher, of Sacramento.

Mr. Graham is the only survivor of the construction engineers who supervised the building of sections of the road east of Sacramento. He is 88 years old and remembers almost as well as if it were yesterday the events of the late '60's when an army of Irish and Chinese laborers, working with pick and shovel, one-horse dump carts and wheelborrows, built the roadway for the tracks over and through the granite ridges of the Sierra Nevada mountains. After the line was built, he became the first resident engineer for the section of newly-built track from Colfax to Reno, which position he held until 1881.

Mr. Wilder started work for the engineering corps during 1866, when final locations were being made for the line over the mountains, and later was a machinist in the Sacramento shops. He retired from railroad work in 1920, after 54 years' service. He is now 80 years old. Mr. Bowsher was foreman of telegraph construction during the time the railroad was being built. It was he who connected the wires to the spike and hammer so that the telegraphic signal could be flashed over the country when Governor Stanford drove the "last spike" at Promontory, Utah, on May 10, 1869, completing the first transcontinental railroad. He is now 89 years old and retired on pension in 1911, after 43 years with the railroad.

Also present at the ceremony were the following descendants of the Judah family who are now residing in California: Mr. and Mrs. Floyd S. Judah, grandnephew, president of the Peck-Judah Company of San Francisco; Mr. and Mrs. Henry R. Judah, grandnephew, and daughter Miss Janus Judah, great grandniece, Santa Cruz publisher; Mrs. Sarah E. Judah, grandniece, and son Theodore, great grandnephew, of San Francisco; Mr. and Mrs. Chester L. Roadhouse, grandniece, and daughter Miss Katharine Roadhouse, great grandniece, of Davis.

"The story of the conception, promotion and construction of the railroad eastward out of Sacramento is a soul-stirring California romance, second only to the discovery of gold, and Judah is the moving genius in the whole story," stated Mr. Kirkbride in his speech. He told how Judah had come to California in 1854 at the age of 28 years as chief engineer of the Sacramento Valley Railroad, the state's first rail line; how he became obsessed with the idea of a transcon-

Have We Your Correct Home Address?

EMPLOYES who are not receiving the Bulletin through the new system of mailing direct to the homes, should advise their supervising officer of their correct home address. Every effort is being made to keep the mailing lists corrected as changes in address occur, and prompt report of your new address will be of great assistance. Extra copies of current and back issues of the Bulletin are always available and will be supplied on request.

tinental railroad and persisted in working in its behalf in spite of ridicule and discouragement; how he discovered a suitable pass and carried his arguments to Washington with so much conviction that Congress enacted the legislation which made it possible to build the road; how he "sold" his idea to the Sacramento merchants who later became famous as the "Big Four"; and how he contracted fever and died in New York when only 37 years old.

"When we consider his accomplishments in his brief span of life, we must concede his genius and energy," concluded Mr. Kirkbride. "The men and women carrying on the work that he started have determined by their voluntary contributions that there shall be erected a memorial to his memory. It shall be erected here in the city where he dreamed his dreams and where his dreams became realities. It shall be of eternal granite removed from the Sierra that he labored over, received inspiration from and loved so well. The stones comprising the monument are to be unhewn, symbolic of the ruggedness of the Sierra Nevada and of the unspoiled and unflinching, rugged character of the man. May this monument survive throughout the ages—an inspiration to all those who come and go."

Conductor's Help to Passenger Makes Friend for Company

Frank Gomes of Kearney Park, Calif., in a letter to Superintendent Rowlands of the Western Division, expresses his appreciation of an accommodation extended him by one of our conductors. Mr. Gomes says:

"I take great pleasure in bringing to your attention an act of courtesy extended to me by Conductor George V. Musser. En route to Oakland, I boarded train No. 55 at Herndon, Calif. When Mr. Musser came through to collect the fare, I put my hand in my pocket and discovered I had left my pocketbook at home. After some conversation, and as your company does not accept personal

checks, Mr. Musser, instead of putting me off the train as he had a perfect right to do, very kindly took my check to himself and advanced the fare from his own funds. As I was on my way north to Humboldt County on business, an interruption to my journey would have been highly inconvenient, and I think it only right that I should let you know how much I appreciate Mr. Musser's courtesy."

Heavy Spring Lettuce Movement Ends in Arizona
By Leroy Magers, Correspondent, Tucson

The spring lettuce movement on the Tucson Division has terminated after three busy weeks in getting some four thousand cars loaded and started on their way to the eastern markets. There were 3160 cars loaded in the Salt River Valley. Mesa was the center of activities for cars loaded east of Phoenix, while those loaded west of Phoenix were handled from Phoenix. Assistant Superintendent Goodfellow, who made his headquarters at Mesa, rode herd over the switcher and hauler crews in that district, and Trainmaster Ed Wheeler was kept on his toes west of Phoenix.

This movement was not confined to the Salt River Valley entirely, however. Yuma shipped 526 cars out of the Yuma Valley, Eloy had 57 cars, Coolidge 55 and Gila 11 cars. This is the first time lettuce has been shipped out of Gila.

About 20,000 acres were planted to spring lettuce at these several points, and the crop would have been larger had the weather not turned very warm during the early part of April.

Retiring Towerman Receives Gift From Co-workers

When, on March 31, William F. Byers ended his day's work as towerman at Fruitvale, he also ended his service with the Company, being retired on pension. As he was preparing to leave the tower for his home, he was somewhat surprised to see a number of towermen and signal maintainers gathered outside, and he was still more surprised when their spokesman, on behalf of the gathering, stepped forward and, with a few words of congratulation, presented him with a handsome bill fold. Upon opening the bill fold, Mr. Byers found that it contained a handsome sum of money contributed by his fellow workers as a token of esteem and a parting gift.

Mr. Byers was so overcome by the unexpected gift that he was unable to make a fitting reply at the time, and desires, through the columns of the *Bulletin* to assure the doners that he is deeply appreciative of the gift, as well as the kindly spirit of friendliness which prompted it.

Mr. Byers served continuously as towerman at Fruitvale for twenty-four years.

Chapter Three
The Sacramento Division Before 1895

In 1864, the Central Pacific Shops were located at Sixth Street and D Street next to a swamp that was given the name of Sutter Lake. The three buildings and the swamp were the beginning of a facility that eventually employed nearly five thousand people.

During the first twenty-five years of the Shops' growth, there was no electricity. Gas lights were common, as were fire concerns related to them. Using electric power to run heaters during the winter didn't yet exist.

The large amounts of water from two nearby rivers and the marshy area of Sutter Lake were health concerns due to flooding in winter and stagnant water during late spring and early summer. Illnesses from poor water quality and rotting debris in motionless water were challenges to all residents of Sacramento.

The railroad work during the pre-1895 period was done without hard hats, two-way radios, flashlights, steel-toed boots, or goggles with un-breakable lenses. The Southern Pacific continued to make larger locomotives, assemble longer trains, and handle more car loadings, despite the lack of safety and communication tools that in the twenty-first century we consider necessary for this work.

Even though the early operations of the railroad seems primitive compared to one hundred and fifty years later, shipping goods by rail for three hundred miles was a huge improvement over loading the same amount of freight onto a cart to be pulled by horses at one sixth the speed.

The following two essays and nine articles from *The Bulletin* describe the earliest operations of the Central Pacific and Southern Pacific railroads. The articles from *The Bulletin* are:

"Romance of the Sacramento Division," August 1920

"Our Loco Has a Leak," March 1920

"Old Times Recalled at Sacramento," July 1919

"Sacramento Employees Recall Days of '49," July 1922

"Rough and Ready – The Sacramento Division of 1870," August 1920

"Hand Brake Setting was an Art," September 1920

"When the Central Pacific was Young," May 1920

"Highly Toned Rail Travel – Eating Stations, Silver Palace Cars, Reading by Candlelight," October 1922

"Back in the Good Old Days on $85 per month," April 1927

❧ ❧ ❧

Lighting—From Steam and Belts to Electricity

The innovation seen at the Sacramento Shops began nearly forty-five years before Henry Ford produced the Model T and ended about twenty years after Neil Armstrong walked on the moon. Electricity was used for the first time at the Shops in 1889, following a twenty-five-year period of growth with no electricity. It wasn't until the mid-1850s that Sacramento and its rail yards saw a series of developments that led to wide spread use of electricity at the Shops.

In 1854, the Sacramento Gas Company was formed near the intersection of the Sacramento and American Rivers. Work on a gas plant began that year, but progress was delayed by high water until the summer of 1855. On December 17, 1855, the first lights using gas shined in Sacramento. When construction was underway for railroad buildings in 1867, the rivers acted up again and undermined the granite foundation of the gas plant. To shore up the foundation, the railroad was asked to bring cobblestones and granite from Rocklin, about fifteen miles north of Sacramento. The new rocks served to stave off the floodwater, and the railroad eventually used similar rocks for their own foundations. The following year, 1868, the railroad bought gas from the company and there was gas powered light for shop work.

The local gas industry continued to evolve in 1872 when a new gas company was organized: the Citizens Gas, Light, and Heat Company. Coincidentally, the purchasing agent for the Central Pacific Railroad (CPRR) was elected president of the new gas venture. He and a few other CPRR officials owned land on the Sacramento River near S Street where they built a new gas plant. Within three years, the original gas plant that was used in 1868 was shut down, leaving the railroad yard with no gas—as no piping existed to transport gas from anywhere else. Resourcefulness showed itself within the railroad when management bought and renovated the property of the old gas plant and used it to make their own gas. Though it seemed like a good idea at first, the need for repairs became a nuisance, and it didn't last very long. The railroad sold some of the hardware for scrap in 1885.

In 1886, when electricity was being produced elsewhere in the country, the railroad bought two generators operated by a large steam engine that was already being used to turn machinery in the saw mill, cabinet, and pattern shops.

Because Sacramento was at the junction of two large rivers, getting power from hydro generators on the Sacramento River or up the American River was in the mind of entrepreneurs such as Horatio Gates Livermore who, in 1895, initiated the building of the first powerhouse to serve the Sacramento area.

The events that led to the first light from his Folsom facility began in the mid-1860s when he planned to use water power to operate a saw mill near Folsom. He started to build a holding pond for logs that had been cut higher in the foothills. Expenses mounted with increased labor costs, and two decades passed before state officials agreed to provide prisoner labor for his project. But, during the delay, Livermore died and his two sons continued the effort. Their idea was to forgo the idea of water wheels and to use the water's force in a different way: hydroelectric power. By 1895, their plan came to fruition with the completion of a powerhouse with four 750-kilowatt, 800-volt, 542-ampere generators built by General Electric. Each weighed more than 57,000 pounds. The water reached the turbines and generators through four eight-foot-diameter penstocks. On July 13, 1895, the Folsom Power Plant began operation twenty-two miles east of Sacramento. Electricity was delivered to Station A in Sacramento which was at the corner of Sixth and H Streets.

After receiving the wonders of electricity from Folsom, Sacramento celebrated with a "Grand Electric Carnival" on September 9, 1895 that included lights strung along downtown streets and on the State Capitol. Of course, the railroad made arrangements to receive power, and by the end of the year, small motors had been installed and lights were in the master mechanic's office.

As with the implementation of any new technology, there were problems. The learning curve of the Sacramento Electric Power and Light Company included the moment when one of the generators failed and damaged the armature. A recurring problem was sand, silt, and small gravel finding their way into the penstocks, which damaged the turbine blades. Turbines were replaced as often as two times per year.

The Livermores' faith in water-powered electricity persisted, even after many problems, and eventually led to the construction of another facility in 1897 just downstream of the first one.

Other competitors constructed powerhouses to take advantage of the flows in rivers from the Feather southward to the Tuolumne. One of these companies was the California Gas and Electric Company, which operated the Colgate plant on the Yuba River (one large watershed to the north of the American River).

The overall success of hydroelectric power for lighting, industry and public transportation overwhelmed the capacity of the Folsom complex, which led to the Livermores buying power from the Colgate powerhouse. By 1903, the California Gas and Electric Company had acquired the Folsom facility along with other power plants in the foothills. In 1906, the business was reorganized to form the Pacific Gas and Electric Company, which operates today.

The lower powerhouse, downstream of the original plant, was shut down in 1924 due to problems that persisted from its inception. The first facility, with its beautiful stone and brick construction, ceased operation in 1952 and is now part of the California State Park system.

❧ ❧ ❧

Eucalyptus Trees and Malaria

In the late 1800s, and up until 1906, the area surrounding the Sacramento Locomotive Works contained somewhere between 1000 and 3000 eucalyptus trees— a species of fast-growing tree that had been brought to California by the Australians during the Gold Rush. They no doubt added beauty to the Shop area, but had more likely been planted there as a response to the high number of citizens affected by malaria.

Malaria was not known in the 1870s as a disease transmitted by mosquitoes. Prior to 1897, a common belief was that the symptoms spread from person to person via the air. It was also believed that eucalyptus trees cleansed the air thereby reducing the spread of the disease.

The concern regarding malaria was serious enough to cause the California Board of Health, in the Third Biennial Report (1874-1875), to publish a section called "Malarial Fevers and Consumption in California." The report stated, "Although the properties of the [eucalyptus] tree, in regard to the prevention of malarious disease, have in all probability been exaggerated, still there is strong evidence that it does exert an advantageous sanitary influence." The report also referred to a lecture made by Professor Bentley before the Royal Botanic Society of England. He said,

> The first and most important influence which this tree exerts, and that which has brought it more especially into notice, is its power of destroying the malarious agency which is supposed to cause fever in marshy districts; from which circumstance it has been called the 'fever-destroying tree.' It is in this respect commonly regarded as being serviceable in two ways; first, by the far-spreading roots of this gigantic tree acting like a sponge, as it were, and thus pumping up water and draining the ground; and secondly, by emitting odorous antiseptic emanations from its leaves.

Though it may have been a coincidence, the Central Pacific Railroad planted the eucalyptus trees on the grounds of the Sacramento Shops one year after the report was published.

In 1897, Ronald Ross, who had studied at the Royal Colleges of Physicians and Surgeons in England, found the malaria parasite in the stomach tissue of a mosquito. Ross continued his work and proved the role of mosquitos in the transmission of malaria parasites in humans.

(The discovery was not dismissed as trivial. In 1902, Mr. Ross was awarded the Nobel Prize for his malaria work.) This made perfect sense: Sacramento was situated in an area prone to high water due to the nearby confluence of the Sacramento and American Rivers. Drainage was minimal and the resulting stagnant water served as the home to mosquitos.

This new information, and the tree's tendency to burn ferociously due to oils in the leaves, caused the sentiment to change regarding the trees' existence at the Shops. They were removed in 1906. The wisdom of this is disputable: While the leaves' aroma didn't help the malaria problem, their consumption of large amounts of water certainly helped reduce standing water. There is one undisputed fact: The eucalyptus trees provided one necessary feature during a hot, still, summer day: shade.

Shown above is Southern Pacific's steam locomotive No. 244 at Bay 18 on the west side of the machine shop. The three-story building housed the railroad's offices. This image dates between 1887 (locomotive construction date) and 1892 when the office building was extended. The machine shop is oriented approximately north to south. The eucalyptus tree is likely one that was planted in the 1870s.

The northeast corner of the building dates to 1868, and an enlargement to the south was added in 1875, which increased the length of the building to four-hundred feet. The new addition had a foundation comprised of one-foot-square redwood timbers on which similar-sized redwood beams were laid crosswise. The south wall of the building was made of wood to allow for further extension. That foresight was rewarded when the building was enlarged again in 1888. This remodel included an overhead monorail to accommodate a crane. Shortly after the introduction of electricity to Sacramento in 1895, two large motors were installed that replaced three steam engines. Leery of the reliability of electricity, the steam engines were kept in working order until being disposed of in 1901.

In 1948, D.L. Joslyn, a former employee in Sacramento, described his impressions of the eucalyptus trees: "When I first went to work in the shops in August 1902, many of those old trees were still around the shops, and the place looked as if it had been built in a forest; and the shade was very grateful.[sic] However, in 1906, orders were given to cut down all trees in the shops, and this order was speedily carried out, a great many of the employees coming over on a Sunday to cut down and take home the trees, as they made ideal firewood for the stoves in their homes." – Courtesy California State Railroad Museum

The Bulletin

THE BULLETIN IS PUBLISHED MONTHLY IN THE INTEREST OF AND DISTRIBUTED FREE TO THE EMPLOYEES OF THE SOUTHERN PACIFIC COMPANY. CONTRIBUTIONS ARE INVITED AND SHOULD BE ADDRESSED TO THE EDITOR, BUREAU OF NEWS, 65 MARKET STREET, SAN FRANCISCO. CALIFORNIA.

VOL. IX.　　　　SAN FRANCISCO, AUGUST, 1920　　　　No. 8.

The Romance of the Sacramento Division

A remarkable story of the inception and development of the father of all divisions; the Southern Pacific's "Alice in Wonderland" trail

By J. D. BRENNAN, Superintendent of the Sacramento Division

THE Sacramento Division is the nucleus of the Pacific System of the Southern Pacific Railroad and the pioneer division of railroads of the "Golden West."

It embraces in its 675 miles all of the old lines and much of the new lines of railroad that have made California famous from ocean to ocean. It is replete with historical events and romance worthy of volumes of word pictures that only an artist can supply.

The first unit of the Pacific System, twenty-two miles in length, was built in 1855-'56. It started at Sacramento and ended at Folsom and was known as the "Sacramento Valley Railroad." Today that father of California railroads is a busy section of the Placerville branch of the Sacramento Division and serves a rich fruit belt and a fleet of gold dredgers that dig millions of dollars of gold annually and cast up mountains of boulders that are crushed into suitable sizes for the foundation of California's unsurpassed highways, furnishing tonnage for this branch line to the exent of sixty to seventy-five cars daily.

In 1865 this branch was extended to the rich mining town of Latrobe, and in 1888 to "Hangtown" (now Placerville), whence there operated twenty-one stage coaches to the richest of the rich gold and silver mines—Comstock Lode, Virginia City, Silver City and Gold Hill.

Theodore D. Judah, chief engineer of the "Sacramento Valley Railroad," naturally gained much prominence as builder of California's first railroad and his services were eagerly sought by such men as Asa Whitney, who, in 1846, conceived the idea of a transcontinental railroad, and Dr. D. W. Strong of Dutch Flat. In 1859, after making a survey for a line over the Sierra Mountains (at his own expense), Judah was sent to Washington, D. C., to present to the Thirty-sixth Congress a proposition for the government to assist in building the "Pacific Railroad," as the dreamers

J. D. Brennan, Superintendent of the Sacramento Division.

of that day called it. History here records a failure.

In 1860 young Judah was introduced by Dr. Strong of Dutch Flat to these stalwarts of pioneer days: Leland Stanford, Mark Hopkins, E. B. Crocker, Charles Crocker, and C. P. Huntington, who incorporated the Central Pacific Railroad January 28, 1861, capitalized at $8,500,000, with a subscription list of but $148,000.

Again the young civil engineer, Theodore D. Judah, visited the Congress at Washington, and to him is credited the work of convincing Congress that government aid was necessary for the tremendous task of going "over the top" of the majestic Sierra, suspended amid storm clouds and surrounded by plangent cataracts.

Once government aid was obtained, Congress, no different then than now, gave painstaking thoughts to the subject of a standard gauge for the railroad, and, not being willing to accept the judgment of young Judah on that momentous question, called upon Abraham Lincoln to decide the point. Mr. Lincoln reported at great length that the gauge should be five feet. Congress, true to form, adopted a gauge of four feet eight and one-half inches.

Directly the act of 1862 became a law, the hardy pioneers of California lost no time and, on January 8, 1863, Leland Stanford upturned the first shovel of dirt at the foot of "K" street, Sacramento, in the real making of California.

Thus, then, was started that masterpiece of railroad, that "Alice in Wonderland" trail of the iron horse! The Sacramento Division over the mighty Sierra, whose problems try the soul of man; whose anxieties prematurely whiten the hairs of men of iron; whose terrors stilled forever the heart of one superintendent as he emerged from his office car in the snowsheds to view the scene of a holocaust, which probably he could not have explained to an exacting superior officer who had never passed through a fire in the Sierra Nevada snowsheds and possibly did not understand the human element that enters eternally into the life of the "Truckee District." Let us mercifully pass by the case of one other superintendent bereft of reason by the baffling problems that confronted him, in the days when but few trains used the iron pathway of today—the same that the dauntless young Judah surveyed—where today the traffic has increased more than 100 per cent.

The Land of Old; the Land of Gold!

Let us continue our story through the country of Bret Harte and Mark Twain; the country of gold and ro-

THE BULLETIN

"THE HUB OF THE WHEEL"

mance; the country as glorious as a June day.

As in most vast undertakings, the originator failed to realize his vision, and Theodore D. Judah died in 1863 from Panama fever at the age of 37 years. S. S. Montague was appointed chief engineer; however, the actual location made by Mr. Judah was followed to Colfax, or Lower Illinoistown Gap, Mr. Montague making a change in the original survey in the vicinity of Long Ravine, again following Mr. Judah's alignment from Cape Horn to Alta, the line from Alta to Summit being located by L. M. Clement.

The first forty miles involved expenditures in excess of the original estimate and the builders found themselves well into Bloomer Cut, between Zeta and Auburn, and up against the hardest conglomerate material encountered up to that time (1864). The stockholders became discouraged on account of the slow progress; the Public proclaimed the adventure the "Dutch Flat Swindle," Dutch Flat being at that time a roaring camp of gold miners, and, according to rumor, the temporary eastern terminus of the railroad.

Gunpowder and Chinamen were the only weapons of combat the road builders had with which to fight the earth and stone through which they had to pass, laid in their path centuries ago by the Creator. Dynamite was yet to be discovered, but, when obtained, proved to be the "Open Sesame." Finally Dutch Flat was reached and the road continued eastward.

At Cape Horn, Chinamen were let down the steep bluffs in baskets to the grade line, where they worked against the solid rock and shale, dangling on ropes from a hundred feet above, with a gap of 1200 feet below into the American River Canyon.

As the work extended easterly from Cape Horn it became, in a construction way, heavier with every mile until, in 1867, they reached the summit of the Sierras and looked with joy over Donner Lake, where the Donner party perished in their ill-fated attempt to cross the Sierras in 1846. The work ahead, while heavy in portions, did not present the difficulties they had passed.

From a report of the chief engineer in 1867, after surmounting the construction difficulties, he faced the snow problem of the Truckee Pass, where more snow falls than in any place in the United States south of Yukon Pass. We find Mr. Montague concerned with protecting the line by snowsheds. This was the commencement of the "Longest House in the World," which today constitutes a single and double track snowshed 156,259 feet in length, twenty-nine and six-tenth miles, topping the Sierra between Blue Canyon and Truckee, and often called "The House Without End." The board measure approx-

The above map shows how the Sacramento Division acts as a receiving and distributing center for traffic radiating in four directions.

imates 100,000,000 board feet and is of exceptionally heavy construction to withstand the great snowfall, recorded during one season as sixty-five and one-quarter feet.

The Hub of the Wheel.

The Sacramento Division is the hub of the wheel of the Pacific System, at the base of the northern district. It is fed from the west and bay region by the Western Division and over it is transported east, west, north and south all tonnage routed by way of the Shasta, Salt Lake and Stockton Divisions.

The division has a total mileage of 675 miles, the Valley lines stretch from Sacramento northerly paralleling the Sacramento River, tapping the centers of the east and west side of the great fertile valley as far as Tehama, with branch lines reaching from these main lines into the important foothill districts adjacent thereto. These Valley lines include 421 miles of track and feed one of the richest districts in the State of California, adjacent to the Sacramento River and its outlying watershed between the Coast Range on the west and the Sierras on the east.

These lines traverse many counties— Sacramento, Yolo, Colusa, Sutter, Yuba, Butte, Glenn and Tehama— which are famed for their productive-

ness as regards lumber, grain, rice, and alfalfa, with their citrus fruit belts and their olive groves; their great bean districts throughout the reclaimed districts adjacent to the rivers, and their fruits, particularly the peach, pear, prune and grape belt in the central counties; likewise their abundant production of walnuts and almonds.

The main line from Sacramento to Sparks, 156 miles in length, along the original Central Pacific, 115 miles of which is double track, traverses Placer and Nevada Counties on the California slope of the Sierras into Washoe County in Nevada. This line, running northeast between the Yuba and American Rivers, feeds the scenic Sierra foothill district between Blue Canon and Sacramento, a fruit belt producing the greatest quantity and variety of fruits anywhere in California.

The main fruit shipping points of this district are Loomis, Penryn, Newcastle, Auburn, Bowman, Colfax. It is over this portion of the division that all transcontinental east and west tonnage is moved through the central gateway of the West.

Yet another seventy-five miles of tracks surrounds the city of Sacramento and leads easterly to Placerville, tapping similar fruit districts in the same foothill zone, the rock quarries of Folsom and Fair Oaks and the timber areas and mineral wealth of El Dorado and Amador counties.

The fruit production of the Valley counties and the fruit belt adjacent to Sacramento has made the Capital City the largest f r u i t distributing center of the world.

It was near Coloma, a short distance above Placerville, that gold was first discovered in California by Marshall, causing the famous gold rush of 1849. Gold and other minerals are still produced in large quantities in this district, and only this month a vein of gold of fabulous wealth was struck by a former railroad man. Where gold has been worked out, orchards are today producing greater wealth.

The Operating Problems.

The physical and climatic conditions of the division create operating problems in variety. Lumber trains originating in Stirling City, at an elevation of 3520 feet above sea level, drop down the thirty miles to Chico on a three and one-half per cent grade and around curves exceeding fifteen degrees, requiring skill and exactness and perfect adherence to rules. It is on this line where the continuous braking resistance causes the slivers from heated brake shoes to fire the trestles and ties. A gravity car patrolman follows each train with a fire extinguisher and water barrels, extinguishing many fires set by the trains descending the grade.

On the Placerville line, the operated maximum grade is 2.2 per cent for thirty miles of the distance, over which are transported lumber, ores, livestock, fruits and hundreds of thousands tons of rock.

From the Sacramento Southern Line comes the asparagus, celery, deciduous fruits, vegetables, vegetable seed and the many products of the delta region. This is where we compete with river transportation and offer a service which results in our enjoying the bulk of the business moving to transcontinental points.

From the east side and west side Valley lines we receive the Oregon products, particularly lumber, and the enormous tonnage o r i g i n a t i n g throughout Superior California. These lines, with their long tangents and easy gradient, have made possible our solution of an economical operating problem through the medium of the ninety-nine-car freight train, with a maximum tonnage of 10,000 Ms. with one locomotive, for 107 miles, between Gerber and Roseville. The exactness required to start and stop and economically operate a ninety-nine-car train is a source of much pride to the division.

With a monthly output of 90,000 cars, Roseville, the largest freight terminal on the division, is the real haven for box cars and other kinds of cars typical of railroads. It is the home of the freight Mallet, a mechanical device combining in one unit the power

VISION AND REALIZATION

The photo at the right is of Theodore D. Judah, the engineer who visioned the first railroad over the Sierra, and who died before seeing his dream come true. At the left is S. S. Montague, who carried to successful conclusion Mr. Judah's plans.

formerly furnished by many engines. One of these monsters, with its 437,000 pounds, outweighs by 50 per cent the entire equipment of eight engines with which the Central Pacific Railway started operations. Roseville is the junction for the north, south, east and west and it is from this point we handle with Mallet engines the transcontinental tonnage over the old, historic line, topping the scenic Sierras at Summit and dropping down to the Nevada plateau at Sparks.

This line takes us over double track, with block signals, from Sacramento to Rocklin, at the base of the foothills of the Sierras. Thence the eastward and westward lines diverge, the eastward twenty-eight miles, Rocklin to Colfax, through eighteen single and double track tunnels, constructed by William Hood, chief engineer during 1907 and 1912. The westward track still follows the old Central Pacific alignment, the two lines occupying the same roadbed again at Colfax. The tracks, in their entanglements through the foothill country, create both right and left hand operation, where the eastward track, with its one and one-half per cent compensated grade crosses over and under the original line.

The increase of traffic, carrying with it a continuous increase in the size of cars and in the weight and power of the locomotives, has had to be met by a like progress in the standard of track and structures, until today

this section of the original transcontinental railway is a leader of American railways. To maintain this standard there has been developed a corps of workmen trained in careful and accurate maintenance of track. Water must be collected in the mountain ravines and stored in ever increasing quantities that there may be no delay to the never ending flow of traffic. Climatic changes require particular care in keeping to the highest efficiency the staff system and signals which make for safety of operation.

We have dwelt upon everyday operation; however, every day is different. Our shed renewals require the reconstruction of over a mile a year; all track work must be done during the summer, making an interesting problem in railway maintenance.

It is likewise so with train operation, for the moment the snow commences to fly the flangers, the rotaries and snow-fighting equipment are a daily and nightly occurrence on that portion of the line outside the sheds, involving many miles of double and single track, and no sooner is the winter over than fire trains are put in operation at four intermittent points throughout the shed territory to protect the investment in this great structure.

Red Mountain, the signal peak, located 7860 feet above sea level, is situated with relation to the shed territory, that the day or night observer may immediately locate and give the

THE BULLETIN

switches and signal apparatus, thereby avoiding delay to passenger train.

John E. McNally, Section Foreman—For discovering brakebeam down under car and signaling train crew to stop and make necessary repairs without resultant derailment and delay to traffic.

William P. Davis, Fireman, Lodi—For vigilance displayed in discovering brakebeam dragging and taking necessary action to stop train and avoid hazard of accident.

Frederick E. Cox, Telegrapher—For prompt action taken when discovering erroneously routed car.

COAST DIVISION

Frank E. Webber, John R. Wilson, John R. Winters, Towermen, San Francisco—For exceptional intelligence and initiative in connection with handling of runaway locomotive.

Wm. R. Wilder, Conductor; Joseph H. Laird and Earl L. Murdock, Brakemen; Wm. H. Rucker, Engineer, and Wm. S. Purcell Fireman—For very valuable assistance in clearing main line after derailment.

Wm. L. Wilson, Flagman, Gilroy—For noticing brake beam and rod down on passing freight train and signaling conductor in caboose.

G. Martin, Signal Maintainer, San Miguel—For exceptional service in clearing dispatching telephone wires at time of wire trouble.

SAN JOAQUIN DIVISION.

Edwin H. G. Bock, Conductor—For displaying close attention to duty and exercising extraordinary vigilance, discovered broken rail and reporting same to the proper official so it could be changed.

A. E. Gallion, Operator—For discovering broken rail in main line, reporting same so it could be repaired promptly and taking immediate action to protect traffic from serious delay by assisting to flag through block.

F. Zuber, Engineer, and E. McDonald, Fireman—For repairs to oil feed line on engine and assisting in clearing track after encountering a slide in Tunnel 7, Train No. 49, minimizing delay to this important passenger train.

J. A. Lafferty, Conductor; W. O. Gernreich, Engineer; A. P. Hughes, Fireman; A. O. Eager, W. B. Cooke and J. E. Chrystal, Brakemen; Bay Point—For assistance rendered in combating fire which menaced several oil conducting pipe lines carried on wooden trestles. Had the fire been allowed to spread to the pipe lines and caused a breakage, no doubt a serious loss would have resulted, and such work is highly commendable.

I. C. Mather, Train Checker, West Oakland—For vigilance in detecting car under load which was billed as an empty. The discovery of this error no doubt saved excessive back haul and delay in the shipment and is much appreciated.

J. T. Arey, Yardmaster; W. V. Hennessy, Engineer; T. J. Cole, Engine Foreman; J. F. Campbell and V. Van Gooden, Yardmen; H. A. Shaw, Fireman; L. N. March, carpenter—For assistance rendered in extinguishing fire which threatened destruction of overhead bridge. It is very gratifying to know employees take such a live interest in protection of company property.

GOLDEN RULE APPLIES TO RAILROADS.

E. McFadden, agent at Davenport, Cal., has sent The Bulletin an attractive booklet entitled "Through the Meshes," the house organ of the W. S. Tyler Company, Cleveland, Ohio, in which an interesting article on the railroad situation concludes with the following paragraph:

"We can't expect the railroads to do well by the country unless they can do well by themselves. Men do not part with their money unless they are reasonably sure of getting it back, and with a profit. The only transaction worth while is one that is mutually profitable. This is the golden rule of modern business, and anyone who tries to take profit out of the hide of the other fellow is riding to ruin."

THE STORY OF THE PIONEER S. P. DIVISION.

alarm of fires. This precaution is in addition to the complete fire alarm system, lookouts and patrolmen within and on top of the sheds. The operating difficulties of moving trains over a single track staff territory is apparent, and, as a shed fire burns at the rate of 100 feet a minute, time saving is therefore the essence of efficiency.

What Stevenson Said.

When blizzards and slides affect the line all efforts are concentrated on the task of keeping the line open for traffic. Summer and winter, night and day the work goes on—the freight is moved and the passenger speeds upon his way, little realizing the army of men who have made possible the comfort and safety of his journey, but we trust he will say as Robert Louis Stevenson said:

"We changed cars from the Union Pacific to the Central Pacific line of railroad. The change was doubly welcome; for, first, we had better cars on the new line. The cars on the Central Pacific were nearly twice as high and so proportionately airier; they were freshly varnished, which gave us all a sense of cleanliness, as though we had bathed."

Tonnage trains of 5000 Ms leave Roseville with Mallet engine ahead and a helper engine cut in twelve or fourteen cars ahead of the caboose and make the first lap to the snow line of the Sierras to Colfax, thence operating along the double track alignment of the original Central Pacific from Colfax to Blue Canon. This section of second track was completed in 1914. In this construction operation was simplified through building Tunnels Nos. 33 and 34, which eliminated the hazards encountered at old Cape Horn, one of the historic construction features of the early builders. Trainmen on their first trip through these tunnels dubbed the change "The Panama Canal."

"Longest House in the World."

Blue Canon, at the eastward end of the double track, finds our train at an elevation of 4693 feet, encircling with its forty-five car length this station well known to every traveler on our line. Here our train heads into what the enginemen term "The Longest House in the World"—the snowsheds. From Blue Canon to Summit (elevation 7017 feet) twenty-six miles of single track on a maximum grade of 2.2 per cent. The helper is cut out at Summit and train descends the 2 per cent grade into Truckee and thence on double track to Sparks.

This mountain district, over which a monthly tonnage of 150 million gross ton miles is handled, is as unique from an operating standpoint as is the history connected with the route. Iron rail of the early days gave way to light steel, which in turn gave way to the heavy rail, ties and rock ballast to

withstand the ever increasing weights of power and equipment which also necessitated an increasing dimension of tunnels.

Train dispatching on the Sierra Nevada Mountains is done by telephone; the dispatcher is in direct communication with each staff office. Our staff system is the safest, surest and quickest means of dispatching trains that has yet been devised.

The growth of the Sacramento Division is reflected by the following comparisons: When the Central Pacific started operation of transcontinental trains it had a total equipment of eight locomotives, eight passenger coaches, four combined mail and baggage cars, thirty box cars and thirty platform cars, all of small dimension. This would provide equipment for only one passenger train and one freight train of today. Combined it could be hauled over the Sierras with one Mallet engine. An army of 4,500 employes is required to serve the 260,000 people residing on the main line and branches of the Sacramento Division.

Old records show the payroll in 1867 was approximately $6750 per month, while the division payroll today for one month is $675,000, which would total in a year an amount equal to the entire sum for which the Central Pacific Railway was originally capitalized.

And yet we are growing!

HERE'S A NEW WAY TO CATCH YOUR TRAIN.

An automobile and an airplane were pressed into emergency service by Dining Car Conductor W. Shields on Train 49 from Los Angeles to San Francisco recently when he found himself left at Fresno. Shields was securing change in the ticket office when his train pulled out.

A chase of two miles in an automobile proved futile. Just as the conductor was about to give up he saw an airplane preparing to ascend from a nearby field. He hurried over and explained his predicament to the pilot and a woman passenger.

"Hop in," they said, "we'll get you there!"

The airplane raced over the train. Shields waved frantically.

The engineer peered out of the cab window and responded but he did not recognize the conductor, nor did he slow down. The race ended at Chowchilla, where the train stopped and Shields boarded the diner in time to attend the breakfast service.

The life of stamp pads is indefinite if properly cared for. They should be scraped periodically with a penknife or any other instrument for the purpose of removing dust and dirt and then re-inked, using only enough ink to make clear impression with stamp. Old stamp pads should be turned in in exchange when obtaining new.

BOY, GET A PLUMBER—OUR LOCOMOTIVE HAS A LEAK

But That Was In The Early Sixties Before The Sacramento Shops Were Founded. Now Monthly Payroll Is $350,000 And Building Locomotives Is Only a Side Line

By D. S. WATKINS
(Superintendent of Shops, Sacramento, Cal.)

"Boy!"

"Yes, sir."

"Run across the street and tell the plumber our locomotive is leaking."

"Yes, sir."

The above colloquy would sound rather strange if uttered today in the great shop plant of the Southern Pacific at Sacramento with its thousands of busy workmen; its monthly payroll of $350,000; its tall chimneys pouring the smoke of a vast industry toward the California sky. But there was a time, 1863 or thereabouts, when the Central Pacific Railroad had but one engine, and when anything went wrong with it help was obtained from the machine shop and foundry in front of which the original locomotive was assembled, much to the edification of the natives.

The first engine on the Central Pacific Railroad, "Governor Stanford" No. 1, was set up on I Street in the city of Sacramento in front of the shop of Goss & Cambert in the year 1863. The depot at that time was on Front Street, where the present freight shed is located and the main line turned onto I Street, as far as Sixth Street, then ran on Sixth Street to what was known as the B Street Levee. This track was later run due north from the depot, going around the present location of the shops onto the B Street Levee.

The first shops of the Central Pacific were located in October, 1863, at what is now Sixth and H Streets. They were moved a year later to Sixth and D Streets, and consisted of a small roundhouse with a machine and blacksmith shop and small car shop.

The present roundhouse was built and finished in 1868. The permanent shops, which have since been extended considerably, were occupied in September, 1869. Since that date various additions have been made. The capacity of the machine shop was doubled in 1872, and another section added in 1888. In 1904 its capacity was again doubled by adding on another bay which comprises the present erecting shop.

In the early days the firm of Goss & Lambert, who had a machine shop and foundry on I Street, did considerable work for the Company, and in the latter part of 1868 the firm was bought out by the railroad, which moved the machinery into new shops.

Engines Known by Names.

The early engines were known by name as well as by number. The

No. 1 was the "Governor Stanford," and had cylinders 16 inches in diameter by 22-inch stroke and had a total weight of 28 tons.

The No. 2 was the "Pacific." It was a 16 x 24-inch engine and had a total weight of 30½ tons.

The No. 3 was the "C. P. Huntington" and had cylinders 11½ inches in diameter by 15-inch stroke. This engine had only one pair of drivers, 54 inches in diameter, and had a total weight of 10 tons. The number of this engine was later changed to Southern Pacific No. 1 and was exhibited in the P. P. I. E. alongside of a Mallet Consolidated engine weighing 218 tons. This engine has now found a resting place in the plaza between the Sacramento depot and the railroad bridge, where it may be seen by passengers traveling over our line.

The No. 4 was known as the "T. D. Judah," and was the same make and size as the C. P. Huntington.

The No. 5, called "Atlantic," was a Mason eight-wheeler with 15 x 22-inch cylinders and weighed 29 tons.

The No. 6 was the "Conness." This engine was a Mason ten-wheeler with 17 x 24-inch cylinders and weighed about 35 tons. It was later known as engine 1509, then as engine 2000, and is at present doing good work at the State Prison at Folsom.

A great deal of work was put on the earlier engines in the way of embellishment. The jacket bands were of brass and were kept polished, also the handrail brackets, and often the handrails were of this material. Sometimes the dome and sandbox were of sheet brass, and as all these parts were kept polished by the fireman they presented a very attractive appearance.

Sartorial Splendor.

The painter also had his innings, and vermilion, green and other colors tended to improve the looks of the motive power. On the sides of the tenders they often painted pictures which corresponded to the names of the engines. The "Gorilla" had a picture of that animal painted on either side of the tender, while the "Black Fox" and "White Fox" and others were also illustrated in paint. Some very artistic work was often put on the sides of headlight case, which was considerably larger than the ones at present in use and offered a vantage

ground for the work of the artist. At the present time our locomotives are very plain and severe; brass is not used for ornamentation but only where its use is more practical than iron or steel. A plain coat of black paint is considered sufficient and no decoration is allowed.

When the Central Pacific was first completed and ran from Sacramento to Ogden wood was the universal fuel. The Sierra Nevadas were covered with pine forests and from these forests fuel was procured. All engines for a great many years were wood burners. They were later turned into coal burners as wood became scarce and coal was more plentiful. Now all locomotives owned and operated by the Southern Pacific Company burn fuel oil.

Shop Site a Swamp

The present site of Sacramento shops was originally a swamp and filling at first was only prosecuted to take care of the needs at the time. The swamp was filled up from year to year as more ground was needed until the last of the railroad's property was entirely reclaimed last year. In the earlier days this filling was all done with teams. Later the dirt was brought in on cars loaded by hand, then steam shovels were brought into play to load the cars. During the last ten years suction dredges have deposited the material.

From a small beginning, employing only a very few men, Sacramento shops have attained the proportion of a small city, and constitute by far the most important industry in the State's capital.

As a result of the determination of the early owners, who were a long way from the source of supplies, to do a great part of their manufacturing, foundries, both for making gray iron and brass castings, and a rolling mill rounded out the complement of shops. We are now finishing up a steel foundry which is equipped with a six-ton Herroult electric furnace.

The payroll of the shops has increased from a few hundred dollars to $350,000 per month and the output of the shops has likewise grown.

In addition to the repairs made to locomotives the shops have built and put in service during the last two years thirty-six locomotives. A large number of flat, box, gondola and automobile cars were built during the war period, and we hope that more of them will be manufactured here. We also believe that the service given by the locomotives built in the company's

These photos tell their own story. One was taken in 1872 and shows the beginning of the Sacramento General Shops and the marsh which was later reclaimed. The other picture shows the major portion of the modern plant with the new steel foundry in the middle distance. The new frog shop was too far to the left for the camera to include in the one picture.

shops has been such that the placing of future orders for this equipment on the shops will be seriously considered.

In addition to the work of repairing and building locomotives and cars an immense amount of other work is done at Sacramento. Track work, from the repairing of picks to the building of the most complicated crossings, is taken care of in our frog shop, which was erected two years ago. The building of this shop relieved the machine shop and blacksmith shop to a very considerable extent and made it possible to go ahead with the construction of locomotives.

Only 40 per cent of the payroll is used on repairs to equipment, the balance going into general work on different lines. Much locomotive, car and track material is produced at this point and shipped to all points on the system.

What the Shops Produce.

During the year 1919 the following work was accomplished at the Sacramento General Shops:

18 new locomotives constructed.

25 new locomotive tenders constructed.

371 locomotives overhauled.

833 passenger cars overhauled.

17,115,200 pounds gray iron castings manufactured in the foundry.

59,077,545 pounds of finished bar iron manufactured in the rolling mill.

71,434 car wheels manufactured at wheel foundry.

3,545,494 pounds brass casting manufactured at brass foundry.

641 track frogs manufactured at frog shop.

32,967,970 pounds of billetts, slabs, tracks and other bolts, angle and tie plates.

The above list represents only a small part of the activities of the Sacramento shops, and with the increasing business of the company and the demand for greater movement of freight and passengers we look forward to the doubling, if not the tripling, of our output in the near future.

SOUTHERN PACIFIC TO ENTER PORT OF HOUSTON.

L. J. Spence, traffic director for the Southern Pacific with offices at New York, recently announced to a gathering of prominent business men at Houston, Texas, that this company will enter the port of Houston with a steamship service if Houston wishes it, and that we are now building at a cost of $4,500,000 three vessels suitable for the trade. There will be two types of vessels. One will carry 4000 tons of cargo at nineteen-foot draft and the other 600 tons at a twenty-one-foot draft.

The Houston "Post" comments editorially upon the matter in part as follows:

"The decision of the Southern Pacific to enter the port of Houston, announced by Traffic Director L. J. Spence of New York, marks another important step in the development of the port. The company is having constructed at a cost of $4,500,000 three vessels for the Houston trade.

"This announcement is welcome for several reasons. It brings to the port one of the greatest and strongest corporations engaged in water transportation, co-ordinating its rail and water facilities, thus assuring for the port opportunities for trade expansion that

scarcely any other immediate connection could give.

"Not only that. It removes the last possible cause for friction between the Southern Pacific and the commercial interests of Houston and assures that the ancient friendship between this city and its greatest transportation interest will not be impaired or broken in the future.

"Houston owes much of its progress and commercial consequence to the Southern Pacific. Its lines radiate from this city in eight directions, making it the most important distributing point of the system between New Orleans and San Francisco. In appreciation of the great part the system has played in the development of Houston the commercial interests of the city have lavished a vast patronage upon it. It has dominated both inbound and outbound freight and passenger traffic. Its officials and employees, all told numbering into the thousands, form a considerable percentage of the population, as well as of the social and business life of the city."

In men whom men condemn as ill
I find so much of goodness still,
In men whom men pronounce divine
I find so much of sin and blot:
I hesitate to draw the line
Between the two where God has not.

                        —Joaquin Miller.

OLD TIMES RECALLED AT SACRAMENTO.

By GEO. DUNN

The first car shop at Sacramento was located at Sixth and E streets, about 1863, and was so used until the quarters becoming crowded or cramped, a more pretentious building was erected near the site of the present buildings, and this has been known for fifty years as the "Old Shop."

The shop at Sixth and E, after being abandoned as the car shop, was used as a drying house for lumber until about 1883, when it was destroyed by fire.

The site of the present shops was selected about 1865 or 1866, and the present buildings erected under the supervision of Arthur Brown, civil engineer, with Joseph Wilkinson as assistant. In 1868 the shops were ready for occupancy, and the old shop was therefore abandoned and thereafter used as a storeroom for finished lumber for the freight department.

The early years of the car department were confined to the building and repairing of flat and other freight cars used principally in road construction, and not until 1874 was any attempt made to construct other than this class of cars; in that year, however, an order was placed by the directors of the company with the car department for the construction of twenty-four first-class passenger coaches. They were built under the supervision of Benjamin Welch, master car builder, J. A. Seamen, general foreman, but more particularly under the immediate care of Victor Limey, cabinet shop foreman, and James Young, foreman of construction. They were considered at that time, both from design and workmanship, as neat and well built coaches as any railroad company had in service.

And this can also be truthfully said for the builders, that while perhaps coaches with much more elaborate furnishings, and, therefore, to the eye much more attractive have since been built, none have given longer and better service than the twenty-four new coaches; a name which still clings when any reference is made to them by the older employees of the car department (and there are a few left, the writer being one of them.).

One of these coaches was converted into a private car for the late A. H. Towne, who carried the title of superintendent. Of course, it was a modest, inexpensive coach, but perfectly in keeping with the taste of the modest, able and very much respected gentleman who occupied it.

In 1876 the car department was called upon to furnish a lot of bobtail horse cars for use in San Francisco, and in this particular branch of car building proved itself adequate to the demands made upon it by furnishing cars of this type equal to those built by any of the Eastern shops specially equipped for this class of work.

About this time there were also built a large number of coaches known as "Oakland Seventh Street Locals." They were built with a very light framework, it being desired to eliminate weight as much as possible, and so rigidly was this idea adhered to that the builders were somewhat doubtful as to whether safety had not been sacrificed to the elimination of weight. But the years of daily and very strenuous service to which these coaches were subjected proved conclusively that the draftsman (the late Fred Schnauss) had fully and correctly worked out the difficult problems as to strain and strength of material.

In 1883 the "Stanford car," for Mr. Stanford's personal use, was built. No expense was spared in its construction, consequently the very best materials the market afforded were used, and the best mechanics obtainable were employed in its construction. Every convenience the most fastidious could desire was provided, and the compartments consisted of parlor, bedrooms, dining-rooms, porter's room and kitchen.

The parlor was finished in rosewood and the balance of the car in mahogany, and as every detail had received the very closest attention it necessarily follows that when this very beautiful private car was finished it was pronounced by disinterested inspectors to be the acme of coach construction.

The building of all classes of cars needed by the company followed in rapid succession, and among others were a large number of cable cars built in 1886 for the Market Street and Hayes Valley Railroad in San Francisco, demonstrating beyond the shadow of doubt that the employees of the car department, officered by able and efficient foremen, were equal to any emergency or demand made upon them.

EX-SOLDIERS AND SAILORS ENTER TUCSON DIV.

By H. A. MACLACHLAN
(Superintendent's Office, Tucson Division)

Recently the Tucson Division has been returning to service quite a number of discharged soldiers and sailors, also many who had not worked here previously. The following are old employees who have returned to the service: Miles Baughn, F. A. Ragin, A. R. Provencio, E. E. Elias, M. G. Cochrane, W. P. O'Neil, J. H. Locker, E. E. Peterson, Wm. A. Acunna, L. Y. Pironi, T. E. Pryor, B. F. Pounds, Joe Oatis, H. F. Stapp, F. B. Romero, H. L. Miller, H. A. Moores, J. Almada, F. Urguides, W. B. Horton, M. Quichus, G. R. Michaels, T. A. Derr, J. J. Lurchan, E. R. Menitz, W. H. Kulsmire, R. B. Cook, J. A. P. Strout, Roy Williams, J. W. Lumley, A. B. Roark, L. M. Saunders, I. C. Allen, C. Bouchet, T. E. Gilliland, F. O. Elias, M. L. Mourey, A. H. Isaacks, C. O. Bartlet, H. F. Wills, A. E. Barral, D. L. Duke, J. B. Kelly, F. S. Whitson, J. B. Kitchens, A. C. Williams, A. C. Woods, H. S. Hall, T. L. Vanleer.

In addition, the following returned soldiers and sailors have entered our service for the first time: Brakemen, 12; boilermaker helpers, 3; boilerwashers, 1; batterymen, 2; callers, 1; clerks, 5; carpenter helpers, 7; firemen, 1; machinist helpers, 3; painters. 2; pipefitters, 1; signal maintainers, 1; switchmen, 1; signal helpers, 1; wipers, 3; water service helpers, 4; truckers, 3.

During the early part of June four Mexican extra gangs were released from this division, three being shipped to the Shasta Division and one to the Salt Lake Division.

On May 12th the end of double track was extended to east end of Polvo and put in service, and new signals operated between Stockham and Polvo.

New members have been appointed on the Safety Committee: E. P. Allen, assistant signal supervisor, succeeds Mr. Board; Mr. C. J. Greenwell succeeds Mr. E. Chamberlain; Mr. T. Smith, section foreman, succeeds Mr. P. Grimley; Mr. E. M. Neely, engineer, succeeds Engineer Sarrells; Mr. C. W. Littlefield, fireman, succeeds T. M. Miller; Mr. L. W. Harter, yardman, succeeds S. B. Homer; Mr. J. F. Brown, agent Casa Grande, succeeds J. H. McClure; Mr. B. M. Montgomery, conductor, succeeds W. C. Lancaster; Mr. H. F. Brown, brakeman, succeeds T. W. Donnelly; Mr. R. D. Hampson, W. S. repairman, succeeds F. Lauth; Mr. Ted Goff, B. & B. foreman, succeeds Mr. Jerome; Mr. Robert Cockburn, assistant boilermaker foreman, succeeds Mr. McAllister. In addition, Mr. W. H. Williams, machinist; Mr. C. J. Feeley, car repairer; Mr. E. R. Beck, boilermaker, have been appointed to serve as members.

Assistant Signal Supervisor Sellick, after several years' service without a vacation, is taking a month's leave of absence and visiting in the East.

During the month of May and first ten days of June, this division handled 2650 cars of cattle, this being the largest stock movement this division has handled for a number of seasons. All this was handled without a hitch or delay, and without a cross haul of a single empty car. Cattlemen co-operated with chief dispatcher and car distributor and made this showing possible.

The Garbanza crop from Sinaloa, Mexico, is beginning to move, and this business will amount to about 4000 tons, most of which is moving to Cuba and Spain. This movement will continue for about thirty days.

The Salt River Valley cantaloupe crop has started to move. This business will require approximately 500 cars to handle. In addition, the Yuma and Toltec territory will ship about 300 cars of cantaloupes, which is the largest crop of this kind ever moved from this country.

The mines are opening up again and bullion is moving to New York via Sunset Gulf from the Globe district and Cananea, Mexico.

THE BULLETIN

SACRAMENTO EMPLOYES RECALL 'DAYS OF '49'

Grouped around the famous old "C. P. Huntington" are several whose efforts made the Sacramento celebration a success. On the engine, from left to right are: A. D. Williams, Gen'l Sup't Motive Power; Tom O'Connell, Pensioned Engineer; J. Uren, Pensioned Foreman; J. Lonegran, 1st Engineer on "Huntington;" Wm. Norton, Ass't Gen'l Yardmaster; W. J. McGee, Fireman. In front are: H. C. Venter, Gen'l Foreman; H. E. Diggles, Chariman "Days of '49 Celebration;" T. F. Bellhouse, Roundhouse Foreman; G. A. Knoblanch, Gen'l Foreman; T. Ahern, Sup't Sacramento Division; C. E. Speer, DF&PA; C. C. Fisher, Trainmaster; J. S. Spelman, of Sacramento; B. W. Spanger, Engineer; C. L. Seavey, Sacramento City Manager; B. J. Byrne, Sup't Sec.; Miss Vivia Hemphill and H. M. Smith, press representatives. BELOW at left are the Sacramento employes dressed in "Days of '49" regalia. Insert is of J. H. Dyer, General Manager, Southern Pacific, at right; and C. E. Virden, Pres. Sacramento Chamber of Commerce.

SOUTHERN Pacific participation in Sacramento's "Days of '49" celebration May 23 to 28 contributed materially to the success of this event, one of the most distinctive ever staged by an American city.

Employes of the Company, the men wearing the now world-famous whiskers of pioneer days and the girls and women attired in the quaint costumes of the period, joined with the rest of the citizens of the city. The mining town of '49 was revived and built upon Southern Pacific land, and a big feature of the the ancient locomotive "C. P. Huntington No. 1," under the direction of Thomas O'Connell, first fireman and now retired Company employe at Sacramento, and J. E. Lonergan, 81 years old, former Central Pacific engineer and now wealthy Philadelphia manufacturer.

The whole city lived for the week in days agone. Picturesque stage coaches replaced street cars. Log fronts on places of business, red shirted, bewhiskered waiters in all the restau- rants, bewhiskered bell boys in hotels, tall-hatted, silk-vested gamblers of the mining camps, miners with high boots and shooting-irons and throngs of girls and women wearing pantalettes and hoop-skirts lent color to the scene.

Pioneer Engineer on Run

The old "C. P. Huntington" engine, after having been on exhibition for many years, was fixed up by General Shops to operate under its own steam over a schedule made by Super-

Fair employes at Sacramento as they appeared for the "Days of '49" Celebration. Top row—Signa Kimball, Alice Swinney, Alice Beavers, Pearl Douglas, Miriam Ruhkala, Edna Maltby, Magdaline Sonntag, Maud Wright, Marie Williams, Adaline Rosebush. Bottom Row—Helen Taylor, Maurie Fisher, Clara Cramer, Nellie Walters.

intendent Ahern and District Freight and Passenger Agent Spear, from Sacramento to Brighton and return, hauling on a flat car, fixed up as an observation car during the six days of the celebration, 2,864 passengers. The old engine nobly responded and no difficulties were met with in the way of engine failures, such as sometimes unfortunately occur to her larger improved relatives of the present day.

Although the only free transportation honored was that held by our worthy pensioners, many Southern Pacific employes to whom the riding of trains is not a novelty, paid their fare and availed themselves of the opportunity of making this trip. The General Manager was required to pay fare for himself and party and one of our high Traffic officials who endeavored to slip by without payment of fare wa promptly ejected from train as an example to others.

Employes in Novel Dress

For many days before and during the celebration, the local offices looked like a scene from one of Bret Harte's novels, the fair stenographers and comptometer operators blossoming out in hoop skirts and crinoline, with bonnets and accompanying paraphernalia, while the male employes—who could —grew beards of various styles, wore red shirts, high boots, etc., and presented a truly ferocious aspect. Visiting railroad officials and friends from other locations had difficulty in recognizing their acquaintnces.

The sand lot next to the station was given over to the mining camp and the town grew up seemingly over night, like the camps of old, into mixture of dance halls, general stores. and replicas of conditions in the early days.

The Whiskerino Club and contest for the longest whiskers mentioned in the press all over the country had many Southern Pacific employes, and it is worthy of mention that J. B. Frees, rodman in the engineering department, won second prize for the longest

whiskers grown from March 17th to May 28th, out of a field of over five thousand participants.

Shop Men Build Float

Shop employes built a float, under the supervision of Master Car Builder Hall, representing mountains and streams with characters to take the part of miners panning gold and throwing out souvenir nuggets to the spectators, which participated in the allegorical pageant. This float was followed by about 500 employes garbed in their various costumes and whiskers and received second prize for the largest turnout and representation of the period.

The increased travel on account of the celebration required operation of many additional trains each day all of which were handled without complaint or personal injury, demonstrating the ability of the S. P. employes to render one hundred percent service and at the same time actively participate in the festivities.

Mr. Lonergan is now wealthy and is the president of three large corporations, namely, the California Vineyard Association, the largest tokay vineyard in California; the John M. Lonergan Company, manufacturers of safety appliances, and of the H. Brinton Co., manufacturers of knitting machinery for making hosiery, underwear, neckties, etc.

The old pioneer has always had a warm spot in his heart for the Southern Pacific Company and Mrs. Lonergan, who accompanied him here, said that she believed that no service was to great for him to perform for the Company, so warm was his affection for his associations of many years ago.

Early History Told

In speaking of the Huntington and his connections with the earlier years of railroad life in California, Mr. Lonergan said:

"My first position here was as machinist in the shops of the Central Pa-

cific Railroad, but being an engineer, I was soon sent on the road, and the little engine Huntington was one of the first engines that I ran after coming out west.

"I never ran the engine for commercial purposes, except switching cars in the yard. The Huntington was used for taking officers of the road out for inspection of the road, and for special runs for the officers of the road, never having more than one car attached.

"The engine was used on all Divisions out of Sacramento. At that time one Division was from Sacramento to Oakland, and another from Sacramento to Marysville, and the Hill Division was from Secramento to Truckee.

"The Huntington had a speed of about forty miles per hour on level ground, which was considered excellent in those days.

Rough and Ready was the Early Day Motto

Railroading a half century ago on the Sacramento Division of the Central Pacific took brawn as well as brain

By T. R. JONES, former Supt. Sacramento Division

I was employed as a telegraph operator on the Sacramento Division of the Central Pacific Railroad on May 1, 1872.

In Sacramento at breakfast time on that date I had no more idea of becoming a railroad man than of becoming a tragedian.

As frequently happens in the life of a young man the aim and ambition of his future life is changed by the unexpected meeting and remarks of another person and he is sent in another direction. Probably it was opportunity knocking at my door.

The position of postmaster in the Assembly of the State Legislature, which I obtained through the political influence of a friend, had become no more by the session coming to an end.

I was looking for employment. An application I had made to the manager of the Western Union office had been turned down and I was trying to get in touch with "Van." He was J. L. Vandenberg, superintendent of the A. and P. Telegraph Company, to whom I intended to apply for a job. "Van's" office was in his hat. If you could find that you would find him in. He seldom removed his office from his head. I heard he had a large tumor on the top of his head. That maybe was the cause. He was constantly on the go between Oakland and Ogden looking after wire trouble and was hard to locate.

On J Street that morning I met a State official to whom I mentioned my intentions. He informed me that he was well acquainted with Colonel J. W. Bowen the superintendent of the Sacramento Division of the Central Pacific, and that he had just come from the depot. While there he had overheard the dispatcher inform Colonel Bowen that the division was short two operators. I immediately went to the depot at Front and J Streets.

It was a long rambling one-story wooden shed-like structure nearly a block long. At its north end, next to the bar and lunch counter, were two offices, one for the superintendent and the other for the telegraph office, both with their doors open to the public.

I found Colonel Bowen a very affable man and after stating my business was immediately turned over to the chief dispatcher, or rather the dispatching force, which was Ned Hartwell.

It took him about three minutes to

Responding to our request, Mr. T. R. Jones, formerly superintendent of the Sacramento Division, has written an article for THE BULLETIN *giving his interesting experiences in the old days of railroading on the Central Pacific. These reminiscences are so well written and comprehensive that the article will be given in three installments, the latter appearing in the September and October issues. Mr. Jones has furnished a vivid picture of the early days and* BULLETIN *readers will no doubt read this and the subsequent issues with pleasure.*—THE EDITOR.

hire me, take my name, write out a pass and direct me to take the train to Rocklin to work there as night operator.

I did not have to make out a personal record, furnish a tintype, undergo a medical examination, nor state my experience as a telegrapher. I suppose it was taken for granted that I could fill the place or I would not apply for it.

Made Good on Nerve.

Some years later, when superintendent of the division, I sometimes admired the nerve I must have had to apply for a job as a railroad telegrapher. I was little more than a lad in my teens; there were signs of an incipient mustache on my upper lip that has since become an array of gray bristles; I had never been inside of a railroad telegraph office; never had heard the "27" of a train nor the "13" to a train order, and did not know such things existed; I had traveled less than 500 miles on railroad trains and had no idea of the methods by which they were run, and my experience as telegrapher was limited to a few years in a Western Union office in a small mining town, where an average of a message a day would make a big month of business. I frequently made six dots instead of five for a P, which is the attribute of a "plug." But I found out later that expert handling of key and sounder did not count for as much as reliability. The operator always on deck and answering his office call promptly stood a better chance of holding a job than the expert who could handle "A. U. B.," as a press report was then called.

I was on hand when the train arrived early in the afternoon. I em-

phasize the term the as used by Dispatcher Hartwell, for it was the only passenger train going East daily.

Only one passenger train each way was run daily over the division. This was composed of mail, baggage and express cars, sleepers, smoker and coaches. The sleepers were run ahead of the coaches. All the cars were painted yellow.

The two passenger trains were through trains, but made stops at all stations and did all the local work. Engines assigned to these runs were, as compared with the Mallet of today, pygmy eight-wheelers. Two crews were assigned to these trains between Sacramento and Truckee. Up one day and down the next. The second-class passengers were then called emigrants and rode in emigrant cars, somewhat better than stock cars, attached to the rear of through freight trains Nos. 5 and 6.

Bar Most Popular Spot.

All freight trains carried passengers in their cabooses. The depot at Sacramento had a long bar and short lunch counter. A large number of passengers got off on the arrival of the train, most of whom went to the bar to get a drink while a few purchased and ate sandwiches.

The probable cause of this was the passengers had dinner, as the noon-day meal was then called, at the Lathrop eating house. As the west-bound passengers took dinner at the Roseville eating house before arriving at Sacramento, this station was not a regular eating house, hence the limited accommodations. It was managed by Jack Biderman, a political friend of Governor Stanford, and like a number of others was probably given a profitable privilege on that account.

While waiting for the train to start an amusing incident occurred.

On account of its animosity and the personalities published against the company and its directors, the Sacramento "Union" was not allowed to be sold by newsboys at the depot. A five-board fence along the platform extending to I Street kept them off the north end of the depot while Special Officer Fred Burke with a heavy cane patrolled and kept them from the south end and from climbing the fence, which they frequently did when he was out of sight. The passengers were anxious to buy and read the paper and I saw a couple of them beckon to "Hash," Harris Ginsberg, afterwards an attache of the superintendent's office, one of the news-

32

THE BULLETINTHE BULLETIN

boys looking through the fence for a paper. He could not resist the opportunity to make a dime or two so clumsily climbing over the fence, he took a chance. He had just made the sale when Fred Burke came in view, moving northward.

"Hash" made a dash for the fence, which attracted Burke's attention, and he made a rush too. Just as the largest and heaviest part of "Hash's" anatomy rose up to go over the top board it received a resounding whack from the officer's cane that caused a roar of laughter from the onlookers.

I got on the train when the conductor, Captain G. T. Witham called "All Aboard" and standing on the coach platform saw him gracefully step on after the train was in motion. The last person to speak to him as he signaled with uplifted hand the engineman to start was a young man named Harry Breckenfeld, who came out of the superintendent's office with a letter, which he handed to the conductor with a few words of information. As I saw the conductor hand this letter to John A. Muir, the agent at Rocklin, on arrival there and heard him say there was a night operator on the train for him. I supposed the letter referred to me, hence my recollection of the incident.

I learned afterward that young Breckenfeld, then in his teens, was the chief clerk, or in fact the whole clerical force, of the superintendent's office. That he made up the payrolls, handled the correspondence (in long hand), conductors' reports and supplies; kept a record of every car moved on the division and besides being a telegrapher performed numerous other duties essential to the business at headquarters. The force there then consisted of a superintendent, clerk, train dispatcher, a day and night operator, a janitor who was ex-officio messenger boy and a night watchman who fired up the wood-burning stove in winter for the night operator. Captain Witham was not in uniform. He wore a slouch hat and a gray suit of clothes. He had side whiskers and a military demeanor that well fitted his title. He seemed to know every man on the train by his first name and where he was going before he took up his ticket. The upper pockets of his vest were bulging with cigars of different sizes and brands. He refused to accept a dozen or more offered him by the passengers, who saluted him as "Cap," giving as a reason he had enough to last him until he returned from Truckee.

No Ban on Smoking.

I afterward saw one of the other "old reliable" passenger conductors pass through a coach while taking up and punching tickets while holding a lighted cigar between his teeth, which he did not remove to either expectorate or flip off the ashes or talk while in the coach.

I learned subsequently that Captain Witham on account of his courteous

manner, strict attention to duty and his success in getting his trains over the road on time, stood high in the estimation of the officials of the company and the traveling public.

I went on duty at 6 P. M. to be off at 7 A. M. I received no instructions from the agent, he probably thinking I would not have been sent there if I needed any, and I, in my ignorance and egotism, did not think I needed any. Fortunately for me only one train passed through Rocklin that night and I had no train orders to handle.

When the conductor of the train handed me the "soup ticket" showing No. 8 arrived and departed on time with a stated number of cars I wired as a message every word from beginning to end on the ticket to Sacramento and received in return the unsatisfactory comment of "Ha ha! Where in h—— did you come from?"

It was the duty of the night operator at Rocklin to lock and seal and make a report to the freight auditor at Sacramento of the cars of merchandise arriving on the way freight from the East when it came in after 6 p. m. If it arrived before that time the work was done by Dick Moore, the office boy—now a veteran passenger conductor. The second night I was there the train came in late, so, with the ladder, seals and sealing iron, I essayed to do the work.

Just as I reached and prepared to seal the first car an engine coupled on the east end and away went the train to the other end of the yard. I saw it stop, and, thinking it was going to get away with the cars unsealed, I hastened after it. Just as I reached it away it went back to where it had been taken from.

I went after it again and then I was informed it was being switched. When it was made up I obeyed instructions to lock and seal all cars that did not contain lumber or wood. I did not take a list of the box cars I sealed, but, knowing that every one except those loaded with lumber or wood had been sealed, I took the yardmaster's record of the train and made the report for the freight auditor from it. Unknown to me the yardmaster had added about ten flat cars loaded with granite to the train list and I duly copied their numbers and reported them as locked and sealed. Had I known then that the box cars carried even and the flat cars odd numbers I would not have made the error. When the next day the freight auditor returned the report for correction and with it a sarcastic letter I had an opportunity to tell J. A. M. how little I knew about railroad business. He kindly and patiently took me in hand and gave me such instructions how to correctly perform my duties that I made no further errors and had no further trouble. Shortly afterward I was transferred to Truckee. On this, my first trip over the division to and my sojourn at Truckee, I learned more from observation of the practical operation of trains, from contact and conversation with my fellow employees in

the different departments of the work than I before had considered it possible for any one man to know and that railroad service was a serious and arduous vocation.

When the Central Pacific was being constructed it was short of funds. To save the expense of erecting depots at the important stations a plan was devised to let others do it. To a number of popular caterers was given the privilege of building the depots and maintain therein a barroom and eating room and then partition off, usually at one end, rooms for the company to use as a ticket and telegraph office and a waiting room. This privilege was given at Roseville to J. R. Watson and a former French chef, Louis Bulens; at Rocklin to Dana Perkins Sr.; at Pino to James Loomis; at Auburn and Colfax to Curley & Mahon; at Alta to Senator Bonvard; at Blue Canon to L. G. Peterman; at Summit to Jim Cardwell and at Truckee to John F. Moody. The depots at Alta, Blue Canon, Summit and Truckee were also hotels. I have understood this arrangement in subsequent years caused the officials of the Southern Pacific Company much trouble when they tried to close up the barrooms and stop the selling of liquor at the stations. No written deeds or leases existed, as the agreements were verbal ones made by Charles Crocker, superintendent of construction, with the depot owners. As possession is nine points of the law, no legal means were found to remove them. Finally, one by one, most all of them were burned down and the company, refusing to allow them to be rebuilt, solved the problem.

The wood sheds west of Blue Canon were located, evidently by design, at a proper distance east of these depots, so that while the engines were being wooded up the passenger cars stood in front of the depot. Thus the owners had an opportunity for the passengers to loiter and refresh themselves in the barrooms, while away the time until the engines were ready to start and not kick at the delay. Everybody seemed satisfied and we who remember the old-timers of California and Nevada who traveled on the trains in those days know that an oasis on the line of the road could not be of too frequent an occurrence.

I mention this condition so that the long stops at these stations on my trip to Truckee will be understood.

Leaving Rocklin in the afternoon the first stop was at Pino, a few miles east. The name of this station has since been changed to Loomis on account of its similarity on way bills to Reno, causing overs and shorts of freight. The Sacramento engine was wooded up here. The wood in the Pino shed was cut in that vicinity and hauled up by teams to the shed.

(Mr. Jones' article will be continued in the September number.)

THE BULLETIN

Days When Hand Brake Setting Was An Art

Regular contests were held by train crews to determine the strongest brakemen on the run

By T. R. JONES, Former Superintendent Sacramento Division

In the August issue of THE BULLETIN *Mr. Jones told of his experiences as a telegrapher in the early days. In this article he relates other interesting details of the way trains were moved over the mountains and the means used to meet the many obstacles nature had thrown in the way of the pioneers.*

At Colfax a large number of passengers got off to take stages for Grass Valley and other Nevada County towns, while a large number entered the depot. Here they were smilingly welcomed by big Dennis McCarthy. He knew nearly everybody by his first name and everybody seemed to know him. Busy as a bee, he was passing out glasses and bottles and taking in the money. From his appearance and the way he was greeted I considered him the biggest and most popular man I had seen on the division.

Alta was the next stop, where the passengers had supper and the engines were wooded up. "Senator Bon" appeared to be as popular as was Big Dennis. Lou Bonvard, then a curly-headed youth wearing a military cap, son of the Senator, now a veteran telegrapher at Sacramento, was an operator there.

Alta, during the season, was a summer resort for a bevy of school-marms and he had something of a Don Juan reputation. I heard he was occasionally laid off for going on trains to Sacramento without getting permission from the train dispatcher and a pass. After being introduced as one of the C. P. family by Conductor Marshall I had supper, for which no charge was made because I was a C. P. employee traveling. After supper I went up to the head end to view the work being done there. I saw half a dozen passengers cheerfully helping the crew to wood up. They were throwing the sticks on the tenders faster than the firemen could tier them, and one asked for help from the wood pitchers. I volunteered, and like all volunteers I soon found myself more in the way than efficient.

Barney Kelly, the engineman, invited me to finish my trip riding on the fireman's seat and to ring the bell when necessary. I enjoyed my first ride in the cab of a locomotive as a novelty I had never expected to experience. So many things interested me that it seemed no time at all before we entered the snowsheds at Emigrant Gap. From there the ride became so monotonous and I inhaled so much smoke that on arrival at Summit I went back to a coach.

Welcomed Because Sober.

On arriving at Truckee I reported to James W. O'Brien, the day operator, now a veteran of the Salt Lake Division. He was anxiously awaiting me. I was sent to take the place of the night operator, who had on the previous night kicked a lighted lamp off a table and nearly caused the office to burn. I think on account of my being sober Jimmy gave me a hearty reception. As I expected to stay there temporarily he invited me to share his bed. He slept in a room back of the office. I think the bed was composed entirely of blankets. When he got out in the morning and turned back the blankets that had covered him I was ready to get in under them and pull them over me. As he was immune to Truckee temptations and sports he went to bed early, soon after I got up. I don't think the bed was made over once while I was there. The blankets were not disturbed more than was necessary for us to get in and out, and they were never unoccupied long enough to lose their human body warmth.

Now, meeting the trainmen and other employees of the Wadsworth as well as the Sacramento Division, who loitered, when off duty, about the depot hotel and telegraph office, I soon became aware they were loyal to the company in their work, ambitious and polite.

These were the days when the engines burned pine wood and were wooded up en route by the crews; when the trains were controlled by muscle and hand brakes; when the link and pin coupling was in use on both passenger and freight cars; when the rear brakeman with his red flag and torpedoes was the only block system; when oil as a fuel, air as a braking

---

FIFTY-FOUR YEARS IN S. P. SERVICE—HIS STORY.

In the October number of THE BULLETIN *will begin an article by J. O. Wilder, a veteran Southern Pacific employee, who has fifty-four years of service to his credit. His fellow workers recently held a reception in his honor and gave him an easy chair to appropriately celebrate his retirement from active service. His article contains much that will illustrate the spirit of adventure and accomplishment which characterizes so many of California's pioneer railroad men.*

---

power, the automatic coupler and block system were undreamed of on the division, yet the employees were more contented and less critical than they are now.

How Wood Was Stored.

The wood burned in the engines was pine, cut in two-foot lengths. About seven cords could be piled on a tender. On an eastbound freight train between Rocklin and Truckee the engines would have to be wooded up five or six times; from three to five cords thrown on at a time. The woodsheds were placed at convenient places along the division, mostly at sidings, but several were on the main line. They were all filled in September and October with a winter supply. About 100 flat cars fitted with racks were in constant use and Chinese gangs on work trains did the loading and unloading of these cars. The principal points of supply were near Emigrant Gap and Tunnel 13. Long V flumes were built by contractors into the canyons near the line of the road and the wood was floated down them and piled alongside of spur tracks. The longest one of these spurs was called Champions, named after the agent at Truckee, who was also a big wood contractor.

Fires occasionally destroyed some of these woodsheds. After a fire got a start it was impossible to do anything that could stop it, and it burned for hours. It was always an unfortunate occurrence when one of these big woodsheds burned at the beginning of winter after they were filled, as they sometimes did. A fireman had to be well developed by nature with muscular arms and a back to throw twenty to thirty cords of wood into the firebox on a trip, as they often did.

A source of annoying delay in the movement of trains occasionally was from the netting in the smokestack of the engines developing holes through which they began "throwing fire" and thereby endangering the safety of the sheds. When the netting was all right small sparks, not dangerous, were thrown out so the dangerous large sparks were not always quickly seen. Some disastrous fires occurred that were attributed to this cause. When reported as "throwing fire" the engine was cut out at the first siding and the train was either reduced or held for needed power. As the sidings, except Summit, would hold but one train vexatious delays and inconvenience resulted.

When one of these occurred we understood that A. J. Stevens, superintendent of motive power and shops, would make the capable master mechanic, Menzo Cooley's hair stand on end and he was likely to become bald-headed at an early age.

Fires Always a Menace.

Fire was the arch fiend that the division had to contend with. Not only were all the station buildings and sheds built of lumber, but all the trestles with timbers. The right of way in many places was thickly covered with inflammable brush and small pines, while forest fires, starting several miles away, fanned by a north wind or a still south breeze would sweep up the canyons and mountain sides, causing a desperate fight to save the sheds and woodpiles. Since then the trestles have become fills, the bridges are of steel, the Red Mountain lookout and fire alarm system in the snowsheds have been established and the right of way has been cleared. Fires, small and big, during the summer and autumn months totaled a heavy loss.

Another cause of delay was when an engine got a tender load of green wood, as unseasoned wood was called. As steam pressure went down it had to stop and "blow up," and so continue a spasmodic movement until the next woodpile was reached.

The brakemen while in Truckee mainly discussed their probability of promotion to be conductors. They seemed to think it was a sort of a lottery where luck was more of a factor than worth. There was no seniority rule. When a conductor laid off, was disabled or fired and another had to be made, the most available caboose brakeman was advanced and a man had to be in the right place at the right time to have promotion lightning strike him. The brakeman who had a reputation of muscle that could set a hand brake so tight it could not be let off by others without using a club, and on the other hand, the brakeman whose reputation was favorable to his ability to let off any brake any other brakeman could set was regarded with envy and admiration. These "arm-strong" men stood in the estimation of the twisters as do prize ring champions with fight fans.

I found there was an ambition with the passenger crews to be considered the best brake handlers and train stoppers on the division. The station baggagemen were the judges. They wheeled their trucks each day to a certain spot ahead of the arrival of the passenger train. If the stop was made with the baggage car door opposite the truck it was a perfect stop, but if the truck had to be again moved it was bum. The crews were as anxious to obtain and keep up the best record as the volunteer fire engine companies were of holding the fox tail for getting water on a fire first. The freight crews were ambitious to be considered the fastest in wooding up their engines and thereby making time.

No such thing as a claim for over-time was thought of. A trip was a trip, no matter how many hours it took, and to ask to tie up on the road for rest would have been considered as show-ing a lack of stamina. Crews coming in late were frequently ready to return without asking for stated hours of rest. When an emergency came and things were moving wobbly, everybody got in and did their best to regain a normal condition.

Formalities Were Few.

We night operators, when we received a "flag and hold" order, after its "13" and "O. K.," would light a red-globe lantern and place it in front of the office to stop the intended train. We wrote our train orders on ordinary clip paper. The original copy the conductor and enginemen signed, and after we got "O. K." to it we made copies on clip paper for those who had signed it. Comparison was made by the conductor and enginemen reading their copies aloud and all checking word by word with the original.

Only the through freight trains, Nos. 5 and 6, were scheduled to run on Sundays, and we had a dull night then. To relieve the monotony and keep awake the "Owls" in the snowshed offices played checkers. Each office had a checkerboard with the squares numbered and the games were played by tapping on the wire the numbers of the squares moved off and on. We who were not playing could follow the game by moving on our board as the moves were wired. Many interesting games were thus played. Special trains were infrequently run and were a sort of an event when they did. Usually a "23" message addressed to "all concerned" was sent the day before announcing the time of its expected start and stating its kind, whether silk, tea, etc. The purpose of the message was to advise section foremen, work trains and shed gangs likely to be away from telegraph offices so that they would not have the track up or otherwise delay the movement of the special. Subsequently all train orders given to special trains ended with the admonition, "Whistle at all curves and obscure places." It can be imagined the number of times an engineman, obeying the order, would toot on the Sacramento Division, yet this was necessary to stop, if possible, handcars being struck.

Side tracks like telegraph offices were few and far between, compared to what exists today on the Sacramento Division. One cause of delay and annoyance to trains was the Cisco siding, then the only one between Emigrant Gap and Tamarack. It held only twenty-two of the company's thirty-foot freight cars, two engines and a caboose. Occasionally, on account of freight trains having foreign cars of different and longer lengths in their trains the dispatcher would be taken by surprise and the trains meeting there would not clear. They were then compelled to saw-by. This would cause a delay, consuming a lot of time and causing a lot of profanity.

"Fixing Up" a Train Order.

I learned at Truckee that I was the beneficiary of an undeserved popularity among the trainmen. About a week after I began work at Rocklin, No. 8, the night freight train, left Blue Canon about an hour late. The conductor, after directing the caboose brakeman to drop the train at a certain speed an hour, not having a pin to pull en route, laid down on his bunk and fell asleep. It was due to arrive in Rocklin at 4 a. m. Shortly after 2 a. m. Jerry Creedan, the night yard-master, who was taking a nap on the office counter, awoke with a start, as we heard a locomotive whistle coming from the east. "What's that?" he exclaimed. I did not know. I had not heard Colfax, the only night office between Rocklin and Blue Canon, report anything special coming west, so Jerry told me to ask Sacramento. Fortunately for all concerned Sacramento did not respond. Jerry went out to investigate. He returned shortly and in a hoarse whisper said, "Never mind. It's all right. It's No. 8." Shortly afterward, I heard Jerry and the conductor conversing outside the office door, and Jerry saying, "He's a new man. He's all right."

Not thinking it strange I should be discussed I gave his remarks no attention. At the proper time the conductor handed me his "soup ticket," showing No. 8 arrived and departed on time, which I "27'd" to Sacramento. The caboose brakeman had obeyed instructions, for after the train had made up its lost time he kept the steady drop until it arrived in Rocklin about an hour and a half ahead of time. I, on account of my inexperience, did not know that there was anything wrong with the train arriving ahead of time. I learned that I now stood high in the estimation of the trainmen as being all right, and a man who would not give a trainman away or tell all he knew. I had noticed that Job Calderwood, Fred Turner, George Crocker, Tom Allister and other freight conductors had a kindly greeting when we met, but I did not know the reason why. A wreck near Summit one night tied up the road. The westbound passenger train was held at Truckee. Attached to it on the rear was Governor Stanford's private car. I did not know of this as the outgoing conductor had not yet registered the train.

In the morning, while I was on duty, Governor Stanford, president of the Central Pacific, with his private secretary came into the office to ascertain the cause of the delay. I did not surmise at first that the Governor was an official of the company, but it was easy to see the secretary was one. The Governor thanked me in so appreciative a manner that it would seem I had done him a great favor to give him the details of the accident and answer in a civil manner his questions, as I would have done for any one else entitled to know.

He remembered me when, several months later, I met him in the dispatcher's office at Sacramento, where I had been unexpectedly transferred to work as night operator.

No wonder we employees all loved him.

(The End)

THE BULLETIN

THE DAYS WHEN THE CENTRAL PACIFIC WAS YOUNG

Business Directory of the Early Seventies Speaks Enthusiastically of "The Most Perfect Highways Yet Conceived by Man," and Details Their Extent in California

The February issue of the Bulletin mentioned the first locomotive that was set up outside the Central Pacific shops at Sacramento in 1863 and the growth of our general shops at the capitol city to their present size and importance.

Now there comes from J. R. Hockett, chief clerk at Oakland pier, a copy of the Pacific Coast Business Directory, 1871-73, reflecting in its advertising columns, reading matter and maps, transportation conditions as they existed in that stirring period of early California life.

Not the least interesting page in the volume is that which contains a map of the old California Pacific Railroad showing its western terminus at Vallejo, whence its line ran through Bridgeford, Fairfield, Vaca Junction, Dixon, Davisville to connect at Sacramento with the Central Pacific. A branch line ran north from Davisville to Oroville with a projected extension to Chico. The terminal railroad from Vallejo to Oakland was not yet constructed. Connection between Vallejo and San Francisco was made by steamship.

The enthusiasm with which everyone in those days regarded even the primitive facilities which the railroad pioneers were able to provide is apparent in the editor's review of railroad enterprises in the state. The railroad was hailed as "an ease and comfort to man"; an "achievement of the age"; an "aid to development," etc.

Viewed in the light of the present conveniences of modern railroad travel and the extent to which service has been extended to all portions of the Pacific coast since that time, the following description of the railroad situation of those days is both interesting and enlightening:

(From the Pacific Coast Business Directory)
(1871-73)

The progress of railroad building on the Pacific coast, within the past few years, is a pleasing assurance of the extraordinary resources of the country, and gratifying evidence of its continued and increasing prosperity.

These are the most perfect of highways yet conceived by man, the strong iron bars forming a smooth and uniform track over rocks and marsh alike, crossing rivers, plains, mountains and desert sands, with no hindrance to the whirling wheels nor fatigue to the powerful machine, which, like a living monster, drags his mighty train along, with a speed unattainable by other means. No roads of history, ancient or modern,

exhausting the resources of an empire at the command of a despot to construct, can compare in efficiency to the two simple iron rails, scarcely observable as they lie upon the plain.

This grand consummation of genius and science, in this most practical age of the human race, has now come to aid the development of the country, and give ease and comfort to man, and is now regarded as indispensable to progress. Great minds appreciate the rising importance of railroads, and with strong will and untiring energy are working like giants in their construction.

To the Sacramento Valley Railroad is usually accorded the honor of being the first constructed west of the Rocky Mountains; but this property belongs to the Cascade Railroad. The Sacramento Valley, however, was the first in California, and the first for a number of years in importance. This was constructed in 1855 and 1856, from Sacramento to Folsom, a length of twenty-two and a half miles, and at a cost of $1,100,000—a very large sum, considering the level country over which it was made.

The Sacramento Valley and Placerville Railroad is an extension of the first, running from Folsom to Shingle Springs, in El Dorado county, a distance of twenty-six miles.

The San Francisco and San Jose Railroad, fifty miles in length, between the two cities whose names it bears, was a great advance in the construction of this noble means of travel. Within the last two years this road has been extended to Gilroy, thirty miles from San Jose, bearing the name of the Southern Pacific Railroad. It accommodates the passenger travel, and carries the mail for the southern part of the state; stages branching from the terminus to Visalia, in the valley of San Joaquin; also to Watsonville, Monterey, Los Angeles and intermediate points. This will eventually form the terminal section of the southern railroad crossing the continent, but whether the thirty-fifth parallel route, or the thirty-second, remains to be seen.

The great achievement of the age has been the construction of the Pacific Railroad. This, in California, has borne the names of Central Pacific and Western Pacific, but these are now merged in one. The Central Pacific now extends from the borders of San Francisco Bay to Ogden, in Utah territory, a distance of 880 miles, there connecting with the Union Pacific, 1,032 miles to Omaha, and at

that point with other roads, to every part of the East and to the Atlantic seaboard, making a continuous line across the continent of 3,305 miles to New York. This is one of the noblest works of the age and is one of the great measures in fulfilling the destiny of the United States. This company now owns the Sacramento Valley Railroad, the Southern Pacific as far as constructed, the San Joaquin Valley Railroad and the California & Oregon. The former roads, known as San Francisco and Oakland, connecting San Antonio with the San Francisco and Oakland ferry, and the San Francisco and Alameda, from the ferry to Haywoods, are now combined with the Central Pacific; and the California Central from Junction to Lincoln, and the Yuba Valley from Lincoln north a few miles, now constitutes portions of the California & Oregon Railroad. The Central Pacific also has a branch, fourteen miles in length, from Niles to San Jose, where it joins the Southern Pacific.

The San Joaquin Valley Railroad, joining the Central at Wilson, is completed to the Stanislaus river, and is a most important adjunct to the main branch.

The California & Oregon has been rapidly extending during the past year and is now completed to near the head of the great Sacramento Valley, opposite Red Bluff, 135 miles from Sacramento City. This is a most important road, passing through a fertile agricultural section, and destined to join the Oregon Central Railroad, now building south from Portland, and, when both are completed, making a continuous line of 630 miles in length.

The Los Angeles & Wilmington is the only railroad in the southern part of the state. This has a length of eighteen miles and connects the city of Los Angeles with the port of San Pedro at Wilmington.

The California Pacific Railroad, with its branches, is next in importance to the great transcontinental road. This gives the shortest passage from Sacramento and the northern part of the valley to San Francisco. Its main line is from Vallejo to Sacramento, sixty miles in length. The first branch is the Napa Valley Railroad, leaving the main line seven miles from Vallejo and extending to Calistoga, thirty-one and a half miles from the junction. At Vaca station, a railroad four miles in length connects with Vacaville. At Davisville, an important branch, forty-two miles in

Photographic reproduction of railroad advertisements in business directory of early seventies. Note illustration of our crack Overland Limited.

length, leads to Marysville, passing through Woodland and crossing the Sacramento river at Knights' Landing. At Marysville, connection is made with the Northern California Railroad, of twenty-nine miles in length, to Oroville, thus giving a continuous line of 140 miles from San Francisco, including twenty-three miles of steamboat connection from Vallejo.

A large number of railroads are proposed, and the construction of many of them will be entered upon without great delay. The principal of these is the Northern Pacific, to be built from Lake Superior to Puget Sound, making the second in importance in the United States. Several branches to the Central Pacific, leading to the different mining districts of Nevada are proposed. The Stockton & Copperopolis and the Stockton & Tulare railroads have been projected for several years, but no progress has been made in their construction. Lines are proposed from San Jose to Alviso; from Gilroy to Watsonville; from Salinas City to Monterey; from Wilmington, via Anaheim, to San Bernardino, and,

in the north, from Vallejo to Sonoma, Petaluma, Santa Rosa and Healdsburg; also, others of less prospect of early construction. Surveys have been made recently for a railroad between San Francisco and Sacramento, by the most direct route from Oakland north, crossing the Straits of Carquinez at or near Benicia, by ferry or bridge. This is undertaken by the Central Pacific Company, and will be an important addition to that road.

TOM FASSETT LOOKING FOR WEATHER TRAFFIC.

T. A. K. Fassett, head of the Train Service Bureau at San Francisco, acquired a new title the other day. A director of a motion picture company located at Truckee called on the long distance telephone and was referred to Fassett.

"Hello," he said, "this is Bert Foster of the Alaska Motion Picture Company at Truckee. What time is that eastbound storm due here?"

"What?" stammered Fassett.

"The storm," yelled the movie man. "Heard there was a storm headed this way and I'm anxious about it. We've been waiting here a week to take a snow scene and there isn't enough snow to fill a salt shaker."

Fassett looked out the office window. Not a cloud marred the calm blue sky.

"Sorry," he shouted over the telephone, "there's no storm routed over the Southern Pacific today. Try some other road."

Now they call Fassett the "weather traffic agent."

SMILE MOTTOES.

If you can't do anything else for your country, you can smile.

Travel with those who smile, not with those who grunt.

Don't smile just to show your teeth, smile to show your heart.

You don't have to smile. You don't have to have friends.

When in doubt—try a smile.

THE BULLETIN

'High-toned' Railroad Travel Fifty Years Ago

Eating Stations; Wood Agent's Job; Manning the Hand Brakes; Queer Passengers; Silver Palace Cars; Reading by Candle Light; Old Guide Books

By CHARLES B. TURRILL
Former Southern Pacific Employe

IT seems only the day before yesterday when trans-continental railroading was a new thing. During the past fifty years many changes have come. But back about 1870 we thought the highest point of comfort and luxury had been achieved. The completion of a railroad across the continent called for many new things in railroading. During the past half century improvements have been increasing annually but at a much slower pace than was necessary during the first year or two that the old Central Pacific line to Promontory operated as a part of the first transcontinental railroad.

By 1871 the general problems had been pretty well worked out. The snowsheds on the Central Pacific made winter travel possible, although the old style snow-plows were often called out from their stables at Cisco and other points.

Old wood burners puffed up the steep grades. Long piles of wood corded along the track, sometimes for miles, were objects of interest to us travelers as, from the open windows, we watched the entire train crew sweating as they tossed cord wood into the tender to carry the train a few miles further up hill.

The position of Wood Agent was an important one in the early days of the Central Pacific and even well up into the '80s. To him fell the responsibility of seeing that there was always an abundant reserve supply and to be aware of where additions could be had. There was a clever trick practiced in placing crooked sticks in the pile so that a cord might not always represent full measure and it was the duty of the Wood Agent to see that the Company was not swindled. Oak wood was brought along the line of the old Western Pacific all the way from McConnell's to Lodi. Pine was less liable to fraudulent piling.

Hand Brakes Meant Work

Train crews had more strenuous lives in those days than now. The whistle for brakes might be heard at most unexpected times. Then all hands to the brakes—hand brakes too, for that was before the day of Miller couplings and long ere air brakes were dreamed of. The setting of the brakes quickly saved many a disaster on curves and grades. Sometimes when the signal for brakes came and the train seemed, to our over-wrought imagination, to be gaining dangerous speed on the curves coming down from the Summit, the timid turned pale through the dust on their faces and did not breathe comfortably till the train stopped. Then we went out and looked around to see what the matter might be. What curiosity there was in watching the brakemen working around a hot box, cooling it off, and packing it so as to get through! We passengers took a personal interest in the running of those trains. We gained much information which we never personally put to use.

We also took some interest in our fellow travelers. We took long strolls through the coaches and, if fortunate, sometimes had a chance of sitting on the fireman's box in the cab of the locomotive or possibly riding with an accommodating baggageman. Thus we beguiled the tedium of a long trip. As we passed through the coaches we had pleasure in watching the Chinese in the smoker, in their curious garb, with their strange baggage. Once the writer was particularly attracted to a Chinese miner on his way to a distant "camp" from San Francisco. He had a precious thing in his lap that was not wrapped. It was one of the old style round eight-day clocks, loudly ticking the time away. That Chinaman took especial pride in his charge.

Indians Enjoyed Rides

Then, as we crossed the dusty plains of Nevada, what a constant interest there was in the Indians sitting on the dusty platforms riding from station to station along the line, enjoying the novelty of locomotion, unmindful of dust and heat, happy in the free rides.

There were no vestibules in those days. The blind-baggage was almost always full of aborigines and even occasionally other platforms as well. What a pleasure we white travelers had sitting on those platforms as we traversed stretches of road through scenic attractions. This pleasure was frequently increased through the necessity of being helpful in holding our arms around the waists of young

The above is one of the earliest type of passenger coaches used in the West. This car was one of the first to come through from the East after the two railroads were connected at Promontory point, it arriving in Sacramento in Governor Stanford's special train the evening of May 12, 1869. The picture was furnished by D. L. Joslyn, Sacramento Shops, from the original belonging to James Hall, Master Car Repairer, Sacramento Shops.

38

THE BULLETIN

ladies also interested in scenery. We sat in proper way in the coaches but on the platforms duty prompted the support of our arms. While there might have been inconveniences in travel in the olden days there were compensating advantages—and some tunnels.

There were no buffet cars in those days. Neither were there diners. Even the "peanut butcher" had not developed into a general merchant. When we started on a long journey we carried in our baggage a few good books, a guide book (then very common) and a deck of cards. Oh, yes, sometimes a pocket flask also. There might be an accident and some stimulant needed. We also usually carried a lunch basket.

Problem of Eating

Those were the days when trains tried to make eating stations on time. But under most favorable conditions the intervals between meals were very variable. If we had our own basket we felt secure above all inconvenience and took advantage of the "twenty minutes" for meals as a time for promenades up and down the platform, while the car cleaners were sweeping and dusting the coaches.

It has been stated we carried guide books in those days. The making of such compendiums of intelligence was an important industry. Many of those books are now rare. Probably the scarcest is "The Alta California Pacific Coast and Trans-Continental Rail-Road Guide." The first issue (of 1871) is full of most important information. On the Central Pacific eastward, trains bore uneven numbers. The "Express" No. 1 left San Francisco (by way of Niles and Lathrop) at 8 a.m. No. 3 "Passenger" left San Francisco at 4 p.m. by the same route and arrived at Sacramento at 9:35 p.m. No. 5 was a "Mixed" train as far as Sacramento, leaving S. F. at 6:45 p.m. and arriving at Sacramento at 7:40 the next morning. From thence eastward, leaving Sacramento at 9:00 a.m., it was a "Passenger" through to Ogden. From Sacramento a "Mixed" train, leaving at 9:00 p.m. ran through to Ogden. Such was the service eastward. The "Western Division" extended from San Francisco to Sacramento; the "Sacramento Division" thence to Truckee; the "Truckee" to Winnemucca; the "Humboldt" to Toano; and "Salt Lake" to Ogden. It may be of interest to recall that "All trains west of Ogden are run by Sacramento time."

Early Guide Books

Referring to this old "Alta California Guide Book" let us glance at the preface:

"But two Guide-books for the great Trans-Continental journey have hitherto been published. They both begin at its eastern end, and both are wanting in much-needed information about California—the place of greatest interest on the whole journey." (There was boosting even half a century ago!)

The preface continues: "Hitherto overland travelers going East from the Pacific coast, if they used a Guide Book at all, had to begin at the end of the book and read backwards—an infliction that involved much trouble, and gave but little information. Moreover this grievance would be felt, in an intensified degree, by all travelers coming from the Australian Colonies, China and Japan." Trans-Pacific business was sought and obtained.

We must bear in mind that the Guide-Book before us was prepared in a measure for use by Trans-Pacific passengers unaccustomed to American methods, but also for our own people, many of whom had not been "back to the States" since the earlier days of steamship or stage conveyance. Therefore we find very explicit "Hints to Travelers." Which are interesting to quote.

"Before our readers purchase their tickets for the trans-continental journey, or any other, a few practical hints will be of service to them.

"How are you going to travel? Let us help you to answer the question. There are three classes of Railroad travel, viz: first, second and third. To everyone who can afford it, we say travel first class. To those who cannot, we say, do not sacrifice your money to your pride. Second class cars are attached to every express train, and go through as quick as first class cars do, but third class cars are attached to freight trains only and take from seventeen to twenty days from San Francisco to New York. Pay the difference and avoid the tedious ride, if you can.

Food on the Way

"There are excellent hotels at convenient distances along the road. At these, meals can be obtained for 75 cents in coin, or a dollar in currency.

"Emigrants with large families, and others, who cannot afford to pay these prices, reasonable though they be, carry a basket of provisions with them. This can be replenished at any of the principal stations along the line. At these stations milk, and hot tea and coffee can also be obtained. It is unnecessary for us to specify what articles of food the basket should contain, but we would suggest that a corner of the basket should be reserved for comb, brush, towel, soap, tin

Back in the early days of Western railroads there was no such thing as a dining car. During the day small tables were placed between the seats in the parlor car on which passengers might spread their own lunches. The above unusual picture, taken about 1870, shows the interior of one of the early Silver Palace sleeping cars converted into a diner. "Wet goods" are much in evidence, also it is interesting to note the colored table spread, candle lights, cuspidors and the prevailing styles in dress.

THE BULLETIN 9

cup, small basin, sponge, hand-mirror, tooth-brush, etc., etc.

"Only first class passengers can engage berths in the sleeping cars. The second class cars are seldom full, and passengers taking a blanket or plaid can generally make themselves comfortable in the cars at night."

Candles for Light

Those were the days of the old "Silver Palace" sleeping cars. What palaces they were! Not as heavily built as the few "Pullmans" which occasionally found their way here with private parties. How brightly the large candles shown from their points of vantage in the ceiling. What a comfort it was to slide up the little panel between the windows, light the candle inside and read until the porter thought it time for us to retire! How those porters regulated our movements! How some of us managed to sit up a little later and enjoy those side candles! What was the tip? Never mind that, for the porters were human, too.

And the "coaches" at night! What an interesting journey through them we could have while that porter was "making up" our section, or on our way to the "sleeper" in search of accomodation had we boarded the train at a way station! What a study was afforded on that journey as every sleeper had taken a different uncomfortable position in search of comfort! What a relief was the fresh air as we crossed the platform between cars! That was fifty years ago but the memory of odors seems everlasting.

Our old guide book has mentioned "excellent hotels at convenient distances along the road." We who traveled only cared for the dining room features of these, at Sacramento (Lathrop came later in the game), Alta, Humboldt, Battle Mountain, Elko and Ogden. When we came back from "the East" the eating house at Colfax supplied our breakfast. We would rush from the train in our hungry earnestness to find a place and to help ourselves from the food that had been rushed on the table when our engineer had whistled our approach—at least some of the food was hurried on before our arrival.

In the letter files of the Passenger Department destroyed in the fire of 1906, was a tissue copy of a diplomatic letter from that grand old Passenger Agent, T. H. Goodman, to the proprietor of one of the Nevada eating houses in which it was stated that many of the passengers had complained of the coffee served by him. Mr. Goodman wrote a good letter and had the knack of making his meaning perfectly clear. He realized the difficulties that beset a "hotel keeper" on a Nevada desert and tried to help him out by sending him a full account of the method practiced in making coffee at the Palace Hotel in San Francisco. The coffee probably improved in quality as that was the only matter of culinary instruction that went from the General Passenger Department.

Terminal Yard To Cost Millions

Construction Started on First Unit of Extensive Southern Pacific Facilities at Los Angeles

CONSTRUCTION of the first unit of the gigantic Southern Pacific freight terminal at Los Angeles is now under way. This unit will consist of modern classification, receiving and storage yards and will cost approximately $300,000.

The entire terminal plan calls for the expenditure of several millions of dollars and includes extensive classification and receiving yards, car storage yards, delivery tracks, repair tracks, stock resting corrals, mammoth fifty-stall roundhouses, powerhouses, drill tracks, and many other improvements for the efficient and expeditious handling of large volumes of freight.

The new terminal yard will extend along the San Fernando Road and the Los Angeles River for a distance of more than two miles and will be one of the most extensive and comprehensive freight terminals in the country.

Freight traffic in and out of Los Angeles has grown in the past few years to such an extent that during the month of July the Southern Pacific alone handled in their yards there more than 99,000 cars, the greatest in their history.

It is estimated that 51,000 cubic yards of grading will be necessary for the first unit now being constructed. The unit will contain 47,000 feet of trackage to be laid with 90-pound rail. There will be six long tracks in the receiving yard and twelve tracks in the classification yard.

A modern 150-ton railroad track scales, capable of handling any mobile load and costing $10,000, will be a feature of the new yards. Another feature will be a system of compressed air lines throughout the yards for testing trains. This air pipeline system and compressor will be installed at a cost of $10,000.

Some of the outstanding features of the entire terminal yards plan are: a huge 2000-car capacity receiving yard of 26 tracks for handling incoming freight trains; a 19-tracks delivery yard; a 600 car classification yard of 25 tracks for breaking up and classifying incoming freight trains; modern stock corrals for feeding, watering and resting livestock; an electric powerhouse and illuminating system for the yards; 2 huge 50-stall roundhouses for switch engines; caboose tracks; drill tracks; corral tracks, water tanks and elaborate repair yards.

Beginning operations in Los Angeles in 1873, when the population was less than 10,000, the Southern Pacific's growth has increased until it is now handling nearly 100,000 cars a month

in the local yards, this being the figure for July.

Geo. W. Corrigan, Division Engineer of the Southern Pacific's Los Angeles Division at Los Angeles, will be the engineer in direct charge of the new construction work of the terminal.

VISITORS AMAZED AT WORK OF SACRAMENTO SHOPS

Thirty-five members of the Pacific Railway Club were recently the guests of the Southern Pacific Company at the Sacramento General Shops. The visitors included representatives from several other railroads including, the Western Pacific, San Francisco-Oakland, Sacramento-Northern, San Francisco-Sacramento, Northwestern Pacific, Market Street Railway and also the Railroad Commission of California.

Arrangements for the reception of the visiting railroadmen were made by a committee headed by A. D. Williams, Superintendent of Motive Power. Luncheon was served at Hotel Land, the Club's President, J. N. Clark, Chief of the Southern Pacific Fuel Bureau, presiding. The guest of honor was John E. Lonegran, one of the first engineers on the famous old "C. P. Huntington." In an interesting address D. S. Watkins, Superintendent of the Shops, told of the remarkable growth of the shops and the quantity and quality of the work being produced. Short addresses were made by representatives of the other roads.

Following the luncheon the party was divided into groups of five and each placed in charge of a Southern Pacific official who acted as guide through the shops. "The visitors were amazed," says W. S. Wollner, of the Northwestern Pacific, who arranged the trip, "by the stupendous scale with which work is done in the Sacramento Shops and expressed their appreciation of the kindness of the Southern Pacific and its officials in making this interesting and entertaining day possible."

They Sailed on It

A burley old skipper and his mate entered a restaurant and ordered dinner. In a few minutes the waiter brought two plates of thin, watery looking soup. "Hi, me lad, but what is this?" The waiter informed him that it was soup. "Soup! By gum, Bill, just think o' that. You and me been sailing on soup or! our lives and never knowed it afore."—Exchange.

In former days fools blew out the gas. Now they step on it.

The history of dining on trains started when all the passengers in a car shared the same cup for drinking water and continued to the era when dining was an elaborate event with special dining ware and white table cloths. The service provided to passengers could be enjoyed off the train too. During the 1910s, ice cream was available outside the Southern Pacific station in Sacramento. – Courtesy California State Railroad Museum

SOUTHERN PACIFIC BULLETIN

When derbies were the rage and be-whiskered chins were much admired there were a few future greats in the personnel of the Sacramento Freight Station force. Names of most of those in the picture, taken some time in the 90's, are identified in the article below.

Back in the "Good Old Days" of Sacramento

By Miss M. J. White *Sacramento Freight Office*

THE Smith Brothers of cough drop fame have nothing on some of the boys of the '90's, judging from the many varieties of chin decorations in the above picture of the Sacramento Freight Station clerical force taken some thirty or more years ago.

C. C. Cooley and L. W. Robitaille, fourth and eighth in the back row, were new in the service at that time, and are the only members of the group still with us. Little Rufus Harmon, at top, is now one of the truly big men with the Western Pacific in San Francisco. First in the middle row is John T. Skelton, popular Native Son and ex-coroner of Sacramento, really a handsome man in spite of the Bolshevik camouflage. Next to him is Emsley E. Griffin, retired, whose classic, smooth-shaven profile peered at us from the August Bulletin. Sixth in that row is P. A. Fitzgerald, editor of the Yolo Independent, famous for his views on the eighteenth amendment. Third, fourth and sixth in the front row are ex-Agents H. J. Kilgariff and C. J. Mills, and Mamie Chamberlain, stenographer. These three have passed into the great beyond. Sixth in the top row is J. S. Watson, deceased, who established the first agency at Red Bluff. W. B. Lawson, well known representative of the E.M.B.A. at San Francisco, is eleventh in the top row. E. L. Sharp, telegrapher at Sacramento, is tenth in the middle row. The last

member of the group is M. E. Gates of San Francisco, who was agent of the Sacramento Valley Road in the early days.

Our veterans have furnished us some interesting information about the cost of living in the '90's. In those days "Nobody worked but father." On a salary of $80 or $90 a month, a man could raise a small family—America's great middle class majority. Rent for a five-room cottage averaged $12 or $15 a month; eggs, 15c a dozen; potatoes, 75c a sack; flour, sugar and other foodstuffs were proportionaly low compared with present prices.

In the matter of clothing, "Father" could get a good ready-made suit for $15 and a derby for $2.50. He wore the latter in all seasons, usually not discarding it until brim and crown were ready to part company. "Mother" usually bought two hats a year, a winter felt or a summer leghorn. She trimmed the former with ostrich tips and the latter with a wreath of roses or daisies. Her hats rarely cost over $3.50 and she often wore them the second season; she hadn't much time to gad. Shoes for little Willie and small sister Susie averaged about $1.50 for the stout kind for school and $2.50 for the Sunday pair with patent leather tips. Susie could have piano lessons at 50c each, and the family and neighbors considered her an accomplished player when she

could rattle off "The Maiden's Prayer" or "The Black Hawk Waltz."

Standards change with the times. Textiles seem to have been noted for their enduring quality in those days. We remember "Sandy" Smith, that dear old scot now on the retired list (we never knew him to tell a lie), once whispered to us confidentially that a certain other clerk in the office had worn the same pair of pants for over 20 years. They were fine tweed of a nondescript color, and, like Mary's faithful lamb, always went everywhere with him. However, in justice to their owner, we must say that he finally received an inheritance and promptly took the pants out of active service. We eventually lost sight of him.

LARGE SHIPMENT OF AUTOS OVER LONG HAUL

The largest single shipment of automobiles from Indianapolis was hauled over the long haul from New Orleans to Los Angeles during the early part of March. The special train of 25 big automobile cars loaded with Marmons valued in excess of $2,000,000 was received from the Illinois Central at New Orleans. Arrangements for handling the big shipment were made through the office of General Agent Lyon Liston at Indianapolis by W. M. Young, traffic manager of the Marmon Motor Car Co.

Chapter Four
Recollections from 150 Years of Railroad Work

Construction of the Shops began in 1863 on twenty acres of filled-in slough donated by the City of Sacramento. During that wonderful year in United States history, the Central Pacific (CP) built two buildings: one for materials and car repair and the other, containing one forge, for locomotive repair. In 1864, another car shop and a new locomotive building were added. From 1877 to 1922, the area of the property grew from 50 to 145 acres.

During the period from 1863, when the shops opened, to 1878, the number of employees increased from 15 to 1,700. But, despite the overall historic success of the railroad, employment did not increase continuously. In 1883, payroll was 25 percent lower than five years earlier. By 1898, the number of employees was back up to three thousand. An extraordinary business slowdown accompanied the 1907 depression and Bankers' Panic during which all employees were laid off temporarily. Tough times returned again in 1914 when the facility was running only six to seven days per month. Soon after, however, support for WWI necessitated a workforce of nearly 3,500 in 1920. During World War II, seven thousand people aided the cause.

Prior to about 1880, work rules were rather flexible. As with any business, there needed to be a procedure for dealing with rule violations. One method used was the "Brown system" which was named after the General Superintendent of the Fall Brook Railway. His philosophy was to not automatically suspend an employee for an infraction. Instead, there were demerit points entered into one's personal record. This action was taken because suspension or dismissal were hard on both the employee and

the railroad operations. These "brownie points" could be reduced after a period of good behavior. The cause and effect of a violation were considered valuable information and it was summarized in "bulletins" as a way to spread the lessons learned from an incident.

By the twentieth century, job assignments, overtime, and time off were based on seniority. In addition, when layoffs were needed, senior workers could "bump" those with less seniority to lower paying positions.

At the turn of the twentieth century, shop workers were paid once a month in gold and silver coin when the pay car arrived. In order to reduce the chance of a robbery, the train operated on no pre-determined schedule, but employees were given two days' notice. The railroad cared about the spending habits of their employees. Personal debt was not looked upon highly, and money owed to local merchants could be withheld from one's pay. The Southern Pacific began to pay salaries with checks in 1905.

In 1926, the February issue of the Southern Pacific *Bulletin* honored an employee who had served the railroad for over fifty-six years. He experienced a half-century of changes in working conditions, pay, strength of the local and national economy, labor rules and equipment. Mr. Trott witnessed the growth of the Sacramento Shops from fifteen employees in 1863 to 1,700 in just the

first fifteen years. During his tenure, he adjusted to fifty-seven years of changes.

One of the benefits to historians is that publishers

of early editions of *The Bulletin*, in 1920 as an example, were able to hear first-hand experiences from railroad employees who worked during the very early days of the railroad. In addition to the thoughts of Mr. Trott, *The Bulletin* was able to interview J.O. Wilder who provided descriptions of early surveying and construction. His accounts were published in the October, November, and December issues of 1920. Those articles follow.

Another long-time employee of the railroad was Roger Trewick. Mr. Trewick's last day was December 31, 1926 after fifty-seven years. His experience was honored in the February issue of *The Bulletin*. The issue stated, "Trewick's record has probably never been equaled by any railroad man in the west."

The editions of *The Bulletin* that cite the experiences of Mr. Trott, Mr. Wilder, and Mr. Trewick follow:

"Fifty Years at the Throttle," February 1926

"My Half Century of Railroad Service," October 1920

"An Engineer Who Railroaded for 57 Years," February 1927

Fifty Years at the Throttle

By ERLE HEATH, Associate Editor

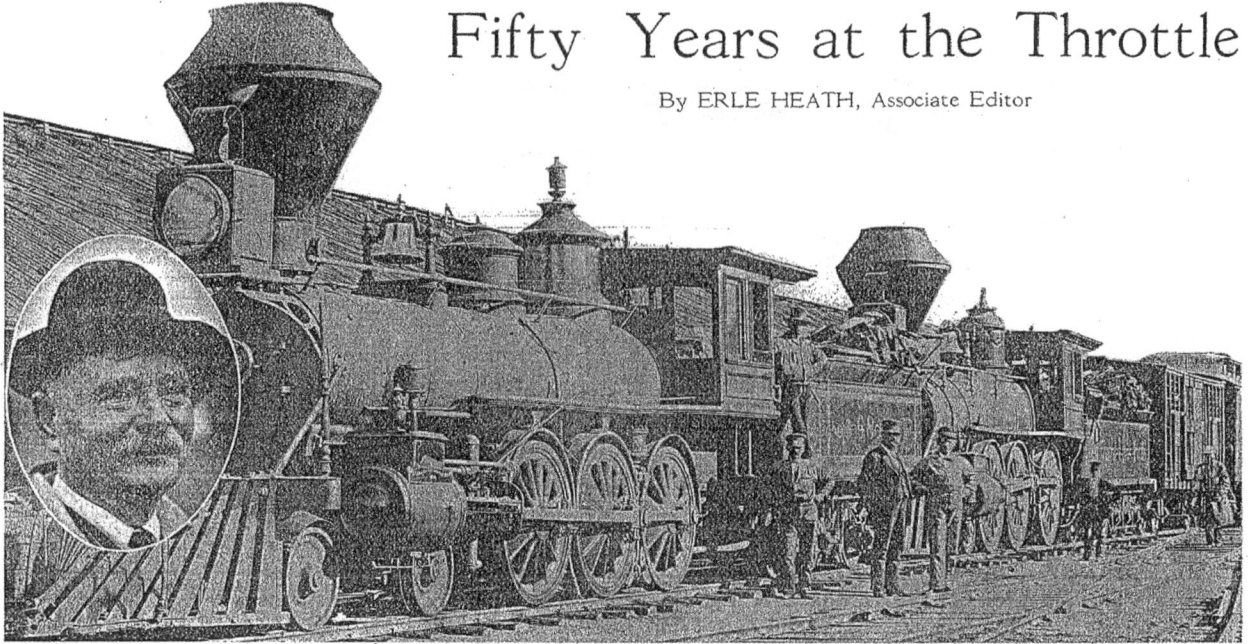

This picture was taken fifty years ago just a few months before Charlie Trott became an engineer. He is shown standing in the cab of the old Peoquop No. 68. Below him are Engineer Theo. Burt, Conductor Joseph Calderwood and Engineer Chas. Clark. The car in the rear of locomotive No. 198 is not a cattle car, but a ventilated fruit car of that day.

WHEN Charles C. Trott, locomotive engineer on Sacramento division, stepped down from the cab of locomotive No. 1453 at Sacramento roundhouse December 30 after bringing train No. 55 in from the round trip to Gerber, he brought to a close a career of 56 years 2 months service on Southern Pacific Lines. All of this service was spent on Sacramento division and for half a century he was in the cab as an engineer. This is believed to be a world's record for continuous service in this branch of railroad work.

Engineer Trott now takes a place on the honor roll of pensioners. His long service record is exceeded by that of only one living veteran, Patrick Sheedy, formerly superintendent of motive power at Los Angeles, who was with the company three months longer than Trott. Only two other engineers, Charles H. Ball and James Jefferson, were in the service more than fifty years.

Seventy years have been kind to Trott. He looks to be many years younger. As he sat in the cab of No. 1453 waiting for the signal from Conductor Wm. Schwab to start on his final run, his hand was as steady and his eye as clear as most of the younger engineers.

Not Anxious to Quit

"It wasn't what you would call a moment of joy," said Trott. "I had been working for the railroad since I was fourteen years old. Riding an engine was just to my liking and I didn't look forward with a great deal of pleasure to leaving the job. It's going to be hard to get used to a new routine, but I expect something will turn up to keep me from 'going stale.'"

All along the line from Sacramento to Gerber the news passed that Charlie Trott was making his last run. Charlie was known everywhere. He had been on the choice valley run for sixteen years. At every stop, from the first at Roseville until he got back to Sacramento in the evening, he met with congratulations and well wishes. This last trip was no "dinky" run, such as some veterans finish their active service on. In spite of his age Trott was still capable of handling any run on the division. On his last round trip he made the 251 miles in 7 hours 40 minutes running time, including 31 stops, four of which were from 10 to 12 minutes each.

Next morning Charlie was back at the roundhouse to "close up" his affairs. To Fireman "Mike" Gleason he gave his gold cap badge with the two stars given him for his fine work in fuel oil saving. The coat he wore when on duty went to Engineer "Spark Plug" Pelham, and his hat to Engineer "Buddy" Jeardeau. Another hat was given to Hiram Ford, helper who sets wedges, and a pair of gloves to "August" Bryant, air brake inspector.

A Tribute

By A. D. WILLIAMS
Sup't of Motive Power, Sacramento

CHARLIE TROTT is a man who faithfully and efficiently performed the task assigned him; one who keenly enjoyed his work and one who could be depended upon in every emergency. It is expected and hoped that he will long enjoy the rest to which these many years of active service entitle him.

Younger men in the service like to have a keepsake from a retiring veteran. Engineer Al Brown will have Charlie's old locker at Gerber.

Pasted on a window in the office of Roundhouse Foreman Bellhouse is a letter from Trott thanking his fellow workers for the large, overstuffed chair presented him Christmas Eve by a committee of veterans still in the service. The gift carried a message of camaradarie born of years of friendly association. Trott's letter is one of heartfelt appreciation.

Started When a Boy

Charlie Trott went to work as a boy for the Central Pacific at Rocklin. His parents had moved there from Volcano, Amador county, Cal., a few years after Charlie was born December 10, 1855. Rocklin was a busy railroad point at that time. It was here that young Trott saw his first locomotive. The little, puffing wood burners seemed wonderful to the boy. They were so bright and shiny in their gay colors and brass trimmings. He persuaded his father to let him quit school so he could become a railroad man.

Master Mechanic M. W. Cooley, who is now 84 years old and living at Santa Barbara, put Charlie to work in November, 1869, cleaning the little engines that were so fascinating to him. That was just a few months after the first transcontinental railroad was opened for travel. Thomas O'Connell, retired engineer living at Sacramento, who was a fireman then, likes to tell how he first remembers seeing Charlie Trott scampering to work at Rocklin with a bun in one hand and holding his breeches up with the other.

A little over two years passed before Charlie realized the first step in his ambition. He was given a seat in

Trott, seventy years young, sitting in the cab of No. 1453, the locomotive used on his last run out of Sacramento.

a locomotive as a fireman. It was one morning in April, 1872, that he started out of Rocklin with Engineer Jim King on the old "Bison," locomotive No. 57. But his first trip was destined not to be a long one. The locomotive went about ten car lengths and was derailed because a stub switch had been left open. Young Trott then had his first experience at a real hard job. There were no injectors on the locomotives in those days and no way of getting water into the boiler when the locomotive was standing still. Charlie had to pull the burning cord wood out of the fire box, stick by stick. He admits it was some task.

A year later he was given a switch engine in Rocklin yard and was an engineer for several months. Then he went back to firing. He also worked as machinist helper for a while.

Burn 20 Cords of Wood

"There was nothing easy about a fireman's job in those days," recalls Trott. "The engines were all wood burners and on a round trip from Sacramento to Truckee about 20 cords of wood had to be loaded on the tender and in turn shoved into the firebox. There was no time to loaf for the fireman. He could always be busy polishing the engine. I mean the whole outside of the engine, not just the brass trimmings. Some engineers would run their hand over the frame to see that all dirt and grease was gone. I have known firemen to be called back to the roundhouse to clean off some spot of dirt the engineer or foreman found.

"I don't remember ever being called back. I wasn't perfect in my work. I made mistakes then, just as in the

years since, but I have always tried my best to do the right thing. A job worth having is worth taking care of. That has been my theory. Back there in the early days I took care of the engine cleaning just as earnestly and conscientiously as I have watched my engine and tried to follow orders and rules since I started pulling trains over the main line.

"Locomotives certainly have grown in size and power in fifty years. Why, the old 'Bison' had 18-inch cylinders, four-foot drivers and weighed about 35,000 pounds. Today the most modern Southern Pacific locomotives have 29½x32-inch cylinders, five-foot 3½-inch drivers, weigh 640,200 pounds and are ten times as powerful as some of the locomotives of my early days. What an improvement it was when air brakes came into use. Still, the early day brakemen were very skillful and they 'pulled down' a freight train so that it could be stopped almost as smoothly as is done today.

"One of the funniest experiences I had while firing was with Engineer Provo, a Frenchman. That is, it seems funny to me now, but at the time it was far from humorous. There were no water glasses or injectors then. The pump worked by means of a connection to the cross head. Provo would invariably forget to shut off the pump when we struck a grade, with a result that water and mud would spout out over the engine. 'Well, Charlie, you better get out and wipe 'er off,' he would say. It was humiliating to me to have to do this clean-up job while out on the line. Other firemen joked me about it. I transferred to another engine the first chance I had."

Charlie was 21 years old when pro-

moted to engineer in April, 1876. For several months he had a switch engine at Rocklin. Then he was given the locomotive "Dragon" No. 74 and went out on the main line in road service. He alternated in helper and switch service at Truckee and Rocklin until April, 1882, with exception of a couple of months work at Sacramento shops late in '79 when business was slack. Illness caused him to take out a leave of absence in 1885. He returned to his engine at Rocklin in July, 1886, and remained in continuous service on Sacramento division until retired.

"We used to have our own engines," says Trott, "and we were mighty proud of them. We took care of them like they were persons. Fact is those little engines often acted like they were human. Sometimes they had to be petted and coddled before they would finally settle down to real work. Others were so mean they simply wouldn't get up steam for some engineer but would purr right along for another fellow. Of course we got used to our engines and knew their peculiarities, so probably it isn't odd that they didn't want to work for a stranger. A man got to love his engine like his wife and family; maybe some engineers loved their engines best.

"My first love was my engine. But in '81 I was married and the little engine dropped to second place in my affection. My wife was almost as fond of railroading as I was. We raised three children. My two boys, Charles Byron and Samuel Edgar, served their apprenticeship in the shops at Sacramento. They now have good positions as machinists in the automobile industry. Since my wife died about ten years ago, my daughter, Grace, has been my housekeeping companion.

Lost His Own Engine

"My last engine, before the company took over the motive power, was No. 1377. I say, 'before the company took over the engines,' because we all felt like the locomotives were our own. When an engineer was off duty, his engine was off duty, too. Naturally we hated to lose something to which we were so attached. Of course, the old arrangement caused a lot of power to be uselessly tied up.

"Some fellows have an idea we engineers did just about as we pleased in the old days. But we had inspections then that would make the youngsters today fairly gasp. Take, for instance, 'Ben' Smith, road foreman of engines during the '70's from El Paso to Ogden, and D. J. Brown, who was for years foreman of the roundhouse at Rocklin. All the old-timers remember these men for close inspections and strict discipline. Just the same, they were fine, likeable men.

"We never knew when we were going to pick up 'old Ben.' It might be at a wood pile or water tank. The news soon spread when he hit the division. Enginemen and trainmen passed the word along to be on the lookout for Ben. But even with this warning he had a faculty of showing up most unexpectedly. His eyes were keen as a cat's. If there was anything wrong with an engine from the top of the stack to the bottom of the tender 'old Ben' saw it right off. We used to say he could see a loose nut half a mile away. The first thing to catch his eye when he got in the cab was the steam gauge. Maximum boiler pressure was then 125 pounds. If the gauge registered one pound more we sure heard about it.

"After Master Mechanic A. J. Stevens died Ben resigned and went to Towles Station to live. One day a few years later when I was waiting for the staff at Towles, 'old Ben' came up to my engine. 'How many pounds you got, Charlie?' he asked. Two hundred pounds, Mr. Smith,' I answered. 'Look out, Charlie, or you'll sure get blown to the devil,' the old man warned. Of course, maximum pressure had increased since he was on the line, but he couldn't imagine any engineer safely handling more than 125 pounds.

"Throwing the Babbitt"

"I don't suppose there is a veteran engineer who ran over the mountains that doesn't recall many and frequent sessions with Roundhouse Foreman Brown. 'Throwing the babbitt' was the offense he took particular exception to. If the call boy followed us home with the message that Mr. Brown wanted to see us, it was almost a cinch he had been looking over our engine and found that the crank pin had gotten hot and the babbitt melted away from the brasses. This act was inexcusable; no need to make explanations; he knew we were just careless.

"I respected Mr. Brown's point of view, but just the same I knew it was almost impossible at times to keep from melting the babbitt. So I planned to get out of being called on the 'carpet.' I managed to get from the storekeeper a block of babbitt and several sheets of emery paper. During a layover at Truckee I could babbitt and reset the brasses, or polish off the pin if there was evidence against me.

"We used to have some fun with Mr. Brown later when he went out on the road again as an engineer and 'threw' more babbitt than any of us. He laughingly admitted that it couldn't hardly be helped. I got a letter from him a few days ago. He is living at Napa and is 88 years old."

Known Everywhere

Probably no employe on the Pacific Lines is more widely known than Trott. Having spent a lifetime on Sacramento division, which was the first link constructed on the company's lines, he is known to the oldest men in the service. From among his friends have come many of the company's officers. He knew J. H. Dyer as a boy at Colfax before the present general manager was a brakeman for the Central Pacific. As a young man he worked at Rocklin with "Pat" Sheedy, former superintendent of motive power. While Assistant General Manager T. Ahern was working his way up from his first job on the section, Charlie Trott was one of his closest friends.

"Not long ago," said Trott, "I had to appear in court as a witness. The judge and the jury seemed startled when I answered I had been working for the Southern Pacific 56 years. I thought my word was going to be questioned, when the judge turned to me and said, 'That must be a fine company to cause you to stay with it so long.'

"Well, we have got a fine company. I am going to receive a liberal pension that will be a big help in making the days to come more comfortable and pleasant. I am perfectly satisfied. I gave faithfully of my services and I have been given a 'fair deal.'"

Among Trott's cherished possessions is a personal letter from former President Taft thanking him for the "painstaking care and skill" with which he handled the President's special train in December, 1911.

May Go to Europe

"I don't know what I'll do, now that I am out of a job," he answered Superintendent W. L. Hack. "I can't seem to get out of the habit of figuring when I am due out again to Gerber. I catch myself studying my watch every little while. I think I'll have to get away from Sacramento to keep from wanting to wander over to the roundhouse. I've always wanted to go to Ireland and England. My mother was Irish and my father was English. John Wright has been back there since he was pensioned and wants to go again. Albert Brown wants to go, too, soon as he is pensioned. Maybe he and I and Tom Newton will join John on a trip back to Belfast next summer.

"This job of being a pensioner may not be so hard to hold down after all."

SEVERAL OFFICES MOVE TO NEW BUILDING AT S. F.

A newly equipped kitchen and commissary, also ticket office and waiting room for auto ferry patrons, and space for several departmental offices have been established in the building formerly occupied by the San Francisco ferry post office. This building now becomes Annex C to the General Office.

The new commissary and kitchen, which supplies the restaurants on the Company's ferry and river steamers, include many new catering facilities in the equipment and will be a model of its kind.

New location of the auto ferry ticket office and waiting room will greatly facilitate the handling of auto traffic and add to the convenience of patrons. It will be a few weeks before the new office is opened for service.

Steamer Department offices under Captain C. F. Heath are now located in the new building, as is also the revaluation department and the time inspector's office.

Time Inspector S. A. Pope has added to the activities of his department in handling the maintenance and repair of the 1600 clocks used on Pacific Lines. Clocks in need of repair will be given immediate attention by his department instead of by authorized local jewelers.

The largest office in Annex C is that of the district auditor, a newly created position held by C. Peterzon. This office combines the accounting work from divisions and stores in the San Francisco bay district.

Charlie Trott is shown here receiving the congratulation of Superintendent W. L. Hack; and also enjoying a few leisure moments in the comfortable, overstuffed chair, a gift from his fellow workers. The little locomotive No. 57 was the one Trott started firing on in 1872. The veteran engine was rebuilt and renumbered several times and was scrapped about 1905 after being in Oakland local train service. It is interesting to note the old time tallow oiler on the steam chest; water pump under boiler worked by an eccentric; counterweights on drivers secured by bolts; wooden brake beams only on rear tender truck; wooden tender frame and wooden cab with fine cabinet work; elaborate lettering and beautiful scroll work in gold leaf; brass mud guards over drivers and brass strips on tender and engine.

My Half Century of Railroad Service

The dramatic story of a retired mechanic who saw our first railroad grow into the present Pacific System

By J. O. WILDER

WHILE I would prefer to leave the service of the company as I entered its employment, unheralded, yet I feel a sense of duty to my co-workers with whom I have been associated for fifty-four years which impells me to relate some of my experiences as a pioneer railroad man. I have noted the leaps and bounds in the progress made from the days of the "Dutch Flat Swindle" to the actual carrying out of the gigantic plans of the "Big Four."

The courage, zeal and determination of the great Central Pacific leaders was an inspiration to all who came in contact with them or realized the magnitude of the task they had set about to accomplish. They were real Americans, and with true Yankee push swept aside all obstacles and pushed up those rugged and almost impregnable mountains.

Having witnessed and marveled at the gigantic undertaking, coupled with the rapidity with which the work was carried forward, I cannot say too much in praise for such leaders. And I am thankful for the small part that I have had in the undertaking.

It is a pleasure to write the names of Leland Stanford, Charles Crocker, Mark Hopkins, S. S. Montague, L. M. Clement, Arthur Brown, superintendent of bridges, and last but not least that wonderful superintendent of construction, William Strowbridge. The latter had but one eye, and because he could see more with that one eye than many men can see with two he was known among the Chinamen as the "One-Eyed Bossey Man."

I will preface my commencement with the Central Pacific by saying that on May 30, 1866, I left my home in San Francisco, a mere youth with my blankets and carpetsack, with canvas pants, high-top boots, and with seven dollars and a half in my pockets, to find myself among strangers, to take my place in the employ of the Central Pacific. I left San Francisco at 4 o'clock in the afternoon by the steamer "Capitol," that being the only route to Sacramento at that time. Reached the Capital City the next morning.

The usual runners for the hotels were there in full force at the landing, which was at the foot of K Street. One runner, whom I learned years after was William Land, for the Western Hotel, grabbed my blankets, bag and carpetsack, threw them into the Western wagon, and I expected to be thrown in myself, as I had seen runners perform in my home city, but to my surprise said Johnny (that is what they called small boys those days)

At the request of THE BULLETIN, *Mr. J. O. Wilder of Sacramento, recently retired after fifty-four years' continuous service, has written his experiences. The first part of his story is published herewith. Parts two and three will appear in the November and December issues, respectively. As a California pioneer railroader, Mr. Wilder presents a graphic picture of the obstacles faced and overcome by the builders of the old Central Pacific, and of his own connection with the Central Pacific and the Southern Pacific. Loyalty and determination to win against heavy odds,* THE BULLETIN *readers will note, stand out as characteristics of those men who left an indelible impression on the early history of California.*

was told the Western was one block up the street, you may walk. After eating breakfast I went to No. 54 K Street, which was the main office of the Central Pacific at that time. Being the first there I sat on the steps with my blankets and carpetbag and awaited the coming of Mr. S. S. Montague, who was chief civil engineer and to whom I had a letter of reference. He arrived with Fred Steiner, his chief draughtsman.

Life in a Railroad Camp.

It was fortunate for me that Mr. Montague was going up to the front that morning. So with my blankets on my back we walked to the depot, which was on Front, J and K Streets. The train consisted of two passenger coaches, which were about the size of maintenance and work train coaches of today. The baggage car resembled that of a caboose of today, minus the observation top, with side doors. This train was drawn by Engine Pacific No. 2, and was known as Train No. 1. The main line at that time turned into I Street, leaving I Street at Fifth, along the levee to Sixth Street, Sixth to B Street levee, crossing the American River east of what was known at that time as Rabeles tannery.

Our first stop was Rocklin. Other stops were Auburn, Illinois Town, New England Mills, Long Ravine Bridge, then the terminal, Gold Run, which was the end of the track. Here Mr. Montague took his span of horses with light wagon and drove up the dusty road to Dutch Flat to what was known as Chinnie Ranch, now called Alta. Here Mr. Montague was met by Mr. L. M. Clement, assistant chief

engineer. I was invited to dinner with them. Tin plates and cups were used but the food tasted very good to a hungry boy, even if it was my first experience in a railroad camp eating house. After dinner Mr. Montague and Mr. Clement inspected the roadbed, which was then about completed to this point. The tracklayers were working east of Dutch Flat.

After the inspection they started up the line. I was told to ride Mr. Clement's white mule, and what that mule didn't know was not worth knowing. For instance, he knew that he had something of lightweight and seemed to enjoy it, but not so with yours truly, for I had been in the habit of riding on the back of wagons or a street car. While he trotted along with his nose close up to the light wagon I would have given my last five dollars for a pillow. When we reached a watering trough near Emigrant Gap I found myself standing in the stirrups and was glad to get off his back to unloosen the check rein on the horses. I then led his muleship to get his drink. He would drink and then take a sniff at me, as much as to say, "What have I had on my back?" In fact, he had never seen a boy before.

As we passed along up the line of construction every now and then we would hear some gang foreman sing out "Fire," then the explosion, with rocks flying in all directions. Mr. Montague had some trouble with his horses, but to my surprise and delight his muleship didn't seem to care half so much about those rocks as I did. This was my first experience, but was not to be my last. At Emigrant Gap the wagon road leaves the main line and up through a small valley to the top of a mountain, then commences the down grade to Crystal Lake, where we arrived about sundown, which was the headquarters of Mr. Clement, and also the party headed by Engineer McCloud, to whose party I was assigned for a time, doing cross-section work. I remember while with this party, of 200 kegs of powder put in a cliff, called Crocker's spur, which lifted the whole mass to the Yuba River below.

Blazing the Iron Trail.

Pine trees were but pipestems in this crash, and Crocker's Spur became a thing of the past after that shot was fired. It was also while with this party that I saw for the first time a big Piute Indian. He had his squaw with him and he became quite interested in the leveling instruments. After some persuasion Mr. McCloud got him to place his squaw in front of the instrument, turning the big end towards the Indian and he looked in. It made her appear a mile away. He would look in the level, then at his squaw, then finally gave a big grunt and said "Talkey lie" and went upon his way to the lake. There were, at the time of which I am writing, some Indian hieroglyphics on a large table rack on the east side of the lake close to the wagon road. They were worth while looking at, and I don't know whether the general public know of their existence or not.

The only bad feature about these large bowlders was that there seemed to be a rattlesnake's den thereabouts. One evening a Mr. Scuppem, much to our surprise and disgust, came into the cabin with a rattlesnake which he had caught as it was coming out of its hole. He held it behind the head. It was about eighteen inches long and the rattles were in splendid working order. I have never in my life heard so many unkind remarks made to a man's face. Some of the more angry victims of this peculiar prank wanted to throw him into the lake that his ambition for snake hunting might be cooled.

About this time I was assigned to the locating party east of Crystal Lake. The camp was about half a mile east of the town of Eatonville, on the Donner Lake wagon road. This town is now known as Cisco. With my blankets and carpetbag I boarded a fast freight wagon to make another new start on life's journey.

A Bed on the Boughs.

At the new camp I found one man and from his general makeup took him to be the cook, and here was my chance to make friends with him for we were alone among the big pines. He was a German. He asked me where I was from. I told him from San Francisco, which seemed to strike him just right, as he was from there himself. He gave me a piece of bread baked in a Dutch oven; it tasted good to me for I could eat almost any time. The sun was fast sinking in the west when the party returned to camp. I presented my letter to Mr. Guppey from Mr. Clement. He introduced me to the members of the party by saying, "Boys, this is our new backflagman." The party included C. H. Guppey, engineer and transit man; John Currier, leveler; Arthur Ledley, leveler's rodman; Chauncie Brainard, head chainman; John Harding, back chainman; M. N. Denton, stake agent. The cook they called "Haunse." I never knew his right name; and myself and a little dog called "Tip" comprised the party.

I was assigned my sleeping room in one corner of the tent. Mr. Ledley gave me a small hand axe and went with me to find a small fir tree from which to cut boughs to make a mattress for my bed. The ends of the boughs I stuck into the ground, laid my canvas on them, then my blankets and my carpetsack for a pillow. Supper was served in an old log cabin a few feet from the tent. The table legs

INTERESTING PICTURES OF J. O. WILDER

At left as veteran employees remember him; at right as he appears at the conclusion of his service. The interesting picture below shows him in a uniform in which as a youth he served in the Civil War.

four in number, were made from small trees and had been driven into the ground on which rested the boards with benches for seats. The supper that evening consisted of a mulligan stew with good bread, of which I had tasted a few hours before, with dried apple sauce, coffee and brown sugar, eaten from tin plates and cups. Everyone seemed to enjoy the meal. I know I did, for I was always ready to eat.

The cabin was used by Mr. Guppey. He would plan his day's work while others sat about the campfire. I was the center of attraction that evening and was kept busy replying to questions about myself and fighting mosquitoes. The mosquitoes took kindly to me and were very friendly. They were as big as hornets. At bedtime I made a small fire in the tent and smoked them out. They went to the rocks in the Yuba River, wet their bills and returned to sing their sweet songs about the tent. They were tireless workers and laid plans to get in while we were sleeping, which some of them did. It seemed to me that they had me spotted and made straight for me. After a few days and nights one

would have thought I had the chickenpox. I hardly knew myself.

Had the Axes to Grind.

The first thing I heard in the mornings was "Haunse" blowing his horn, the signal for us to get up. I was always the first up, and after putting on my canvas pants, leather boots with hob nails and short spikes in the heels, would go to the river, take a wash to return ready to eat. After breakfast the first morning I was shown how to hold the red and white rod with sharp point on one end, and was then sent to turn the grindstone for Mr. Brainard and Mr. Harding to sharpen their axes before going up the line.

My hands were soft and I was soon changing from one to the other and had blisters on both. I had not as yet fully recovered from my ride on Mr. Clement's white mule, but I stayed with the job. We started up the mountain to commence work where they had left off the day before. We went through underbrush, over rocks, and I was not slow in finding out what the spikes were for in my boots, for they saved me more than once from falling. We picked up the line about two miles east of what was known as Butte Canyon.

Mr. Guppey set his transit on the first peg with a tack in it, I the next one back with a stake driven beside it marked C. T. with red chalk, which I had no trouble in finding. We had not gone more than 1000 feet when the line struck a big pine, which meant it must come down, and it was here that Brainard and Harding came in with their axes. After two hours' work the giant fell, which meant another turn at the grindstone the next morning. We did fairly well that day, which brought us up to about opposite Eatonville.

(To be continued)

The Way Pioneer Builders Met Difficulties

Obstacles in wilderness served only to arouse fight in men who blazed the way for present day rail lines

By J. O. WILDER

NOTHING more of interest happened until we were about a mile above the present town of Cisco. Here we came to a deep gorge where the sides were so steep that we had to use a rope to get down to the bottom. To get up the other side was a problem, but Mr. Guppey was equal to it. He sent Mr. Brainard and "Long John," I mean Harding, to a place where it could be crossed, taking with them their axes to fell a tree across at some point where it could be found. They worked their way up the other side, which was no easy task, there being much underbrush to contend with. The water was deep for that time of the year, and Ledley, Denton and myself were sent to find from whence it came.

At the top of the mountain we found a lake. Those with me thought it was about a mile long and a half mile wide. At the outlet we found a deserted log cabin which at one time had been the home of a trapper. Having found what we went for we returned to the line. On our way back we met a bobcat and shot him. On our return Ledley made his report. Brainard and Harding had come up from the other side, hitched a rope around a tree and were then ready to go ahead with the line. It consumed all of one day to cross this outlet. About one mile east we came upon another stream not quite so deep as the first. In this we discovered a beaver's dam. We did not disturb it.

These gorges we called "Kidds Outlets." They are now known as the Cascades. We were now five miles above our camp, so we spent one day moving to Driver's Creek, about two miles northeast of Tinker's station. This move was a bad omen for our party, for when nearing Summit Valley we met with our first accident. Mr. Guppey fell and his leg was broken. He was sent to Sacramento for treatment.

Explore the Mountain Tops.

While we were waiting for the engineer we visited Soda Springs, also Castle Peak. This was the hardest climb we had undertaken. It was now the first of August, and there was still snow on the north side of the mountain. We were compelled to cross over some of the patches, which nearly cost Mr. Ledley his life, for he slipped and fell and finally brought up against a big rock, but none the worse for his slide. We reached the top, ate our lunch, wrote on a piece of paper who we were, each signing his name, placed it in a bottle, corked it tightly and placed the bottle, big end up, with rocks around it. Here it rested un-

This is the second installment of the experiences of Mr. J. O. Wilder, veteran employee of the Southern Pacific, who recently retired after more than a half century of service. Mr. Wilder told in the first installment, which appeared in the September issue of THE BULLETIN, *of his first duties as a railroad man. In this chapter he relates some of the hardships encountered and overcome by the pioneer builders of California's railroads.*

was found by C. H. Bonty, late chief clerk for the Southern Pacific Company. He and his friends signed their names to the paper and replaced the bottle where they found it, where it has been since August, 1866, and is there to this day so far as I know, for I alone remain to speak for them, as they all have passed off the Iron Trail.

With the coming of Engineer Stevenson we completed the survey to the summit. Here the party was broken up, some to return to their homes, others to Kidder's party in Nevada. I remained with Stevenson and helped him lay out the heading in the shaft of the Summit tunnel. From here I was sent to join Joe Wilkinson at Cisco to lay out the town and sidings. The construction crew was at work on the main line. He would barely get his transit set before some "China" herder, as they called them, would sing out "fire," and he would then have to take up his transit and make a run for a pine tree shelter. I was with him about a week. When this work

was completed I returned to the Summit and Wilkinson went to Sacramento to take charge of the building of the shops, which were then under construction. About this time Mr. Clement moved his headquarters and also Mr. Strowbridge, and I saw much of the latter.

When Chinamen Would Strike.

At times the Chinamen would strike and refuse to take their shifts in the tunnel, but Strowbridge was ever on the job. They feared him in their hearts as much as they did the Chinese devil. He was a fine general. He had a mild but firm way, which was in the form of a pick handle, in dealing with these fellows. He had but one eye, yet he could spot the ringleaders at one glance and would bring his persuader into action and was not particular where it landed, for he was a pastmaster in this line. Inside of five minutes you could not find a Chinaman in camp, and could hear them say "muck-a-high" as they went to their work with Strowbridge acting as escort, for many times he was called upon to settle discords and confusion among the Chinamen.

It was now September. I remember distinctly the nitro-glycerine explosion which took place about this time which cost the mixer his life. He was blown to bits. This accident retarded the work in the tunnels, which were solid rock. It was here that the supply department did some fast work. The delivery was by fast freight wagons and the department was in charge of Superintendent Pratt. He was never known to return short, so in a week's time we had the chemicals to mix more nitro-glycerine, and the new house was ready when it arrived, but

PACIFIC RAILROAD.
Time Table No. 6. To take effect Feb. 14, 1865.

Freight. No. 3.	Mail & Pass. No. 1.	Distance to Sac.	STATIONS	Dis. to Newc.	Mail & Pass. No. 2.	Freight. No. 4.
2 P.M. Dp	6.30 A.M. Dp		Sacramento	31	11.50 A.M. Arr	7.25 P.M. Arr
2.35	6.55	7	Arcade	24	11.25	6.55
3.00	7.10	15	Antelope	16	11.10	6.30
3.25	7.20	18	Junction	13	11.00	6.15
3.45	7.30	22	Rocklin	9	10.45	5.50
4.05	7.40	25	Pino	6	10.35	5.32
4.40 P.M. Arr	8.00 A.M. Arr	31	Newcastle		10.15 A.M. Dp	5.00 P.M. Dp

No Trains will leave any Station ahead of time, unless specially ordered by Superintendent.
Gravel and Extra Trains must keep ten minutes out of the way of all regular Trains.
NIGHT SIGNALS—A light swung over the head is a signal to go ahead. When swung across, or at right angles with the track, is a signal to back up, and when moved up and down, is a signal to stop.

C. CROCKER, Superintendent.

These are the men Mr. Wilder learned his trade with at the Sacramento shops after his experience in the open country with the construction and engineering forces of the old Central Pacific. Wilder is the second from right in the first row. All but one of those who appear in this picture are now dead. Henry Thiel, the tall man in the back row, is the lone survivor.

was built farther away than the old one.

About this time I received orders to join Mr. Cadwalader's party at Prosser Creek. I rolled up my blankets, put my belongings into my carpet sack, had breakfast and waited for the fast freight wagons going that way. I was given a small Spencer rifle and a box of cartridges. I got to Prosser Creek, where I met the party with pack mules and started down the Truckee River. It was a hard tramp along an old trail that may have been the old Emigrant road. We went downstream about fifteen miles, at which place we made camp at sundown. This camp later was known as Camp 24, six or seven miles west of the State line.

Scare in the Night.

The river at this time of the year (September) was quite low. We had passed several parties of Indians spearing trout. All we got from them was a sour look. Farther down we met two more, from whom Cadwalader bought some big fine trout which he cooked for supper, while we unpacked and put up the tent, spread our blankets and gathered wood for the campfire. Supper over we were getting ready to turn in, for it was getting dark. This night I shall never forget, for we were greeted with one yelp after another. None in the party had heard the like before. We had

heard that the Kidder party had some trouble with the Indians and thought they were after us. Cadwalader told us to take our rifles and cartridges and pair off and make for the rocks. Fred King, taking me with him, keeping the campfire between us and the river and keeping a sharp lookout from behind. Here we staid about two hours, with a continual yelp. Finally one of the party fired a shot across the river. This made my hair, which had not been cut for some time, stand straight up, but this one shot won us a victory, for we never heard another yelp and concluded it was some kind of an animal. We later learned that they were called river ottos.

Cadwalader, besides being head of the party, was also chief cook, but he tired of the job, so he put me on the cooking staff. He soon found out that he had made a mistake in his choice. I also found out that I was not much of a cook. He was very fond of hot biscuits, but I was not at my best in that line. I got too much baking soda in them, which made them as yellow as gold. He could not eat them, nor any member of the party, so I lost my job, but did not lose any sleep over it.

Two weeks later there was a man sent to take my place. I returned to the Summit. So the next morning, bidding all goodbye, with my blankets

on my back and Spencer rifle through the ropes, and with my carpet sack, started on my fifteen-mile walk up the Truckee River alone, and as luck was with me did not see an Indian. Now and then a chipmunk would give me a greeting. I walked as fast as I could with the weight I carried on my back. When I came to Prosser Creek I could not find the log, so I took off my pants and waded the creek with part of my belongings and then went back for the balance, got loaded up again and started on my last stretch to the main wagon road. Arrived about noon: sat down and took a piece of hardtack for dinner.

It was my intention, if I had to walk, to make Coburn's station by night. I want to say here that that station is now the town of Truckee. My good luck staid with me, for I was picked up by one of company's fast teams, the driver greeting we with "Johnny, what are you doing here?" I told him that I belonged to the civil engineering corps. He told me to put my things in the wagon and get up on the seat with him. I was happy to have the chance to ride, and more than happy to know I was going somewhere nearer God's country.

We arrived at the summit at sundown. Here I learned that they were getting ready to move the terminal to Cisco. The trains at this time were running to Blue Canyon and work trains as far east as Bear Valley tunnel, which was nearly completed, and the grading 90 per cent completed to Cisco. Then but a boy I noted the rapidity with which these men had pushed the work forward. They had a system like clockwork, even down to Mr. Scovey, who was foreman of construction of culverts and bents for bridges.

Saved the Tipping Rock.

It was his men who later on tried to roll that tipping bowlder from its place on Truckee River, as they feared it might some day roll down and destroy the tracks, but they could get it so far and no farther. And the hundreds of persons who have since visited it, together with the party who now has it housed in, are all indebted to Mr. Scovey, for one of his men wanted to put a shot of nitroglycerine under it and blow it off. He said no, let it stay. I did not see this, but heard Mr. Scovey tell of what had happened. I saw this rock many times after I had been transferred to Camp 5, or I may say Tunnel 13.

This was the best conducted camp I had been in, and it proved to be the last. It was kept clean, all credit falling to Andrew Helmer, who was the camp boss; nor would he allow any "snake bite" in the camp. Mr. William Phelps was the engineer in charge of the tunnel which goes from the Donner Lake side to Cold Stream Canyon. This tunnel is on a ten degree curve and if I remember correctly is 980 feet long. It was timbered with 12x12 studs and the same for arches, eighteen inches apart, and lagged. This work was directed by Charles Fischer, foreman carpenter. He also had charge of the company's sawmills at the lower end of Donner Lake on Donner Creek, where the timbers and the lagging for the tunnel were gotten out and also the bridge timbers. This mill was run day and night.

"Missouri Bill" and His Team.

And right here I would like to speak of a man who was a source of much amusement to me, a tall, lanky person known as "Missouri Bill," who drove the bull team, hauling timbers to both ends of the tunnel. This man could swear by note and never miss one, and he had the voice of two men. You could hear him "cussing" those bulls half a mile away, and the crack from the whiplash he carried. There seemed to be one bull in the team which he liked to pick on, for he would swing with his whip over his head and yell, "Come down around here, Dolly Gal, doggone your ugly skin, I'll knock a horn off you."

Winter was now fast approaching, for the grouse would get on a log and hammer with their wings, and it proved to be a very severe winter. The snowfall was heavy and it was very cold. There was one large snowslide at Strong's Canyon, known as Camp 4. In this camp were two gangs of Chinamen for Tunnels 11 and 12, also a gang of culvert men. The slide took it all, and one of the culvert men

was not found until the following spring. At our camp the snow was so deep we had to shovel it from the roof and make steps to get to the top of the snow. We were snowed in, and our provisions got down to cornmeal and tea. Had it lasted on week longer we would have been compelled to eat horse meat, for there were two hundred or more men in this camp. We broke a road to the main wagon road, where there was a store known as the Donner Lake Postoffice.

There was a complete blockade east of Blue Canyon, and all construction work had to stop with the exception of the tunnels. Between Blue Canyon and Cisco the cuts were filled by landslides, which had to be removed by gangs of Chinamen. Steam shovels were unknown at that time. A push plow loaded with pigiron to hold it to the rails, with three engines behind, would back up and take a run at the snow and keep going till it got stuck, and then back up and take another run. They had to use care not to get the plow off the track, for the roadbed was soft, and in this way the track was cleared to Cisco. This work was in charge of a big Yankee named Nat. Webb, of Sacramento, with Add Vahay, conductor, also of Sacramento, and engines Nevada, Utal, Oneonta, with Pence, Carroll and Mills, engineers, and one of the firemen whom I remember was Jim Ireland.

(To be continued)

E. L. KING NOW SUPERINTENDENT SALT LAKE DIVISION.

Mr. E. L. King, former superintendent of telegraph, has been appointed superintendent of the Salt Lake Division with headquarters at Ogden, Utah, vice Mr. B. A. Campbell, assigned as assistant superintendent of the Coast Division. Mr. King left San Francisco on October 1st, the day the appointment became effective, to take up his new duties. His advancement in railroad circles is another example of the many opportunities that are open to those who begin at the telegraph key.

From a messenger boy for the Postal Telegraph Company in 1891, Mr. King rose rapidly, first as an operator and office manager in various California towns for the Postal and later as a train dispatcher, agent and chief dispatcher for the Southern Pacific Company. He entered the service of the company in 1894 as an operator on the Sacramento Division.

From 1906 to 1907 he was chief dispatcher of the Sacramento Division with headquarters at Sparks, Nev. He then returned to the superintendent's office at Sacramento as train dispatcher, and in 1910 was appointed chief dispatcher of the Shasta Division with headquarters at Dunsmuir. In February, 1914, he was appointed superintendent of telegraph with headquarters in San Francisco.

Mr. Edward Entelman, former assistant superintendent of the Western Division, has been appointed to succeed Mr. King.

LOS ANGELES DIVISION NOTES

Effective September 1, Mr. S. C. McClung, chief clerk, Los Angeles Division, for the past twelve years, was transferred to the claims department at San Francisco. His many friends and associates on the division, while missing him greatly, are glad to see his ability and faithful service recognized by this appointment. As a token of friendship and high esteem in which Mr. McClung was held by all, a farewell gathering of division officials and clerical forces was held in the superintendent's office, at which time appropriate remarks were made by Superintendent W. H. Whalen, a beautiful handbag toilet outfit as well as a portfolio being presented as a remembrance.

Mr. McClung was succeeded by Mr. A. H. Oberg, who for a number of years has served as assistant and head clerk in the transportation department, Mr. E. J. Cleugh succeeding Mr. Oberg.

Mr. T. A. Murphy, at present agent at Santa Ana, is being granted an indefinite leave of absence on account of ill health.

We are undertaking the ballasting of the Imperial Valley branch, also Inter-California Railway.

At Santa Barbara a new concrete platform is being laid at the passenger station.

At San Pedro the Banning Company, which handles freight for the Southern Pacific Company over our wharves, has placed in operation at our lumber slip a locomotive crane which will greatly facilitate the handling of lumber at that point.

All stations which were consolidated with the Santa Fe and Salt Lake Railroad during Federal control have now been reopened by our company, the last to be reopened being Long Beach, on September 15th. With the completion of the new bridge over flood control opening in that vicinity our company is now in a position to handle all traffic the same as before Federal control.

Co-operative bodies will learn by hard experience the same lessons that corporations have learned—that it is necessary to pay high enough salaries to attract responsible and competent executives and managers. At that, the high salary doesn't make a man competent. It takes a lot of business judgment to pick a competent man.—Oregon Voter.

Successful regulation, manifestly, must be elastic enough to enable the roads to earn living wages. That does not mean a constantly changing scale of rates, but it does mean that regulation shall not sentence those properties to long terms of penury. The railroad competently managed should not have to eat, sleep and drink with the specter of insolvency; it should enjoy the prosperity which capacity commands in strictly private industry.—St. Louis Post-Dispatch.

Sacramento Shops Kept Pace With Demands

Rapid development of Southern Pacific marked by loyalty of pioneer employees who made good in early days

By J. O. WILDER

WITH the road open again we were assured of something to eat. The work in the tunnel was progressing rapidly, for we could hear the gangs in either heading 200 feet apart. Engineer Phelps, Fred King and myself were giving grades and centers. As spring approached we got our stakes ready to cross-section from the tunnel up to the head of Cold Stream and down the other side (this is now known as the horseshoe) to the Truckee River.

One morning at 1 o'clock in early May the two headings were broken through. The honor of this event fell to Helmer in the east end and Dow in the west end. These two men were shift bosses, as were Deardoff and Dave Dramer. They worked twelve hours straight, with no Saturday afternoons off and ten days' vacation was unknown. It was just plain work and push ahead and make every move count. I was sent to the summit with a message to Mr. Clement telling him we had broken through. He read the message, got on his white mule and started for our camp. He got there before I did, although I took a fast freight team.

I did not see him on the road, but we will leave that to the white mule, for he could go almost anywhere and had a way all his own in getting there, as he could pick his way up a mountain or go down one as well as a man, and when it became too difficult to walk he would slide. Anyway, Clement and the mule were there when I arrived, which was late in the day, and Mr. Clement spent the night at our camp. I tied his muleship in the stable. Next morning Mr. Clement, Mr. Phelps and the party went up to the tunnel, tried the centers from both ends and found that it was just two inches out. It was a wonderful piece of engineering at that time. So far as I know I am the only one left of the civil engineer corps at Camp 5. Mr. Phelps and Mr. King have long since passed from the iron trail.

Back to Sacramento.

In June we laid out the bents for the bridge across Donner Creek, also the big culvert at the head of Cold Stream. Toward the latter part of the month Mr. Montague requested that I be sent to Sacramento for transfer to the locomotive department. So on the 30th day of June I rolled up my blankets, put my belongings in my carpet bag and bidding them all goodbye I made ready to start. The last one I called on was a man I had learned to love. He was a man well up in years, and like myself, alone. He had always been kind to me. Putting his arms around me he kissed me

This is the third and final installment of the interesting series written by Mr. J. O. Wilder, veteran employee of the company. Mr. Wilder recently retired after fifty-four years of continuous service and his experiences have been read by Bulletin readers with a great deal of interest. During all his length of service Mr. Wilder was only "on the carpet" twice. His recollection of many of the pioneers and his description of the old shops at Sacramento form a valuable addition to the historical data of the Southern Pacific Company.—The Editor.

goodbye with the tears rolling down his cheeks. He said, "Johnny, be a good boy, for we will never meet again," and his words have come true; and it is a pleasure for me to write the name of C. H. Davis, chief cook at Camp 5, and I still have my boyhood regard for this noble hearted man.

Seated on a fast freight wagon and all wishing me good luck I bid Camp 5 my last goodbye. At the summit we stopped to change horses. I stepped in the civil engineer's office to say goodbye to those who were in at the time. On my way down to Cisco (which at this time was the terminal), where passengers took the Pioneer Stage line for Virginia City and Gold Hill and other points in Nevada, I noted the rapid progress the construction gangs were making, one being within a half mile of the west end of Summit Valley, and it was plain to be seen that the iron horse would be there before the tunnel was completed.

I arrived in Sacramento on July 3, 1867. I reported to Mr. Montague, upstairs over 54 K Street, where I had reported to him one year before. He looked me over, told me to go and get a hair cut, clean myself up, get a new suit of clothes, then come back. I bought my outfit from John Silvercrup's store on the northwest corner of Third and K Streets. John Trauback waited on me and my complete outfit cost $25. I went back and could see by the expression on Mr. Montague's face that I looked more presentable. He took me in his buggy to the shops and introduced me to Master Mechanic I. H. Graves, who told me to report for work on the 5th as the shops were closed on the 4th. I was there on time with my overalls, and was told to report to J. L. Gerrish, foreman of the machine

shop. He started me on a tap machine, tapping nuts for bridge bolts. The shops at this time were located at Sixth and E Streets, and seemed to me to be built on a levee. There were four pits for the engines, and G. D. Welch was the foreman. The machine shop and pits were under one roof.

Days in the Old Shops.

The blacksmith shop, boiler shop, pattern shop, paint and car shops were in separate sheds. The foremen at this time were J. L. Gerrish, machine shop; Frank LaShell, blacksmith shop; Jim Hall, boiler shop; I. G. Shaw, patternmaker; Ben Welch, master car builder, with H. W. Seaman foreman. The paint shop foreman's name I have forgotten. The work turned out in this small machine shop was enormous, for the truss rods for bridges were threaded and there were hundreds of them, 2 by 30 inches long; bridge plates, bridge bolts and freight car bolts. Each bolt had a nut put on, and there were thousands of them. We never had to wait on the supply department those days; also the car wheels were bored, the axles turned by Bill Hammond, who run two lathes.

It was about this time the Western Division was being built. This meant more work for the shops, for frogs, switch targets and everything that goes with construction, so an annex was built over the slough, more machines put in and started up. The engine which we called the "donkey" bucked at its load and Billy Moran, the engineer, would pat it on the back with a hundred and enough, but it was not equal to the load, so Foreman Gerrish cut out some of the machines until such time as a larger one could be installed. Inside of ten days we had one with more power. Jim Hall, the foreman of the boiler shop, got the dimensions of the smokestack, had it made and ready to put up. One Saturday night they tore out the side of the shop, installed the new engine and every machine was moving Monday morning. The last time I saw the old "donkey" it was standing on the edge of the slough, a monument to those it had so cheerfully worked for. They, too, have passed out from the shop and were forgotten. So far as I know all the men and apprentice boys who worked in the old shop only two remain in the company's employ, my time-honored friend H. G. Thiel and myself, and while I am speaking about the old apprentice boys I would like to say a word to the boys of today.

In our time at 4 o'clock Saturday afternoon we were told to "wipe the machines." After this was done we were handed a broom and each one had so much floor space to sweep, and you had to make a clean sweep or Gerrish would make a sweep at you with his boot. He was stern yet kind with us boys, and would show us anything we wished to know. There was no nonsense with him. It would not take long for him to tell you in language more forcible than eloquent that the shop was not a juvenile playground, and that you were there to learn. I often think of the advantages you boys of today have been given by the Company, which some of you look upon as a matter of course. I refer to the mechanical schooling given to each one of you alike at the company's expense.

It was not so with us boys. We went to the home of G. A. Stoddard (then a workman in the shops) in the evenings. We furnished our own drawing tools and paid him one dollar per lesson. He was a splendid teacher and later became chief draughtsman for the Central Pacific Company. With the other work we also had the repairs on engines to keep up, for there were some good smashups in those days. One of our engines they never took time to go and look for, as I remember it was engine No. 129, but have forgotten her name, McKay and Aldus makers. The engine struck a rock at Cape Horn and the engineer and fireman saw it in time to jump. Over she went to find a resting place at the bottom of the American River, and is there today so far as I know. I also remember the lightest engine, the William Penn. It was a hook motion and run by a large man by the name of Stockum. She could pull about five old-fashioned freight cars. The engineer was a very large man and took up the whole cab, and without his weight might have been good for one more car.

The First "Overland Train."

"C. P. Huntington" was run by an engineer named Obe Hamlin. The Judah was an extra engine. The engine "Governor Stanford" was in charge of Engineer George Chapman. This was the heaviest switch engine they had at this time, and the heaviest were the ten-wheeled Mason engines, the "Corness," the "Owhye" and the "Idaho." This last named engine had the honor of pulling the first through overland train on what is now the Sacramento Division into Sacramento, with William Mills as engineer. This event I remember as if it happened yesterday, which was either July or August, 1869. The engine was gaily dressed with flags and bunting and the brasswork on the engine glittered like gold. On this train were the first Silver Palace sleeping cars to enter California. We were still in the old shop when this event took place.

The new shops were designed by S. S. Montague, and I. H. Graves with Joe Wilkinson, civil engineer, in charge of construction. The roundhouse was the first of the new buildings to be occupied. Then the car shop next. The machine shop was moved to the new shop in September, 1869. The boiler shop was the last to

move to its present location. I have been connected with the locomotive department for fifty-three years, and have seen the engines grow from the William Penn to the M. C. class. There is one man in the locomotive department to whom I owe much; a better mechanic never stepped into the shop. I refer to Pat Sheedy, present superintendent of motive power at Los Angeles.

This narrative would be incomplete without mentioning the name of one young man who, like myself, started alone. I refer to H. C. Venter, the general foreman of the shops. I have always looked upon him as one of my boys and am proud to claim him as such.

And last but not least it may not be generally known to the apprentice boys of today that D. S. Watkins, superintendent of the shops, is a product of the Sacramento shops, in as much as he finished his mechanical education in them, and there are none in the business to match this gentleman.

In closing, I will say to my superiors that I have none but kind words to offer one and all.

(The End)

APPOINTMENTS.

Mr. T. L. Williamson, roadmaster of the Mina District, Salt Lake Division, has assumed the duties of trainmaster of the Mina subdivision with headquarters at Mina, Nev., according to an announcement by H. W. Wistner, assistant superintendent at Sparks.

Mr. F. F. Small has been appointed chief train dispatcher for the Salt Lake Division with headquarters at Sparks, Nev. Mr. Small succeeds Mr. H. G. Valleau, who has been appointed assistant chief train dispatcher with headquarters at Sparks. He succeeds Mr. H. F. McDonald, who has been assigned to other duties, according to an announcement by H. W. Wistner, assistant superintendent of the Salt Lake Division.

Mr. B. D. Richart has been appointed trainmaster of the Salt Lake Division with headquarters at Carlin, Nev., according to an announcement by F. C. Smith, assistant superintendent of the Salt Lake Division at Ogden.

Mr. D. W. Dower has been appointed signal supervisor of the Los Angeles Division with headquarters at Los Angeles. He succeeds Mr. C. A. Veale, who has been promoted according to an announcement by G. W. Corrigan, division engineer.

Mr. T. A. Allen has been appointed roadmaster of the Deming District, Tucson Division, with headquarters at Deming, N. M., vice Mr. C. Butler, who has resigned. Mr. W. F. Sampson has been appointed roadmaster of the Benson District, Tucson Division, with headquarters at Benson, Ariz., to succeed Mr. Allen, according to an an-

INTERESTING VOLUME BY ROY W. KELLY.

The Bulletin acknowledges receipt of an interesting volume entitled "Training Industrial Workers," from the Ronald Press Company of New York. The author is Roy Willmarth Kelly, manager of industrial relations for the Associated Oil Company of California and director of the Harvard University bureau of vocational guidance.

Mr. Kelly dedicates his book to his father and mother and "to other fathers and mothers whose steadfast practical faith in education makes progress possible."

In an introduction to the volume John M. Brewer, Ph. D., associate professor of education and director of the bureau of vocational guidance, Harvard University, notes that Mr. Kelly's book is written to tell industrial managers and educational directors about the lessons which both school people and manufacturers have learned in shop and factory education, and to show how these lessons can be applied to particular establishments. He brings in review the successful accomplishments in vocational education with the reasons for their success. He points out the need on the one hand for quick training in skill and on the other for the more fundamental education which shall give knowledge of the correct principles back of successful business and for the development of responsibility. Whether the reader wishes to find out the theory back of successful plans or the actual way to begin he will find his answer.

IT IS NOT ALWAYS EASY

To apologize.
To begin over.
To take advice.
To admit error.
To be unselfish.
To be charitable.
To face a sneer.
To be considerate.
To avoid mistakes.
To endure success.
To keep on trying.
To recognize the silver lining.
But it always pays.
—Baltimore Trolley Topics.

nouncement by W. F. Turner, division engineer.

Mr. William Rau has been appointed general yardmaster at Roseville, vice Mr. J. Pfosi, who has been assigned to other duties, according to an announcement of F. J. Berry, trainmaster of the Sacramento Division.

Mr. J. C. Goodfellow has been appointed terminal trainmaster with jurisdiction over the Los Angeles terminal, vice Mr. W. H. Jones, who has been appointed trainmaster of the Los Angeles Division. Mr. Jones succeeds Mr. V. S. Burnham, who has been made trainmaster of the Los Angeles Division with headquarters at Indio, Cal., according to an announcement of A. F. Bowles and C. F. Donnatin, assistant superintendents.

An Engineer Who Railroaded for 57 Years

By Erle Heath, *Associate Editor*

THE first transcontinental railroad had been connected by the "golden spike" at Promontory, Utah, less than five months when Roger Trewick, veteran locomotive engineer of the Coast Division, started his remarkable railroad service as an apprentice in the copper shop of the old Central Pacific at Carlin, Nevada. That was in October, 1869, and Roger had not yet reached his thirteenth birthday.

On December 31 last year, the eve of his seventieth birthday, Trewick made his last run as a locomotive engineer. When he stepped down from the cab of his flag-bedecked locomotive at Pacific Grove roundhouse to receive the congratulations of fellow workers and townspeople, he brought to a close a continuous service of 57 years 3 months with Southern Pacific, a record never equalled in the history of the Company. He has now

taken his place at the head of the list of pensioners, in point of service, of all retired veterans.

Trewick's record has probably never been equalled by any railroad man in the West. There are few engineers in the United States who have had as long a continuous railroad career, and none, it is believed, on roads where the compulsory retirement age is 70 years.

Roger Trewick had become an institution on the Coast Division. For more than thirty-six years, except for a time when he was roundhouse foreman at San Luis Obispo, he had been running an engine on those lines. Young firemen came into the service and received their schooling under him. Their association with him taught them something more than the mechanics of running an engine, and when they went on to their own locomotives they carried with them in-

spirations that made for the highest type of railroad man.

For several years he was division chairman of the engineers' brotherhood. His sincerity and sound judgment gained him the respect of his officers; his loyalty and devotion won him the affection of all his associates.

Tribute to this friendship was paid at two affairs held in his honor following his retirement. As soon as he could get home and into his "party" clothes on the evening of his final run, he was taken by automobile back to Del Monte Junction, where he was the guest at a New Year's eve celebration given by a group of residents of that city; men and women, many of whom had known him since their childhood and had ridden back and forth on his train to high school at Monterey. Many is the time he held the train a few seconds or a minute so some hurrying youngster would

The picture in the upper right shows Carlin, Nevada, as it was in October, 1869, when Roger Trewick, retired engineer of the Coast Division, started his railroad career that continued 57 years 3 months, the longest in the history of the Company. He started work as an apprentice coppersmith in the shop building shown at the right in the picture. The overland train stopped at the station gives a good idea of the observation car and other passenger equipment of that time. In the upper left Trewick is shown in the cab of his locomotive ready to leave Pacific Grove on his regular passenger run. Picture in lower left shows him with members of the train crew after he returned to Pacific Grove December 31, the day of his last run. Left to right are: Fireman R. Williams, Engineer Trewick, Conductor Geo. M. Lewis, Brakeman J. G. Middleton and Brakeman H. Sandstrom. Inset is of Walter Buttle, roundhouse foreman who decorated Roger's locomotive. Trewick is shown with his two brothers in the picture at lower left. Between them the brothers have almost 150 years railroad service. Archie, right, heads the seniority list of Salt Lake Division engineers with 47 years 6 months service, and Harry, center, who is road foreman for the Northern Pacific at Tacoma, has had more than 45 years service with that company.

not be late to school. He knew them all and took personal interest in their welfare. The school train is now a thing of the past, but the young people who used to ride on Roger's trains never forgot him and, when he finished his long service, they were the first to do him honor.

The second affair was a banquet held in the Masonic Club room of the Palace Hotel at San Francisco January 6 when 175 officers, former associates, friends and their wives tendered him a testimonial that carried with it a fine sentiment of regard for his remarkable accomplishment in railroad work and hearty wishes for many more years of happiness. Mayor James Rolph Jr. of San Francisco and Mayor F. L. Sinsheimer of San Luis Obispo were among those present.

Roger Trewick was born January 1, 1857, at Newcastle on Tyne, England. His parents emigrated to Australia when he was eight years old. John Trewick, his father, went on to America soon after and Roger and his mother looked after the family until the father could get settled in the new country. Roger worked long hours in a cracker factory at Melbourne and it was his mother's boast that her little son supported the family. Then came word from John Trewick for them to come to California. After a 63-day trip on the S. S. White Star, they landed at San Francisco and went on to Sacramento by river boat, where the father had taken a job in 1867 as blacksmith with the then newly-organized Central Pacific railroad. A short time later the family moved to Carlin, Nevada, leaving Roger and his younger brother, Archie, in Sacramento to attend school, where they were classmates of Fred and George Crocker, sons of Chas. Crocker, one of the builders of the Central Pacific.

Saw "Golden Spike" Parade

It was while they were at school in Sacramento that the celebration was held on May 10, 1869, the day the "golden spike" was driven at Promontory connecting the Central Pacific and Union Pacific. One of Roger's most vivid recollections is the thrill he felt when the six-horse "carry-all" went by in the parade with ten big Irish laborers who had set a world's record, which has never been broken, in laying ten miles of track in a day.

In the summer of that year it was necessary for the boys to join the family at Carlin. There were no schools in the little railroad town at that time, so Roger went to work a few months before he was 13 years old as an apprentice under Coppersmith Dave Hopkins. His salary was 8 cents an hour. "That wasn't so bad for a kid," Trewick recalls, "and, anyway, I had no idea that the job was going to work into a permanent one— at least not one to last 57 years."

When the prospects of any further school days were set aside, Roger found his boyhood fascination divided between becoming an engineer so he could run one of the proud and puffing little early-day locomotives, or follow in the footsteps of his father as a blacksmith. As far back in the family

Mr. and Mrs. Roger Trewick on the front steps of their home at Pacific Grove.

as there was record someone had always been a master blacksmith. Being the eldest son, Roger decided to keep up the family tradition in the "Master Craft of Hard Jobs" and by the time he was 18 years old he was a full-fledged blacksmith. He must have been a good blacksmith, too, for, in the years after he had left the forge, he was often called on to help in making a difficult weld or use his skill in swinging a hammer.

Shortly after Roger became a master blacksmith, Ben Smith, first road foreman of engines for the Central Pacific, whose district was from El Paso to Ogden, and who was always on the lookout for the most promising young men to train as engineers, "spotted" Roger when on a visit to Carlin. It was through his persuasion that Roger left the shops and became a locomotive fireman on the Humboldt Division in November, 1878. His first engineers were Phil Whickland and Hank Lightner and the first engines he fired on were Nos. 150 and 76.

The fireman's job was no "bed of roses" with its duties of cord wood stoking and brass plate polishing, but in two years he was promoted to engineer and went out on the road with Will Benson as his fireman on the "Antelope," No. 29. He ran this locomotive for almost six years before it had to go into the shops for repairs. Later he had the "Hector," No. 67.

Worked With Sheedy

On the Humboldt and Truckee divisions, which territory now is part of the Salt Lake Division, Trewick was associated with Patrick Sheedy, retired superintendent of motive power, who was then a general foreman at Carlin. Roger claims he gave Pat probably the hardest several hours' work in his long career. A passenger train was stuck in a snow storm near Pecos and could not be reached because of the blizzard. That was in

the days before the big rotary snow plows that now keep the line always open. The Company had bought one of the new type Cyclone snow plows and Roger had been handling it between Wells and Carlin. This particular night instructions came from Superintendent Fillmore to put the Cyclone in shape and rush it to aid of the stranded train. Roger got Pat out of bed at midnight and they worked on the plow until daybreak before it was ready for service. By the time they got to Pecos the train was frozen hard and fast. It took eight locomotives on as many passenger cars to finally break the train through and get it under way. Roger always held a lot of respect for Sheedy for the way he had of getting hard jobs done.

Cleaned Up Carlin Yard

Winnemucca was then a very busy town. There was a great deal of teaming going on and freight was heavy out of that point. Trewick was located there for some time in yard service, doing his own firing. He had a chance to show his ability to handle heavy yard traffic when the regular yardmaster was sick and he was appointed yardmaster for a day. The sidings had been filled with cars for some time, but as soon as Roger got on the job in the morning he mustered all the locomotives available and cleared the sidings as fast as trains could be made up and moved out on the main line. By night time there was not a freight car in the yard. Roger admits his tactics were not exactly orthodox, but it solved a difficult situation and his judgment was upheld in the end.

While on a visit to San Francisco in 1893, he took a trip down the Coast line, construction on which he understood would be resumed soon. The weather was beautiful at San Luis Obispo and he fell in love with the oranges and flowers and sunshine. He asked for a transfer, and, in March, 1894, took a work train locomotive on the Coast Division. The line was being extended south from Santa Margarita and Roger continued at the "front" until the terminus was established at Surf.

During the period of construction work Roger made a friend of a contractor named Hord who later did him a good turn. Neither men had ever met, but Roger saw to it that Hord's water barrel was kept filled with fresh water. Some years later Roger stopped at a horse sale at 13th and Market Streets in San Francisco at which Hord was the auctioneer. Roger introduced himself and was greeted warmly by the former contractor, who gave Roger a chance to buy a fine colt for $50, half interest in which he later sold for $400.

After construction work stopped, Trewick was on passenger runs between San Francisco and San Jose until 1899, when he went to San Luis Obispo as roundhouse foreman. Here he was a jack-of-all-trades and for more than two years was at times a yardmaster, blacksmith, coppersmith, machinist and engineer, as well as su-

(Continued on Page 16)

56

SOUTHERN PACIFIC BULLETIN

(Continued from Page 4)

pervising the work at the roundhouse.

He then went back to San Francisco and handled fast passenger runs between that terminal and San Luis Obispo until about eleven years ago, when he transferred to Pacific Grove.

Ever thoughtful of the comfort and pleasure of others, Roger endeared himself to residents along the Monterey Branch. To many families in the sand-dune region near Marina he was known as the "paper boy," and for years he had thrown off daily papers and current magazines to people along the right-of-way. When some of his friends in Pacific Grove learned what he was doing, they brought papers and magazines for him to distribute to his "customers." Never a Thanksgiving or Christmas passed that Mrs. Trewick did not prepare one or more packages of good things to eat for Roger to give to some of his less fortunate friends out along the line.

On the day of his last run his wayside friends, few of whom he had ever met or knew by name, were reminded by his flag-draped locomotive that this was the last day they would hear his friendly "toot-toot." One elderly lady stood at the crossing near her home and waved a greeting to Roger every time he passed on the three round trips during the day. For ten years he had brought her a daily paper; yet they had never met. On his early morning run out of Pacific Grove he was always careful in using his bell and whistle as little as possible so as not to disturb people living near the tracks. Once he received a check from a woman who appreciated his being so quiet with his locomotive during the early hours. Another woman came near driving her automobile in front of his locomotive at a crossing. He wrote her a nice letter warning her about the crossing. She hunted him up and thanked him for his advice and apologized for being so careless. Friendships developed during the years that filled many hearts with sadness on that last day when Roger's train rolled along down the shore of Monterey Bay to the end of the line.

Had Clear Record

Aside from the length of his service, remarkable features of Trewick's career were the safety with which he guarded his trains and himself and the fact that he never received a demerit mark on his record for oversight or neglect of duty, but finished his long service with 90 days of commendation in his "bank account."

This was called to attention by E. R. Anthony, superintendent of the Coast Division, who presided as toastmaster at the San Francisco banquet given in Roger's honor. "Mr. Trewick's record is a marvelous one," said Mr. Anthony. "Anyone familiar with the operation of trains knows how easy it is to receive a 'Brownie,' and the fact that Roger has been in railroad service longer than any other employe and has never received a demerit, but has 90 merit marks, is something to be mighty proud of."

"Mr. Trewick has made a success of life," continued Mr. Anthony, who has himself served the Company more than half a century and who has known Roger since 1880. "He has achieved a success who has lived well; who has gained the respect of intelligent men and the love of little children. His life is an inspiration and his memory a benediction. His work has been his pride and recreation, and now he has arrived at the pension age without hardly realizing it; almost without knowing he was getting that old."

It is told that Roger was quite a "dandy" in his younger days and had a reputation as a good dancer. Yet one of the only two personal injuries he ever suffered happened on a ball room floor. He slipped and broke his arm while dancing the lancers. Another time he broke his leg while in swimming. He never received a serious personal injury while on duty. He had only two crossing accidents, neither of which resulted in a fatality. He never had a derailment or a break-in-two. A flagman let him past once

Page Sixteen

February, 1927

and he had a rear-end collision with a freight. Only slight damage to one freight car was the result. He had been a good fuel saver and was awarded many monthly certificates in that work. An intervention of Providence once saved him from running into a washout near Riverbank. He was pulling the Shore Line Limited. It was raining heavily, and, when the headlight went out, he was delayed for a few minutes fixing it. This delay probably saved their lives. The section foreman flagged Roger's train about a mile from where the headlight was fixed, and if the train had come along two minutes sooner it would have run into the washout 10 feet deep and 30 feet long.

Across the banquet tables many friends greeted Trewick whom he had not seen for several years. There was a Mrs. Marshall whom he had known as a girl in '77, and Mrs. Tom Neerich, wife of one of his Coast Division firemen. Roger carried her in his arms when she was a baby in 1880. They had not met for 15 years, but at a glance he recalled the name. Then there were Mr. and Mrs. Bob Watson, Mr. and Mrs. Russell Hurd, Mr. and Mrs. Martin, Fred Champlain, and dozens of others.

Newspaper stories about his retirement attracted the attention of his friends all over the country and promoted letters and cards of well wishes

from almost forgotten names and others whom he thought had passed out of this life. A note from Miss Georgie Burgess, a young lady of Guerneville, puzzled him some time until he recalled having given a short ride on his locomotive to a little 8-year-old girl ten years ago.

Mayor Rolph Speaks

Mayor James Rolph Jr. of San Francisco and Mayor L. F. Sinsheimer of San Luis Obispo, chief executives of the two cities that had been terminals for Roger during part of his long career, were among the speakers at the banquet. Mayor Rolph paid tribute to the original "Big-Four"—Stanford, Huntington, Crocker, Hopkins—whose courage and foresight made possible the first transcontinental railroad and the building of the lines which pioneered development of the West. "It is men such as Mr. Trewick," said the mayor, "who have made the Southern Pacific the great institution that it is and who have continued the progress of transportation established by the 'Big Four.' In behalf of the people of San Francisco I thank you for all you have done; I praise you; I honor you."

Telegrams of congratulation to Roger and regret in not being able to attend the banquet were received from a number of officers and friends. President Wm. Sproule wired from Chicago, in part, as follows:

"Distance prevents my accepting invitation to be present at banquet, but do not wish the day to pass without expressing my profound interest in the fact of a continued service of 57 years and 3 months with honor to himself and credit to the Company. I hope Mr. Trewick will have many happy years in which to enjoy the release from work that our pension plan allows, and in which he can look back with satisfaction to his long career of good service, in which satisfaction the officers of the Company under whom he served participate."

"Mr. Trewick deserves any tribute that is within the power of his fellows to bestow," wrote General Manager J. H. Dyer, whom a previous engagement kept from attending the banquet. "He has served the Company faithfully and well. He has accepted and discharged with credit whatever responsibilities have been placed upon him, and he has achieved a record of which he may well be proud."

"I turn over to my friend the honor of being the dean of all retired employes, a privilege I have enjoyed for some time," wrote Patrick Sheedy, former superintendent of motive power at Los Angeles, who was retired in January, 1925, after 56 years 5 months' service. "I have known Roger for more than forty years. As a man, genial and true; as a citizen,

upright and worthy; and as a co-worker, courteous and loyal; he has earned the respect and affection of all."

Many other wires and letters were received from officers, former associates and brotherhood lodges all over the System.

Mayor Sinsheimer called attention to the fact that Roger did not leave his work behind when he stepped from the engine cab, but that he carried the interest of the railroad with him among his friends and neighbors.

Record an Inspiration

First Assistant General Manager F. L. Burckhalter said he felt like a youngster in the service alongside Mr. Trewick. He thanked the veteran engineer in behalf of the management for his years of faithful service and stated that Roger's record should be an inspiration and incentive for the younger railroad men.

Tom Negrich, former engineer on the Coast Division and now a San Francisco attorney, brought several good laughs with his recounting of incidents when he fired for Roger. "I always knew when we were hitting it along right, for Roger brought his jaws down on his chew of tobacco in perfect rhythm with every turn of the drivers, but if we were behind time, how those jaws did work," he said. "Also, it wasn't only at little girls and old women that Roger tooted his whistle."

Assistant General Manager T. Ahern told of the pleasant associations he had with Trewick when he was superintendent of the Coast Division and Roger was division chairman of the engineers.

F. M. Worthington, retired superintendent of the Coast Division, explained the meaning of "Brownies." Speaking from his long experience as a train man and how easy it was to make a "slip up" that would assess a few demerits, he commended Roger on winding up so long a service without a mark on his personal record.

A fine tribute to the part Mrs.

Trewick had taken in making possible Roger's long career was paid by J. A. Christie, superintendent of the Santa Fe's Coast Lines. "It is the faith and encouragement of loved ones and the influence of a happy home that makes the man a success," he said. "When a man's mind is free to think of his work he has every advantage of becoming a success."

Mr. and Mrs. Trewick have been married 44 years. She was Emma Gertrude Linn, the daughter of a Reno justice of the peace. They met while she was teaching school at Winnemucca. "How my wife ever happened to be in such a wild and woolly place I have never been able to figure out," says Roger. They have had five children; three daughters have gone through college and two sons are dead.

Roger has two brothers, Archie and John. Between them, they have the distinction of having a total of 150 years' railroad service. Archie is 67 years old and heads the engineer seniority list on the Salt Lake Division with 47 years 6 months' service. Harry is at present road foreman of engines for the Northern Pacific at Tacoma. He is 62 years old and has been in the service of that company more than 45 years.

Good to Have Friends

It was a few minutes before Roger could collect himself to speak when he was called on at the banquet. "It is good to have so many friends," he said, "and there is something in my heart I would like to tell you, but I can't express it. Except for the number of years I have been on the job, I do not think my service has been out of the ordinary. I have tried to do what was expected of me, and in accomplishing this my friends have been a big help. As Mr. Christie intimated, there is nothing quite so helpful or encouraging as the wife's elbow in your ribs when the caller comes in the early morning. But, in all seriousness, my home has always been a happy one and again Mr.

Christies' remarks are true. It is a noble corporation whose pension system makes this reunion of my friends possible. To tell the truth, I have gotten old and didn't realize it."

Arrangements for the banquet were made by a committee composed of J. A. McCarthy, formerly an engineer on the Coast Division; W. G. Fifield, road foreman of engines on that division; and Lee Hamlin, fuel inspector, also a former engineer on the Coast and at one time a fireman for Trewick.

During the evening music was furnished by an orchestra composed of employes from the San Francisco Freight Station, and Miss Letty Collins, of the Treasury Department, sang several numbers.

Before the banquet was adjourned Mr. Fifield presented Mr. and Mrs. Trewick with a large radio set equipped with all accessories.

Roger has made no definite plans for the future. For the present at least the family will live at Pacific Grove. "Now that I am off the run I find that I really am tired, and am going to take a little rest," he says. We may visit some of our friends and then I am going back to work again. I've got several good years ahead of me and the best way to keep fit will be to keep busy. I regret that my railroad days are over, but I kept the pace, made the run and finished clear, and to me that brings a great deal of satisfaction."

From Trail to Rail

(Continued from Page 11)

bonds upon a certificate from the chief engineer that a certain proportion of the preliminary work had been done, the remaining third to be issued on completion of the 25 miles.

The most important modification was that the act of 1864 authorized the company to issue first mortgage bonds to the same amount as the United States bonds, and provided that the federal bonds should be subordinate to the company's bonds. Until this action was taken the company's own bonds had little market value, as, under the act of 1862, they were but second mortgages on the property.

The act of 1862 granted the Central Pacific authority to build beyond the borders of California if the Union Pacific had not yet reached that far. The act of 1864 limited the construction to 150 miles eastward of the California-Nevada border. This restriction was removed July 3, 1866, and the Central Pacific authorized to continue its eastern way until it met the Union Pacific.[13]

Before Judah left Washington the Pacific Railroad Committees of both Senate and House joined in a testimonial, thanking him for his "valuable assistance in aiding the passage of the Pacific Railroad bill through

Many immigrant trains bringing settlers to the West were handled by Roger Trewick when he was engineer on the old Humboldt Division. Several nationalities may be distinguished by the style of dress shown in this picture which was taken at Mill City, Nevada, in 1886. There were nine cars of emigrants on this train. Mr. Trewick was engineer on the locomotive "Antelope." He is standing in the front row wearing the white sun helmet. To the right of him are: Brakeman L. Jones, Conductor Miles T. Coates, Fireman F. Hammond and Brakeman F. Gillett.

NOTE 19—When this limitation was called to Huntington's attention he said that he was satisfied the Central Pacific could make, under the amended act, such a showing that when the time came the government would remove the restriction.

Chapter Five
National Labor Issue Comes to Sacramento - The Pullman Strike of 1894

During the 1880s, George Pullman created a community at which his famous Palace sleeping cars were built. Pullman contracted with many railroads to operate and maintain the cars. Having a huge portion of the sleeping car business, Pullman also operated facilities in St. Louis, Ludlow (Kentucky) and Detroit. In 1885, at the company's main operation south of Chicago, most of the company's 9,000 employees lived in housing owned by the Pullman Company. This relationship between the employees and the company, which showed signs of friction as early as the late 1880s, continued to fester in 1893 when a national depression developed. By the end of 1893, employment at the Pullman Company was reduced by about 4,500 people and the remaining workers' pay was reduced by 25 percent—though the salaries of executives, superintendents and some shop managers were not reduced.

During the winter of 1893-1894, the workers began to organize. Their action was to join a rapidly growing union called the American Railway Union led by Eugene V. Debs whose resume included a successful strike against the Great Northern Railroad (Mr. Debs eventually ran for President of the United States five times as a Socialist). The Pullman Company, however, forbid any union activity and threatened to terminate anyone supporting the organization of fellow workers. The rumor of company spies led to clandestine meetings, and by the spring of 1894 several thousand Pullman workers had joined the American Railway Union. In early May of 1894, representatives of the employees and the Pullman Company met to discuss the restoration of wages and a reduction in rent. George Pullman and his Vice President Wicks were unyielding and stated that the existing policies were necessary for company survival given the poor national economy. The next day, three members of the employees' committee were fired. During an evening meeting involving the workers and the union, the union urged the workers not to walk off the job. But, the frustration had reached a limit. On May 11, three thousand Pullman workers refused to appear at work while several hundred employees chose to not participate in the walk out. But, even they were not allowed to work when Pullman closed the entire operation with no mention of a date to resume business.

Decisions made in late June were ominous. On the twenty-seventh, Debs announced a national boycott of Pullman cars on all railroads. Within twenty-four hours, a large portion of railroad traffic across the country was halted, and the turmoil that originated near Chicago had reached twenty-seven states, the West and approximately three thousand employees of the Sacramento Shops, one-third of Sacramento's population. The movement of fruit, general freight, mail, and passengers was stopped. The union stated that it had no complaint against the Southern Pacific (SP) and that it intended to move trains that contained no Pullman cars. But, railroad officials still objected to the strike and closed the Sacramento freight offices on June 30 and laid off all employees without pay despite knowing that revenue from shipping a portion of the summer's fruit and vegetable harvest would be lost.

Also on June 30, a Southern Pacific train with Pullman cars departed Oakland for Sacramento. Among the passengers was a group of United States marshals prepared to see the strikers uphold their pledge and not handle any trains comprised of Pullman cars. When the train arrived in Sacramento, strikers disconnected several cars. The train eventually departed, but went no farther than Rocklin.

On the same day, a westbound Overland Express with three Pullman cars was headed to Sacramento. The train was stopped northeast of the station, and the Pullman cars were removed before the train was allowed to proceed to Sacramento. Soon after, the engineer was told to use a yard engine to bring the three cars to Sacramento. But strikers deterred him. Railroad officials attempted to re-route the train to Oakland via the Yolo Bridge but found switches misaligned and the drawbridge open.

Needing outside help, SP attempted to solicit assistance from the Sacramento sheriff. This effort was in the form of a telegram sent on July 1 from the general manager of SP to the sheriff: "This will notify you of twenty-five hundred men who are gathered about the railroad depot in Sacramento. They are in the way and obstruct the operation of the Southern Pacific...In order to prevent riots and bloodshed it is necessary that you disperse the mob from the area."

The sheriff responded, "I will in every way uphold the peace. The men are peaceable and no violence has occurred. If any breaches of the peace occur, I will, of course, suppress them, but cannot interfere with law abiding and peaceable citizens."

The correspondence continued with the general manager replying, "You are not performing your duty by awaiting a breach of the peace. You should act before it occurs...the men...are unlawfully assembled and encouraging lawlessness."

On July 1 or 2, U.S. Attorney General Olney ordered federal involvement based on the fact that some

of the trains carried US mail. He clarified that his order concerning the mail covered not only the mail car but the entire train even if it contained a Pullman car. So, a train with a mail car in the consist was to proceed unobstructed.

On July 2, U.S. Marshal Barry Baldwin and SP Superintendent Fillmore arrived in Sacramento.

On July 3, Baldwin was on a locomotive that was being moved down First Street to couple with the Eastern Overland, which included Pullman cars. After strikers uncoupled the Pullman cars, Baldwin, while running to the depot, was thrown to the ground. After gaining his feet, he drew a revolver but had his arms pinned to his side and was removed from the site by the strikers. Baldwin asked Governor Markham to assemble the National Guard in a continued attempt to move the mail. Guardsmen from San Francisco, Stockton and Marysville were sent to confront the rioters. The troops arrived on July 4, and marched fifteen blocks to the armory where they met five hundred troops from Sacramento. The whole group went to the passenger station where they met two thousand strikers and onlookers. The effectiveness of the military was hindered by the friendship between Sacramento infantry members and the strikers. Company A, one of three from Sacramento, refused their depot assignment. Surely, some pedestrians in the city were thinking of an enjoyable holiday only to find armed troops by the hundreds near the depot.

Brig. General Sheehan, with orders to clear the depot, found the west entrance blocked by the strikers. He addressed the crowd by saying he did not want to use force but would if necessary to assure the passage of the mail. He tried for over an hour to encourage the strikers to move back but they refused. With no success, Sheehan reported the stalemate to Baldwin. The decision was made to shift some troops to the Sacramento and American River bridges. By this time, some troops were suffering from the heat, and many had to be removed for treatment. Baldwin's busy and historic day continued when he called together the leaders of each side to see if an agreement could be reached allowing him to be in control of the depot. The negotiation was successful and a truce was reached at 6:00 p.m. The troops were allowed to seek shade and at 6:30, the troops had supper, and an eerie period of peace prevailed.

Then, on July 10, United States President Cleveland stated that arrest faced the strikers if they continued to obstruct train movements after 4:00 p.m. News arrived that additional troops from the Presidio in San Francisco were en route via ship. With this news, the laborers left the yards with the fear that staying would lead to being accused of treason. During the early stages of the strike, President Cleveland declared a new national holiday: Labor Day.

On the eleventh, the troops arrived and joined the state militia. Together, they took control of the depot and the entire railroad property, which had been abandoned by the strikers the previous day. The expectations of continued hostility led to the assemblage of a hospital tent.

With control of the depot, the troops' efforts shifted to making up a train for San Francisco. The train to be assembled was the same Overland Express that was stopped in Sacramento on June 30. It was comprised of a locomotive, two mail cars, one baggage car, and six other passenger cars. Also added to the train were the three Pullman coaches and a Pullman diner that were in the original train. The train was engineered by Samuel B. Clark, who was one of the oldest and most popular engineers. He departed with twenty-one military guards, including six positioned on the locomotive.

After about an hour, a porter ran into the depot to Superintendent Wright's office with a note from the train's conductor stating that there had been a wreck on a trestle about three miles from the city. Engineer Clark and four soldiers were killed. The engine had fallen off the trestle to land in about six feet of mud and two cars landed on top of it. The fireman had saved his own life

by jumping when he felt the engine tipping. The cause? Spikes had been removed, leaving nothing to hold the rail in place.

This accident caused the passionate allegiance of the resistors to wane, and the public's sentiment toward the strikers weakened also. On July 17, the company provided notice that those who had not participated in violence or destruction of property could return to work on the morning of the eighteenth.

Further investigation into the trestle incident led to a young boy who reported that four men with tools hired him as a buggy driver to the trestle. The boy had remained in the buggy while the four men disappeared. When the men returned, the boy drove them elsewhere. Based on the boy's story, Salter Worden, a union member, and three other union members were arrested. Worden was the only member of the four who was found guilty and his sentence of death was changed to life in prison. He was paroled in 1921 and released from probation in early June of 1927, almost thirty-three years after the train left the trestle. (McGowan 1961, 103)

The photo above was taken in 1894 during the Pullman strike. The view is looking southeast. To the left is the car paint shop, which is parallel to the transfer table on which the railroad car is located. The rolling mill is at the rear center situated east to west. The American Railway Union called for national boycott of Pullman sleeping cars in June 1894. Within twenty-four hours, rail traffic in twenty-seven states was affected. Strikers took control of stations and railway yards in many California cities and significantly disrupted service. The strike delayed the shipping of agricultural products during the growing season in the state's Central Valley. The federal government became involved because some of the trains with Pullman cars carried mail also. Hindering the delivery of mail was an unintended effect of the labor actions. Federal forces were called in during early June to restore the flow of goods and mail. Some troops, however, refused to fire on the strikers. Summer in Sacramento can be exhaustively hot. In early July, some troops, who might have come from the cooler climates of San Francisco, were suffering from the heat and needed treatment. Early in the strike, local favor was with the workers because a large portion of the population was employed by the railroad. Because violence and disruption to businesses continued, citizen support waned. The violence reached a peak when lives were lost during a train derailment west of Sacramento. The derailment was attributed to strikers, and four union members were eventually arrested. The strike ended July 20. – Courtesy California State Railroad Museum

The Pullman Strike of 1894 ended shortly after a fatal train derailment between Sacramento and Davis. The derailment, blamed on union members, signified an increase in violence that was unacceptable to all parties. — Courtesy California State Railroad Museum

Chapter Six
Fire

As the Southern Pacific learned, a fire can destroy shop buildings overnight. A fire at the Shops could be caused by embers from a steam engine, or an accident with a natural gas light, the flame from a torch, or hot pieces of iron near a foundry.

Concerns of fire and the railroad were not limited to the structures of the Shops in Sacramento. At one time, there were nearly thirty miles of wooden snow sheds through which wood-burning locomotives traveled while emitting embers. Even though the locomotive crew could control the burning of the wood in the firebox, fire in the sheds was a major concern.

To monitor fires, there was a viewing point on Red Mountain across the valley from Cisco, ninety-two miles from Sacramento. From that location, the staff could see almost all of the snow sheds from Blue Canyon to the Summit, about twenty miles. Inside the building was a device that looked like a compass. It was used to identify the fire's location on the track. At some locations, the snow sheds were constructed with rolling doors. In the event of a fire, the doors were opened to create a break in the sheds that would ideally stop the progress of a fire. An article in the June 1930 issue of *The Bulletin* described this observing station and is included in this chapter.

The essay that follows is about a large fire at the Shops in 1898. Much of the essay quotes the *Sacramento Union*, explains the effects of the fire on the Shop departments, and notes the lessons learned about fire prevention.

Other articles from *The Bulletin* are:

"Fire Brigade Sets New Speed Record," June 1925

"Better to Prevent Fires Than Fight Them," June 1930

❧ ❧ ❧

The Fire of 1898

There is no doubt that any large facility, of any era, in any business, must have the ability to deal with events that could cause a major shutdown of operations. These scenarios, depending on the location, include tornadoes, hurricanes, floods, earthquakes, and fire. The Sacramento Shops were in a location where floods and fire were possible. Both had their place in the history of the Central Pacific and Southern Pacific railroads.

In 1852, fire destroyed much of the city near J and Second Streets. Certainly, designers of the original Shops in Sacramento were aware of that disaster and were also able to draw upon the hard-learned lessons of eastern railroads. One design feature they implemented was the separation of buildings, as they recognized that the risk of having the foundry connected to the woodworking area was too great. The open land was a fire barrier and not a waste of space. Another feature of the early shop design was the use of brick and corrugated tin as building material. The basic skeleton of the buildings was often wood, however—most likely for economic reasons.

Despite these preparations, in 1898 a major fire destroyed the car machine shop, the brass finishing room, the plating room, the upholstery shop, the dyeing room, the sleeping car equipment room, the planing mill, and the cabinet shop. The cause of the fire was not determined with certainty. But, among the possibilities was an electrical problem. (In 1889, electricity was used for the first time at the Shops when a small steam generator produced enough power for thirty-five lights—a small portion of the lighting needed for operations.)

The fire was well covered by the *Sacramento Bee* in an issue of November 7, 1898 and the account included comments by H. J. Small, Superintendent of Motive Power and Machinery.

The following is the text of the article reproduced as accurately as possible.

FIERCE FIRE AT THE SHOPS

The vast system of shops of the Southern Pacific Company in this city early this morning was threatened with complete demolition by fire. As it was, some of the most valuable property in the shop system was destroyed, entailing an immediate loss of something over $200,000. Five hundred men of the 2200 employed find themselves temporarily out of employment because of the fire. The buildings contained departments that are essential to the running of the shops as a whole, and their reconstruction will doubtless be matter of a very short time.

The Fire Department was called out about 1:30 o'clock [sic] this morning. A general alarm brought all the engines to the scene. Sparks arose in tremendous volume from the burning buildings and were carried by the strong north wind as far over the city as the Post Office, half a mile away. The deep-voiced whistle which calls the men to their labor was shrieking wildly, and its frantic appeals to the men to leave their homes and rush to the rescue of their workshops added an uncanny accompaniment to the roaring flames. From the residence and business portions of the city it seemed that the whole system of shops was being consumed. No passes were demanded of those who sought entrance through the opened gates, for there was no time to tell who were workmen anxious to help save the railroad property and who had been attracted through spectators' curiosity merely.

When the first installment of the flocking multitude arrived, the two-story building, occupied by the car machine shop and upholstering shop, was practically in ruins, although still fiercely burning, while the two-story building adjoining, occupied by the planning mill and the cabinet shop, was a mass of flames. With a terrific crash the roof of first one building and then the other fell in, causing a deafening report, and adding millions of burning brands to the shower that was sifting threateningly over the city. The scene was one of the most intense excitement, heightened by the realization of the crowd that the engine and boiler room was in imminent danger of destruction, the screaming whistle calling attention to that building, with its towering brick chimney, which momentarily loomed up from the flames and smoke that swept around it from the burning cabinet shop, from which the engine and boiler room was separated only by one dead wall of vast strength and thickness.

Had that wall succumbed like those that formed the burning building which it faced, the great shop system would have been struck a heart-blow and no smoke would be found curling from its hundreds of chimneys as usual this morning. Had that wall fallen, there would have been an explosion of the hot boilers which would have rocked the city to its very foundation and blown to storms the fireman who were impatiently yet doggedly playing feeble streams upon the blazing structures. But that wall stood the terrific test as firmly as if it were some impenetrable fortification, and the whistle continued to utter its alarm as if the glass windows of the engine and boiler room were not falling into shattered masses from the surrounding heat.

A feeling of gratification and relief found expression from a thousand throats when it was realized that the engine and boiler room had escaped practically unscathed for, had it gone, the shops would have had to close down for want of motor power. As it is, all the shops outside those destroyed were working full blast this morning, and man would never know there had been a fire unless he was directed to the scene of the conflagration. Once there, however, he would see ruins, which, if compressed, would occupy three-quarters of one of the city's blocks. Machinery which was regarded as the most complicated and delicate known to mechanics was standing up from heaps of smouldering bricks like shapeless pinnacles. The escape of the storehouse, of the round house and of the car shop proper was regarded as most close. As it happened, only the roof and side of the store house nearest the flames were scorched, while the car shop suffered only slight damage at its forward end.

THE ORIGIN A MYSTERY

H. J. Small, Superintendent of Motive Power and Machinery in railroad shops said to a Bee reporter today that the fire this morning would throw about 500 men out of employment; thus affecting about one-fourth of the working force in the shops. Mr. Small said it had not been determined how the fire started. The fire started in the upholstery department, in the upper story, where work had ceased at 5 o'clock Saturday night. The power for running the machinery in that department was supplied by electricity, and it is possible that the fire was started by a live wire.

The watchman employed in that part of the shops has reported to Superintendent Small that he made the rounds of the building where the fire was discovered at 1:10 o'clock and the time register shows that he had made the tour at the hour claimed by him. Then minutes later, or at 1:20, he discovered a bright light in the building, and the light presently burst forth into furious flames.

Superintendent Small says the fire affects car work exclusively. All the iron work, mill work and the work of making trimmings for cars will temporarily stop.

From the upholstery department the flames traveled rapidly to the planning mill and cabinet shop, which contained a vast amount of seasoned timber and a great lot of expensive machinery. The place was soon a mass of seething flames, and was destroyed in an incredibly short time.

Through the heroic efforts of the fireman of the city department and of the shops, the flames were cut off at the boiler rooms and that portion of the shops was saved. Had the flames entered the boiler rooms the fire would have been far more serious than it was, as it would have meant the closing down of every department because of the lack of power. There was such a heavy demand for cars that the shops would have been kept running to their full capacity nearly all winter. Of a necessity the car work must be stopped. This will cause the idleness of painters, upholsterers, planning mill hands and some of the iron workers. There will also be a partial lay-off of men now working in the foundry.

Superintendent Small said he presumed that the buildings would be erected again, but that was a matter to which railroad authorities had, of course, not yet given attention. If the buildings were put up again, he said, there would probably be some changes in the general plans. Mr. Small said that owing to the necessity for cars, it was possible that orders would have to be given to other factories for them.

Superintendent Small, in speaking of the probable loss by the fire said that, at a rough estimate, it could be placed at $200,000. He did not think it would exceed that amount.

Master Mechanic T. W. Heintzleman, Master Car Builder Benjamin Welch and other local railroad officials were early upon the scene of the conflagration, inspecting the remaining walls, with a view to ascertaining whether they were safe, or whether they should be torn down.

The supply of water at the shops was not sufficient to fight a fire of such great proportions as that of this morning. The shops have their own water supply, an abundance of hose and carts and a well-drilled company of firefighters. When the city engines came upon the scene and began pumping from the shop pipes they diminished the power to such an extent that the shop firemen could only get little dribbling streams through their hose. The company's fire trains were put to work and rendered splendid assistance in preventing the spread of the flames. "Had there been plenty of water," said Small, "I am certain that the fire could have been confined in the vicinity where it first broke out. I cannot account for the lack of water, and I was surprised that the water pressure was so light. I have often witnessed fire drills at the shops, and at these drills the streams of water easily passed over the buildings due to the excellent water pressure, without the use of the department's fire engines. That is why I was surprised. There were only three good streams on the mass of fire. Two of these

came from the two department engines, and the other from the railroad fire [engine]. There were probably about a dozen other streams from the surrounding hydrants, but owing to the fact that the water pressure was so light these streams were of little service. In order that the department engines could pump their two streams, I ordered several of the hydrant streams shut off.

"This is the first time that the city's fire engines had to be used for a fire in the shops, for the reason, as I have said, that the hydrant streams, with the shop pumps were sufficient. As it was, the two engines could only pump one stream from each. To save car shop No. 3 some of the hydrant streams had to be shut off to give the engines a sufficient pressure for their two streams, which were played on the car shop. It was a hard fight to prevent the car shop from also being destroyed.

"It was a miracle that the engine house and boiler room that supplies all the shops with motive power were not swept away, owing to the close proximity of the burning buildings. Several times the fire broke into the engine house, and at one time the fire was so hot that the engineer had to leave his post until it subsided. This was the hardest fight. Two good streams protected the engines and boilers until all danger was past. The firemen, at the risk of their lives, fought to keep the fire away from this place. The firemen ran a line into the engine room and another on the roof. They had a thrilling time, for it was not known for some time whether the engine room could be saved. As to the cause of the fire, I can give no explanation. It seemed to me that the fire started about the upholstering department. The railroad fire whistle awakened me, and I was driving to the scene when the bell gave the alarm. When I reached the fire the mill was burning. I had the firemen who were in the mill with the engine streams get out just a few minutes before the roof fell in. It was the falling of the walls, that made the noises which sounded like explosions."

THE GUTTED BUILDINGS

In the car Machine shop, which was entirely destroyed by this morning's fire, could be found all kinds of machinery pertaining to car construction. On the second floor were located the brass-finishing room and the plating room. In the brass-finishing room all kinds of car locks were made and repaired, as well as all the metallic car trimmings. The plating room was fitted up with baths for silver, copper, nickel and brass plating. On this floor, also, was the upholstery shop, the dyeing room, the upholstering store room and the sleeping car equipment room.

In the planing mill, which was completely gutted, many saws and planes were constantly at work. Rough lumber was there dressed so as to reveal all the beauty of the wood.

On the upper floor of this building was located the cabinet shop. It was divided into two compartments. One of these was used for veneering and gluing and the manufacture of general furniture. Much artistic work was turned out there. In the second compartment machine wood work and all classes of bench work were done.

Prior to the fire, the railroad thought it did not need city water and believed that their own supplies, on the banks of the river, were sufficient. An eight-inch pipe from the city water works did exist, however, but was kept closed with an iron gate. The fire was well developed before this gate was opened.

Soon after the fire, there was agreement that a twelve-inch main be installed along with twelve fire plugs on the site. The citizens of the city would vote on a bond measure to fund the infrastructure improvement.

Surely, the question arose regarding the reasons the eight-inch source from the city was inaccessible when the fire started. If this problem had not existed, water would have been on the fire much sooner.

Fire Brigade Sets New Speed Record

Members of Sacramento General Shops fire hose cart No. 5 which set a record of 31 2-5 seconds in a nine man test, starting 50 feet from plug, running out 300 feet of hose, breaking joint, applying nozzle and getting water. Back row, left to right—Fire Chief H. A. Adams, N. Pendleton, A. Sena, G. Toffee, A. Logan (ass't captain). Front row J. Caldroni, J. Battaglia, L. Jorgenson, F. Parisi, B. DeRiso, J. Hernandez and C. Schmitt.

NOW comes the Sacramento General Shops fire brigade with a record of 31 2/5 seconds in getting water through 300 feet of hose, starting 50 feet from the plug and breaking one joint.

Fire Chief H. A. Adams read in the Bulletin about the 33 2/5 seconds made by Dunsmuir fire brigade in a similar test, and was inclined to doubt the possibility of making such fast time. In fact, after the first attempt of his team, he was convinced that the Dunsmuir boys made their record while running down the Siskiyou mountains. After a short period of training, how-ever, the team was ready to make another attempt, which resulted in the new record being established. Following were the conditions of the run.

The course was over a level, bumpy road, with cart carrying 500 feet of standard fire hose with a crew of nine men. The cart was placed 50 feet from the fire plug and on a signal given by the chief, the crew started and ran past the plug, laying out 300 feet of hose, breaking hose joint, applying standard nozzle and getting water in 31 2/5 seconds. Water pressure was slow and the crew waited 1½ seconds for water to respond.

Fire-fighting train in operation. The engine is equipped with pumps and can throw four streams simultaneously in any direction. Two cars of 12,500 gallons capacity furnish the water supply.

Better to Prevent Fires Than Fight Them

Campaign of Prevention Has Been Vital Factor in Safeguarding Timber Country and Railroad Property

IN all the world there is nothing more terrifying than the cry of "Fire." The clang of bells, the wail of siren, and the hoarse cries of an excited crowd send a frightened chill down the hearer's spine such as no other alarm can equal. The thought of fire carries with it a sinister threat of property destruction, personal injury and loss of life that fills one with consternation.

Adequate protection against the menace of this monster of destruction is one of the first considerations of any organized community, and the United States, through its Forest Service, safeguards its timber against destruction by fire by every known means of prevention.

The Southern Pacific, Pacific System, extending over 9000 miles through densely settled communities, small villages, desert country, mountain ranges, dense forests and much unsettled country, has a fire prevention and control problem of immense magnitude, with many unusual features not encountered by the average fire-fighting organization. The Company's fire department operates in seven states, protecting millions of dollars' worth of property, and yet, in spite of the unusual obstacles under which it at times must work, and the vast area of heavily forested country through which the road operates, its fire loss per mile is less than one-half of the average loss per mile of all the steam railroads in the United States.

While its organization is equipped to combat fires wherever and whenever they occur, and constant drilling of fire companies is practiced, in order that they may be kept at a high degree of efficiency, the watchword of the fire department is "Prevention," and much of its effort is expended in the direction of educating employes

to the importance of, as far as possible, making fire impossible. Through the constant exercising of prevention methods over the entire line, the ancient adage that "An ounce of prevention is worth a pound of cure," has been reversed, as in the handling of fires it has been proven that a *pound* of prevention has made only an *ounce* of cure necessary in the large majority of cases.

In the Sierra Nevada, Cascade and Tillamook mountains the railroad runs through mile after mile of densely wooded country. During the summer months there is little, if any, rainfall, and the underbrush dries up and becomes like tinder, presenting a hazard that is extremely acute, and requiring constant vigilance to prevent fire. In the state of California alone, despite the most careful precaution, there are about two thousand

During the dry summer months trains are followed by a motor car, the operator of which is equipped with a pack tank of water with which to control incipient fires caused by friction sparks from car brakes, or by burning articles thrown from the train.

forest fires each year, which burn over more than 500,000 acres, causing timber destruction and property damage amounting to over a million dollars.

There are but two causes for these fires—lightning and man. By far the greater number of these fires are started by human agency, very often due to carelessness and acts of ignorance. As it is estimated that eighty per cent of these forest fires are directly attributable to human agencies, and are preventable, it is apparent that no effort can be too great to bring to the minds of the general public the importance of care in handling fire. Carelessly throwing lighted matches and burning tobacco from cars is a common cause of fire, as they frequently fall into masses of dry grass, smoulder for a time and finally break into flame, which spreads to forests, grain fields and pastures.

As well as safeguarding its own property, the Company, through a sincere desire to minimize these forest fires, is constantly cooperating with the Forest Service. To this end, signs are posted in all trains, requesting patrons and employes to assist in protecting and preserving the forests by not throwing burning cigars, cigarettes or matches from trains.

Doing its own part in prevention work, the Company has equipped all locomotives operating through territory subject to fire hazard with devices for preventing the escape of sparks from the smokestacks. Fire guards six feet wide, are plowed at the edge of the right of way, and the remainder of the right of way is cleared of dry grass or brush by burning at the approach of summer. On the heavy mountain grades where hard application of brakes is necessary, sparks are at times created by the friction. While it is seldom that these friction sparks cause any damage, at the same time they are closely watched. During the summer, when forest and field are dry, and undergrowth furnishes an extreme fire risk, precautions are intensified.

A motor car, equipped with emergency apparatus, follows closely behind each train in the forest country. This car is operated by one man, who carries on his back a pack tank filled with water, with a small hose attached. He carefully inspects the right of way, and any small flame that might have been caused by the passing train, or by burning articles thrown therefrom, is immediately extinguished.

In the Sierra Nevada there are at this time about twenty-three miles of snow sheds, which it is necessary to maintain in order that traffic may move during the winter where a snowfall of sixty-five feet has been recorded in a single winter. These snow sheds are constructed of wood, and, during the dry summer months, constitute one of the Company's greatest fire hazards.

Fire Trains

To protect these sheds, specially designed fire fighting equipment is maintained. Four fire trains are kept in readiness for instant action, one at Truckee, one at Norden, one at Cisco and one at Emigrant Gap. The locomotives of these trains are equipped with duplex pumps with a capacity of 300 gallons per minute. To each locomotive is attached two water cars, with a capacity of 12,500 gallons. The trains are also equipped with 1000 feet of 2½-inch standard first hose, 12 brush hooks, 12 axes, 6 shovels, 2 cross-cut saws, 6 lanterns, 1 portable telephone, 2 chemical extinguishers and 2 fire ladders. There are two moveable nozzles attached to the front of each locomotive, making it possible to throw four streams of water at once, and to throw water from either the front or rear end of the train.

These trains are kept constantly under steam during the summer months, and only require from two to five minutes to get under way. There are no speed restrictions for a fire train—keeping on the rails is the

Fire fighter operating pack tank.

only determining factor in operation. The engines are equipped with sirens, which scream their warning as the trains dash forward to a fire, and all other traffic is sidetracked to make way for the fire trains. With four trains on the mountains, it is possible to send trains to a fire from two different directions, thus fighting the

◀ ━━━━━━━━━━━━━━ ▶

The Cover Picture
By E. J. Whisler

THE fire train located at Emigrant Gap is shown in a demonstration of its ability at fire control. The two nozzles permanently attached to the engine, may be moved in any direction, making it possible to put four streams of water on a fire simultaneously. The train crew: H. Strube, engineer; H. C. Bell, fireman; T. P. Holder, conductor; C. A. Marr, brakeman, and J. L. Kent, engine watchman, appear in the picture.

◀ ━━━━━━━━━━━━━━ ▶

flames from both ends of the fire. It is the proud record of the fire trains that they have never yet had to back away from a fire. The fire losses in the sheds have been measured by the time required to get the fire trains to the scene.

In addition to the specially equipped fire trains through the snow shed territory, ninety-three locomotives in regular train service have been

equipped with fire fighting apparatus consisting of an inspirator and 100 feet of 2-inch rubber-lined hose, so that each engine becomes a fire fighting unit. This allows every train to combat any fire which it may encounter, particularly in an incipient stage.

In the clearing of the right of way, section men use every precaution. Some of them are equipped with pack tanks with which to keep clearing fires from getting out of bounds, and to check forest, field and bridge fires encountered. The pack is also used to wet the sacks with which they fight fire.

Lookout on Red Mountain

At Red Mountain, the highest peak in the Sierra Nevada, a lookout is located for the protection of snow sheds and forest. In the clear air of this altitude, a pair of good eyes and constant watchfulness are all that is necessary during normal daylight conditions, but at night, or at any time when a clear view of the surrounding country is impossible, high-power binoculars and a transit telescope mounted on a fixed pier in the center of a bay window gives verticle and horizontal angel and location on an etched map. So closely has the etched map been scaled that it is possible for the indicator attached to the telescope to reveal within an extremely small range the location of a fire, and also whether it is in a snow shed or in the forest, and, if in the forest, on which side of the shed it is located. The lookout is connected with Norden station by telephone, over which the watchman reports every half hour, day and night.

In addition to the lookout, trackmen patrol the sheds every hour. The sheds are equipped with an electric fire alarm system, ringing into a central office, so that, between the lookout, the patrol and the electric alarms, a fire seldom gains much headway before it is discovered.

To further insure against fire in the wooded country, the Company has provided fifty water tank cars on steel underframes, especially equipped for emergency use. These cars are distributed at the most strategic locations on spur tracks, being first out and

On Red Mountain, highest peak in the Sierra Nevada, the Company maintains a 24-hour "look out" during the dry season. A wide area of timber country is under constant observation from this vantage point and a small blaze is quickly detected. The lookout has telephone connection with the fire train stations. A portion of the snow shed territory is indicated by the dark horizontal line.

thus quickly accessible. The cars are tested once a week to insure that they are always in good working condition. Piping is arranged on the cars for a locomotive to furnish steam for the operation of a Worthington Duplex Pump with a capacity of 300 gallons per minute. Each car is equipped with 500 feet or more of 2½-inch standard fire hose and all tools necessary to make complete fire fighting equipment.

Protecting Bay Property

Second in importance only to the forests and snow sheds, is the fire hazard of the Company's properties around San Francisco Bay, the train terminal at Oakland Pier and Alameda Mole, freight and passenger slips, West Oakland shops and the Company's ferryboats. All of these properties can be reached by water, and a fire tug, the Ajax, has been provided for their protection, being kept on a 24-hour service. The boat is 103 feet in length, has 750 developed horse power and a speed of 10 to 11 knots per hour. It is equipped with pumps that will deliver 2400 gallons of water per minute at 150 pounds pressure. In case of mishap to ferryboats, the Ajax is always available to tow them to safety.

The shops on all of the Company's thirteen divisions are fully equipped with fire fighting apparatus, and the employes are drilled regularly in its operation. Fire companies, ranging in number from one to six, are organized at each division point. Fire drills are held once a week. To make them the more effective, these drills are always surprise calls, so that the men never know when an alarm is sounded whether it is a real fire or a drill.

At large shop terminals, electrically operated fire alarm systems, connected with central stations, and also with the city fire department, have been installed.

Water Supply

At many locations on the Pacific System the water supply for fire protection has been increased, the most important points being Los Angeles and West Oakland, where large values are represented at both terminals in buildings, store stock, lumber and ties. Each new water system has been provided with two 1500 gallons per minute pumps, one steam driven, the other operated by electricity, located in concrete buildings and used for fire protection only. These pumps take suction from a reservoir of 250,000 gallons capacity, and discharge into a network of pipe lines ranging in size from 4 to 12 inches. Fire hydrants have been located at all important points, where our own forces, as well as the city fire department, may attach hose for fire protection.

These new pumps are automatic in operation. Opening a hydrant and

Daylight Is "Dressed in Pearl Gray"

ALREADY the fastest train in the West, the Daylight took its place as the most distinctive and finest train of its kind in the country, when, on April 18, it blossomed forth in a new spring dress of pearl gray paint. In adopting this color for all coaches of the train, from locomotive to observation car, the Company has established an innovation in the exterior appearance of its passenger equipment. Never before has one of the Company's trains been painted in any other than a standard color.

Coincident with the introduction of the pearl gray color scheme, new and improved equipment was added to the famous 12-hour San Francisco-Los Angeles flier. Included in its makeup from now on will be the latest type club car with a separate lounge and smoking compartment for women. This new car is matched in comfort by the bucket seat coaches, reserved seat cars and the observation car. Both dining and lunch cars are carried on the train.

The Daylight Limited was originally placed in service as a weekly feature in 1922, going on a daily run in 1923. Since then it has grown in popularity, not only with Pacific Coast travelers, but with tourists from all parts of the world.

allowing the water pressure on the pipe lines to drop 10 to 20 pounds, will cause one of the large 1500 g.p.m. pumps to go to work. The new water system at Los Angeles cost $83,000, and at West Oakland the cost was $75,000. At Sacramento the fire pump equipment has been reinforced with one 1500 g.p.m. electric-driven pump. The water is taken from two 30-inch city water pipes.

The water supply at these large terminals is practically inexhaustible. The reservoir at Los Angeles is supplied with water from wells and also with city water. At West Oakland the reservoir is replenished with water from the bay. The Sacramento water is from the Sacramento and American rivers.

In addition to water protection afforded by pumps, water barrels and buckets have been placed at every point where there is the remotest possibility of fire occurring. Fire extinguishers, ranging in capacity from 2½ to 40 gallons, have been placed throughout buildings and yards, in outfit and baggage cars, and in all other places where it is deemed expedient to place such protection. Sand barrels and scoops or shovels are placed near oil columns, tracks where locomotives stand, and through roundhouses and other places where oil might create a fire hazard.

Along the line, in territories where

long and high trestles are in use, water has been piped onto the trestle and hose provided to reduce the fire risk. Locomotives serving these districts are equipped with water sprinklers, which the engineer opens before going onto the trestle. By this means the deck of the trestle is wet down, reducing to a minimum the danger of fire from hot bits of metal from brake shoes or sparks from the firebox.

Break Matches in Two

The Company has taken every precaution that human ingenuity could conceive to provide prevention of fire, and to combat it if it does develop. In the last analysis, the prevention of fire in forest and field is squarely up to the traveling public and the employes. The number of forest fires started by lightning is negligible. The balance, or about eighty per cent, are due to human agency, by far the greater proportion of which are started by burning matches and cigars carelessly thrown away. We must educate others as well as ourselves to break matches in two and see that cigarettes and cigars are out before throwing them away.

Fire prevention is not so much a matter of preparation against fire as of eternal vigilance, once preparation is made. After all, the greatest hazards at times lie in the smallest things. Lunch papers cast carelessly aside, balls of greasy waste thrown into corners, fusees and torpedoes allowed to lie around, instead of being kept in metal cases—these apparently trifling things may be the cause of a serious conflagration. Keep the premises clean and absolutely clear, whether it be a large shop, a caboose, a station building, or other property, and you will have removed 90 per cent of the fire hazard. And again, and again, and again—it cannot be repeated too often — break your matches and see that cigars and cigarettes are OUT before they are discarded.

Dogs "Guiding" Blind Persons Permitted in Passenger Cars

The Company's hitherto inviolable rule that dogs on passenger trains would be handled only in baggage cars has recently been amended to provide for one exception. This exception permits dogs accompanying blind persons and acting as "eyes" and guides for them, to be taken into any of the Company's trains carrying coaches; in club cars of Pullman trains; on the upper deck of ferry steamers, and on electric trains.

Blind persons who depend upon dogs to lead them, will be given special permits allowing the dogs access to passenger cars.

Nowadays, nobody cares how bad your English is as long as your Scotch is good.—Ex.

The photo above shows a fire drill at the Sacramento Shops. The view is looking north toward the roundhouse. The boiler shop is to the left, the transfer table in the middle, and bays of the machine shop are to the right. The date is unknown. Even though the exterior of many buildings were brick, the structure inside was often wood. Embers, torches, and solvents were hazards that contributed to fire concerns. Also, when electricity was introduced, the precautionary measures for using this source of power had not been learned. Experience led to the development of codes, guidelines and fire drills. – Courtesy California State Railroad Museum

Chapter Seven
Building Locomotives – Bigger Demand and Bigger Locomotives

The need for larger locomotives grew with the demand for train-delivered goods. Locomotives similar to the Governor Stanford, the Central Pacific's first steam engine, which came to Sacramento in boxes via Cape Horn, weighed only about fifty tons. By 1925, the locomotives being built at the Shops were nearly three hundred tons.

Designers of these engines gave consideration to the efficient combustion of fuels, draft in the smoke stack, effect of load and speed on fuel consumption, superheating of water, and the need to reduce friction caused by the wheel on the rail, wind, curves, and grades.

Much of the early changes to the construction of a locomotive were made due to trial and error because the testing methods of steel strength, valve actions, and forces were very limited. There were many successes, however.

In 1948, D. L. Joslyn, who started working at the Shops in 1902, wrote an article called *Sacramento General Shops, Southern Pacific Company – Pacific Lines* that described the first steam engine built at the Shops:

> As the locomotives were in constant service during construction of the road, they were given severe service, and after May 1869 when the road was completed, a large number of the locomotives were brought to Sacramento and set aside as in need of repairs, or as the old report puts it, they were "set aside in the dead line, awaiting or-

ders." Some few [*sic*] were given new fireboxes in the roundhouse, some of them were completely rebuilt, requiring a large amount of boiler work, and this was done in the old shop, in the Roundhouse and in the Machine Shop. Then, in 1872 the management gave Sacramento the go-ahead to build ten new American Type or 8-wheel locomotives. The No. 173, built in 1864 by J.A. Norris at the Lancaster Locomotive Works for the old Western Pacific Railroad, was brought in shop to be repaired. But being a small engine with 40-inch boiler, and small cylinders, it was decided to scrap this engine. So the first new engine was given the number 173 and was built new from the ground up.

The wheel-making process was a significant activity in Sacramento. The first essay describes this part of the business and the fine tolerances needed to make the wheel ready for service.

Southern Pacific's *Bulletin* extensively covered the building of locomotives in the following articles of this chapter:

"Pacific Type Locomotive Built By Southern Pacific Shops In Sacramento," November 1917

"Large Oaks from Little Acorns," November 1919

"New Switch Engine Made at Shops," September 1919

"Meet the New Southern Pacific #5000," July 1925

"Sacramento Shops Have Built 143 Locomotives," November 1925

"Giant 2-10-2 From Tiny Hayburner Grew," September 1924

❧ ❧ ❧

Wheel Making

For many years, Sacramento was an isolated railroad facility, which required the making of parts, and the tools to make the parts, on site. Consequently, a high level of expertise regarding foundry operations and the physics behind the behavior of various kinds of metals was needed.

The structural engineers that filled a crucial role at the Sacramento Shops for over one hundred years had, as one of their many chores, the design and production of wheels by the thousands. To appreciate the smooth passage of an 80-car train with each car weighing one hundred tons, one must know the small design features measured in fractions of an inch. There are subtle applications of physics to the design of the railroad car's wheel. The wheel must support the car's weight, of course, but also should be designed to minimize wear on the wheel, the rail, and the car's frame. The designers needed to know the actions and reactions that occur every second as the wheel rolls along the track, through switches, around elevated curves and along rails that might have lost their shape and proper function fifteen years earlier.

The first characteristic of almost every railroad car's wheel (some wheels on steam locomotives and high speed passenger trains are an exception) is that the tread of the wheel, which contacts the rail, has a gradually decreasing diameter as measured from the flange to the other side of the wheel so the wheel at the flange has a slightly larger radius. This tapered surface serves two functions. The first function is this: Imagine a car rolling along straight track. As the car shifts slightly to the right, perpendicular to the rail and perpendicular to the direc-

tion of travel, the right wheel slides on the rail to a position of greater wheel diameter. This causes the right side of the car to rise slightly then slide laterally again back to a level position, which on properly gauged rail, is the center of the rails. Considering the high likelihood that the two wheels on an axle do not have exactly the same diameter, a set of wheels *without* the tapered tread would fall toward the side with the smaller diameter creating excessive flange wear on the smaller wheel.

The other function comes into play on a turn. On any curve, the distance the outer wheel needs to travel is greater than the distance traveled by the inner wheel. The differential on an automobile handles this problem with a complicated set of gears yet the design of a railcar handles this problem with no moving parts other than the wheel. As the car moves along the curve, the car slides toward the outer rail and reaches a position where the outside wheel is turning on a larger diameter than the inner wheel. Consequently, the wheel on the outside of the curve can travel a greater distance than the wheel on the other rail given the same number of revolutions.

Another important location on the wheel is called the *flange throat radius*. It is the curved section between the flat tread surface and the flange. An alternate name for this location is the root radius because it is located at the bottom of the flange. The root radius for many years, was smaller than the radius on the inside of the rail. This created a situation where there were two contact points between the wheel and the rail while rounding a curve. One point was between the top of the rail and the wheel tread surface where most of the weight bearing force is located. The other point was between the flange of the wheel and the outer side of the rail. This combination of shapes restricted the movement of the outside wheel in such a way that the flange may contact the rail before the wheel is rolling on the optimum diameter. A costly consequence was additional flange wear.

The crews at the Sacramento Shop no doubt witnessed a variety of wheel wear patterns as the geography

76

over which trains traveled ranged from the curve-after-curve mountainous terrain of the Sierra to the much straighter routes of the Central Valley.

The stress on a wheel is not localized to only the contact point with the rail. The weight of the car exerts forces on the wheel hub and the plate of the wheel (the part of the wheel between the axle and the flange). For decades, the plate was a flat surface on a slight angle outward from the hub. Straight-plate wheels were vulnerable to large stresses during braking because when the brakes are applied, the friction of the brake causes heat on the tread that creates a larger circumference of the tread. But, the heat that expands the rim has a minimal effect on the plate. So, when the tread expands, the plate is stretched and cracks may result. One can imagine the heat generated on wheels during the near 6,500 feet descent along the one-hundred-mile downgrade between Donner Summit and Rocklin. Research revealed that a curved plate reduced the number of defects. A profile of the newly designed plate section had a slight "S" shape. The new geometry enabled the plate to flex and bend slightly while absorbing the stresses of the heated rim. Straight-plate wheels are ten times more likely to fail due to plate forces than curved plate wheels. As of 2009, straight-plated wheels were being regulated off the rails. But, for years the Southern Pacific crews certainly gathered a collection of wheel defects related to the straight plate design.

Wheels with unacceptable wear did not have to be thrown away. Work at the Shops included restoring the correct tread shape. This task was accomplished on very large lathes that could hold and rotate the wheels while the tread was cut. After the lathe work, the wheel was re-installed on the car for use. Because some steel was removed from the wheel during this process, the number of occurrences during which the wheel was reshaped was limited to two or three during the lifetime of the wheel.

Eventually, the wheel was removed for the last time and scrapped. But the material was often recycled to become new wheels. The complete process of melting the old wheels, making molds for new wheels, pouring molten steel into molds, and the controlled cooling of the new wheel was a tough chore performed by the staff in Sacramento.

Over the years, it was discovered that the percentage of carbon significantly affected the usability of the final product. An increase in carbon content increased the hardness. Steel quickly cooled will be much harder than steel cooled less rapidly. So, high-carbon steel that is rapidly cooled is very hard. The Sacramento crews cooled the outer rim of the wheel faster than the rest of the wheel. The result was a hardened surface for rail contact with a less brittle plate and hub.

Wheel plate or web

Wheel tread

Flange throat radius

Wheel flange

Wheel hub

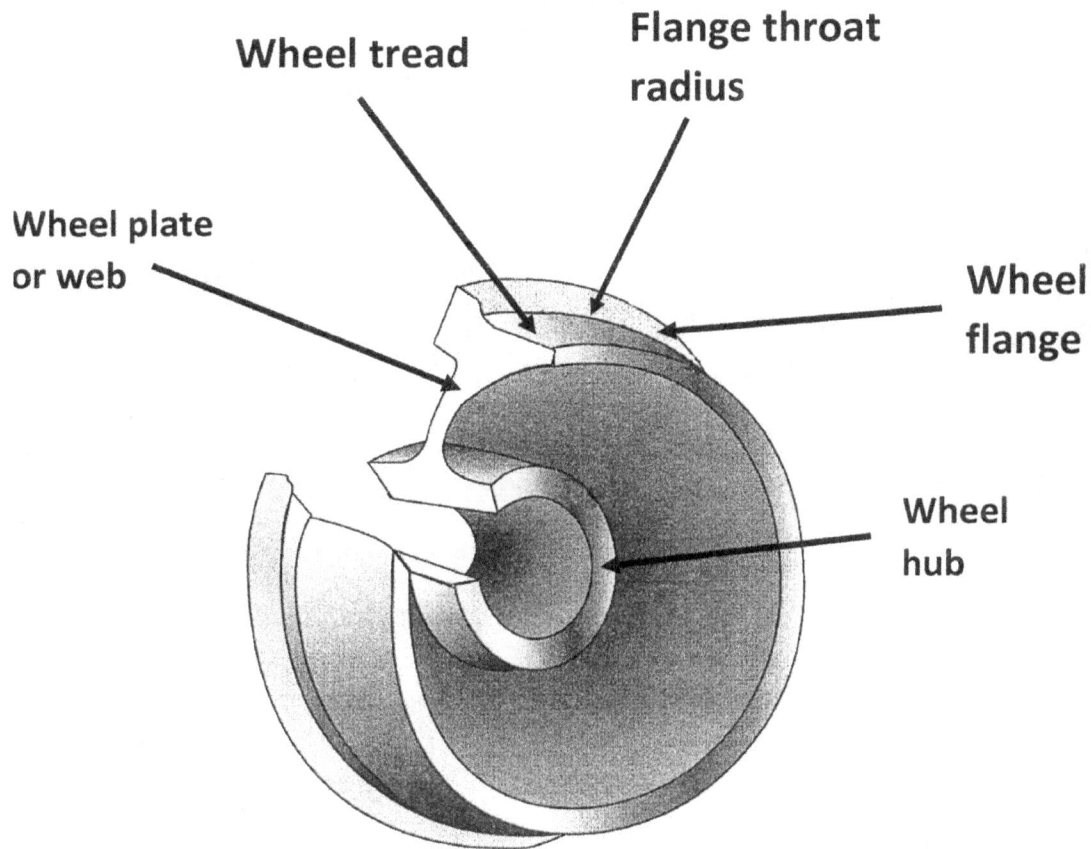

A loaded railroad freight car in 1940 weighed about forty-five tons, was about forty feet long, fifteen feet high, ten feet wide and had wheels with a diameter of approximately thirty-three inches. The rail dimensions varied. A common rail weighed one hundred pounds per yard and was six inches tall. The efficient rolling of the car in regard to fuel consumption and wear has been researched for many years and is a function of about five dimensions of the wheel. These dimensions are measured in fractions of an inch. The shape of the wheel was not the only focus of attention in its design. The composition of the steel affected the brittleness of the wheel and its ability to handle shock resulting from various roadbed conditions and cargo weights.

Pacific Type Locomotive Built by S. P. Shops at Sacramento

This is the Pacific Type of locomotive turned out by the Southern Pacific Company at its Sacramento shops and widely heralded as the "Liberty Locomotive." It has a tractive effort of 31,420 pounds and its weight in working order is 234,200 pounds. In a previous article it was stated that the first ten locomotives to be built at Sacramento would cost about $300,000. Lest such a figure be misleading, it should be stated that much of the material for these engines represented stock already on hand.

Railroad Shops Busily at Work on Equipment to Meet Wartime Needs

Western railroad shops are doing their part to help win the war. With the arrival of delayed material, construction of war equipment is speeding up and a steady flow of cars and locomotives will soon be going forth to meet the ever-increasing demand.

The Southern Pacific is leading the way with a huge program of home construction, all its shops on the Pacific Coast being hard at work to build equipment.

At its Sacramento shops the Southern Pacific has built two consolidation and one Pacific type locomotive, and will have a fourth ready by November 15th. All are the latest superheater type and are part of an order for twenty to be built at home. One hundred and fifty flat cars have been turned out at Sacramento and more are being turned out at the rate of ten a day. The order is for 500, of which 325 will be built at Sacramento and 175 at Los Angeles. Ten freight cars a day will soon be turned out of the Southern Pacific shops at Los Angeles. The program calls for 1500 box cars or 3252 cars of all kinds.

The Southern Pacific shops at Sparks have built and put into commission forty narrow-gauge flat cars for its Nevada lines.

The Oakland, Sacramento and Los Angeles shops have added fifty caboose cars to their accomplishments.

To accomplish this undertaking, extensive improvements were necessary. The Sacramento plant was extended, and many labor-saving devices and much special machinery installed, including a wheel-grinding machine which is able to save thousands of wheels a year.

A giant magnet installed in the Ogden shops is mounted on wheels and with the aid of a crane moves around picking up and depositing heavy metal pieces which would otherwise require human help that cannot be secured. Electric welders will be installed at Sacramento and Los Angeles to supplement the acetylene welders used generally.

The Southern Pacific has opened its new shops at Visitacion, near San Francisco, and is repairing freight cars there. The passenger shops will be opened in another month.

Demand for Steel Plates Promises to Exceed the Supply.

Added emphasis to the need of securing full use of all available locomotives and cars is given by the news that government needs in the way of steel plates will add to the difficulties of the railroads in securing new equipment as well as the repair of that now being put to strenuous use.

Wartime growth in value of United States exports is summarized by National City Bank of New York as:

Section	Year ended June 30, 1914	Year ended June 30, 1917
Belligerent Europe	$1,304,000,000	$4,042,000,000
Neutral Europe	1,061,000,000	2,252,000,000
Other Neutrals	184,000,000	414,000,000
Total . .	$2,549,000,000	$6,708,000,000

Agent Cites Experience in Securing New Business for Company.

An example of how business can be secured for the Company with a little effort is given by E. R. Plumb, agent at Seeley, Cal. While traveling on a Coast train, Agent Seeley chanced to learn that a lady passenger was contemplating an Eastern trip and had about decided to choose some other route than the Southern Pacific. Descending from the train at the first station, Seeley secured folders of the Southern Pacific and the road which the passenger favored. Then introducing himself, he showed the passenger where she could reach her destination a little sooner with only one lay-over and change as against two changes and lay-overs on the other route. The passenger was convinced and accompanied Seeley to the Southern Pacific ticket office when the train reached Los Angeles. There she purchased her overland tickets and expressed her appreciation for the service.

Transfer at San Antonio.

Although the M., K. & T. Ry. of Texas discontinued the use of the joint terminal with the Southern Pacific Lines at San Antonio on September 1st last our Passenger Department is in receipt of advice from them that the automobile bus requires but ten minutes to make the transfer between stations.

All tickets which are routed via the Southern Pacific Lines to San Antonio, thence the M., K. & T. Ry. of Texas, or vice versa, should contain a transfer coupon. However, the present forms of printed tickets may be still used, and in such cases the conductor into San Antonio will issue a transfer check which will be honored by the San Antonio Transfer Company.

"LARGE OAKS FROM LITTLE ACORNS GROW"

"Father and Son" in the Southern Pacific's locomotive family. Photographed in the Sacramento General Shops.

The above illustration is of two locomotives, one of which went into service very soon after the commencement of the building of the Central Pacific and the other which went into service during the late war and was built at Sacramento.

Engine No. 1, S. P., "The C. P. Huntington," was built by Danforth, Cooke & Co. of Paterson, N. J., in the year 1863, was shipped around the Horn on a sailing vessel and went into service at Sacramento in April, 1864. It was known at that time as No. 3, Central Pacific, but was afterwards transferred to the Southern Pacific and known as No. 1, S. P. This locomotive has cylinders 11x15 inches and has a weight on its single pair of drivers of 18,500 pounds. The total weight on the engine is 39,000 pounds. The last work which this engine did was in hauling the Berkeley local. After being taken from this run she reposed at Oakland for some time, when she was set up at the Panama Pacific International Exposition and placed alongside one of our Mallet engines. She now has a little plot of ground all her own in the Sacramento shop yards.

Engine 2371 was built at Sacramento shops and went into service November 24, 1917. She is a 10-wheel passenger engine with cylinders 23x28 inches and has a weight of 174,000 pounds on her drivers and a total weight of 218,500 pounds. She is superheated, and is in every respect an up-to-date modern passenger engine, although not as large as some. This engine was taken into the shop for repairs on July 25, 1919, having made a mileage of 124,881. This mile-age, while it may not be a record for an engine of this class, is nevertheless an excellent performance.

SANTA BARBARA JOTTINGS.

By Agent J. E. SLOAN

The King and Queen of Belgium delighted the people of Santa Barbara during the few days the royal couple were here. The distinguished visitors appeared to thoroughly enjoy their stay, taking advantage of the surf, a hydroplane trip over the beach, motor rides into the country and Sunday mass at Santa Barbara Mission.

The bean and walnut harvests are under way. Prices are somewhat higher than last year, being 34 cents for walnuts and 13 for beans. The lima bean crop is short but the walnuts are normal though the sizes are not so large as last year.

The Gibraltar dam is 80 per cent complete and the last concrete will be poured this month. Completion of this six-year job will give Santa Barbara a magnificent municipal water supply.

Freight and passenger business for Santa Barbara approached a record last month. The recent enlargement of the freight house has scarcely sufficed to care for the increased business.

A municipal landing field for airplanes, several miles of new streets, a new downtown hotel and one hundred new buildings are projects which Santa Barbara expects to see under way this winter.

TRAFFIC NOTES FROM COLTON

At Colton the California Portland Cement Company is engaged in filling an order for 23,000 barrels of cement for shipment to Honolulu for United States Government work.

It is proposed to form a union high school district that will result in the erection of a $150,000 union high school at Colton.

Yardmaster O. D. Guire and Roundhouse Foreman E. J. Shearer are home from a vacation spent at their former eastern home.

A party of twenty interstate deports from the Patton Asylum left Colton October 9 in one of the Southern Pacific combination sleeping and dining cars for destinations in the Atlantic Coast States.

Heavy movement of grapes from the Pacific Electric for eastern destinations continues. Etiwanda, Muscat and Fontana are the principal shipping points.

Red Cross war-time activities have been run on a firm business basis. The same standards will be maintained for the peace program. Help make its institutions financially sound. Answer the Roll Call November 2-11.

Enlist in the fight for better health in your community. Volunteer now at your chapter headquarters to help obtain 20,000,000 members for the Third Red Cross Roll Call, November 2-11.

OAKLAND LONG WHARF SHOAL WELL GUARDED.

Navigation on San Francisco Bay will be adequately safeguarded from Long Wharf Shoals, according to a statement given the press by General Manager J. H. Dyer.

Until such time as the shoals are dredged out by the city of Oakland or other interested parties, the Southern Pacific will maintain there a fog board similar to those now in use in Suisun Bay and various points along the Sacramento River. Upon this board the Government has agreed to install and maintain an acetylene light, flashing at suitable intervals, similar to those used as aids to navigation all along the Coast.

Said Mr. Dyer: "It is not considered that there is sufficient motion of the water at this point to make a bell buoy of any use at the time it is needed. The purpose of the fog board is to afford a support for the light that will be a guide in fair weather and to reflect the sound of the steamers' whistles back in foggy weather, the wings being so arranged that the sound will come back to the pilot house from whichever direction the shoal is approached.

"It is not intended that this board shall perform the functions of the bell, namely, to be a mark for which a vessel shall head in foggy weather in order to locate her position, but that it shall provide a warning to any vessel that comes too close to the shoal. There is plenty of room to the southward and to facilitate following such a course it has been decided to install on Oakland Pier a diaphone, which is the fog signal used by the United States Lighthouse Board for all important stations.

"We are naturally more concerned than anyone else in seeing that adequate protection is afforded our ferry steamers. The installation we have undertaken carries the approval of Captain Heath, our superintendent of ferry steamers; W. H. Rhodes, superintendent of lighthouses for the United States Government in San Francisco, and other experts who have made a thorough study of the subject."

APPOINTMENTS.

Superintendent William Wilson announces that E. D. Leavitt has been appointed assistant superintendent of the Tucson Division, with headquarters at Tucson, Ariz.

Superintendent G. E. Gaylord announces that W. M. Stillman, who has been released from military service, has been restored to the position of assistant superintendent of the Stockton Division, with headquarters at Stockton, vice E. D. Leavitt, transferred.

Dr. O. R. Myers has been appointed district surgeon at Wheatland, vice Dr. M. E. Smith, resigned.

NEW SWITCH ENGINES AT SACRAMENTO.

By H. C. VENTER
(Sacramento General Shops)

The heavy demands for switch power in the past years, due to the steady growth of the vast territory which this company serves as well as the heavier cars being built, calls for modern switch equipment to take care of this business.

There was built at the Sacramento General Shops during the latter part of August the first of ten new modern switch locomotives, and considerable credit is due to the officials and employees for the design and workmanship.

The running gear detail includes several features of interest. The main frames have a width of four and one-half inches and the upper and lower frame rails are secured to the frame tongues at the cylinders. The frames are braced transversally throughout their length. The guide yoke and valve motion bearers are braced to the boiler by heavy wrought iron supports which are bolted to an angle plate studded to the boiler. Driving wheels are fifty-seven inches in diameter over the tires. The valve motion applied is of the Walschaert type, which will give a more equal distribution of steam than the Stephenson design of valve motion. Among the specialties not already mentioned applied to these switch engines is the Schmidt Superheater, W. H. E. T. No. 6 brake equipment with eight and one-half inch cross compound air pumps and Pyle headlight.

The tender is of cylindrical design, having a water capacity of 5000 gallons and an oil capacity of 1100 gallons. Fitted with Miner draft gear with thirty-ton arch bar trucks and simplex bolsters.

Operating with superheater and with 200 pounds working steam pressure and having sufficient weight on drivers, these engines will be capable of handling the long trains now being brought into terminals.

Our shop employees are awaiting with interest the advent of the first engine in the yards for a display of its good qualities.

These engines will be numbered from 1247 up.

In the past month considerable interest was taken in making one of the largest forgings that has been turned out of these shops. The forging is what is known as a strap for the walking beam on the ferryboat Garden City. In order to handle this massive forging, weighing several tons, it was necessary to excavate several feet of ground about the large steam hammer in order to get the part to be forged in line with the die head of the steam hammer. Forger J. Ryan and his assistants are to be complimented for the manner in which this work was done.

The interest which foremen take in their work, as well as for the welfare

STORE EMPLOYEES FORM ASSOCIATION.

By GEORGE R. HIMMELMAN
(West Oakland General Stores)

In the interest of good fellowship and loyalty to the railroad, employees of the Store Department at Oakland have formed an organization known as the Store Department Employees' Association. This includes the West Oakland General Stores, West Oakland Mill, Stationery Store and General Storekeepers' Office, San Francisco.

This realizes the ambition of many who have long felt the need of some means whereby the members of different branches of the department could be brought to a better understanding of each other both in a social and a business way. While the object of the Association is primarily to promote interest in social affairs, its purpose is also to lend assistance to members in personal matters whenever needed. The treasury is supported entirely by the members and a small monthly fee maintains the funds available for benefits and social affairs. To add interest, and as an inducement as well, a generous amount has been set aside for wedding presents. The treasurer has been kept busy lately paying Cupid's bills, for a few members have already taken advantage of this offer.

The executive committee is planning much in the way of entertainment, and from all indications many good times are in store for the members and their friends.

The Association takes no part in wage controversies or grievances of any kind, but is simply a "get-together" proposition on the part of the employees of the Store Department that they may know each other better.

The executive committee consists of the following: S. L. Bouque, acting president; H. B. Montag, secretary; F. J. Cunningham, treasurer; J. F. McAuley, G. Mansfield, F. P. Austin, D. Ritchie.

of the men whom they are in charge of, was shown lately by F. P. Hammack, foreman of the Loco Paint Shop, who designed a sand-blasting cabinet for the protection of his men.

Formerly the operator who did the sand blasting in this department had to protect himself by wearing a hood over his head, which was very uncomfortable. The operators could only stand up to this work for about an hour at a time. With the present cabinet, which is made of a pipe frame covered with canvas, with a large glass in the front, the operator stands inside of this cabinet and does sand-blasting work without discomfort and much faster.

SEEK LOS ANGELES SCALPS.

M. A. Robles, manager of the Southern Pacific ball club at Tucson, directs a challenge to other railroad teams, particularly the Southern Pacific nine at Los Angeles.

Meet New Southern Pacific Type Locomotive No. 5000!

The first of the new three-cylinder Southern Pacific type locomotives to arrive at Sacramento General Shops. Sixteen of these powerful locomotives have been ordered by the Company and will soon be placed in freight and passenger service over the Sierra Nevada and Siskiyou mountains

FIRST of the sixteen new three-cylinder Southern Pacific type locomotives are now in operation.

The first two arrived at Sacramento General Shops the middle of June where they were quickly placed in condition for active service. The other locomotives of this type are now in course of delivery.

The new locomotives, which were built by the American Locomotive Company with the co-operation of Southern Pacific's mechanical experts, attracted general interest on their way to Sacramento from the East. On their arrival at Sacramento they were the center of attention of employes and the leading citizens of Sacramento, including officers of the Sacramento Chamber of Commerce, joined with Southern Pacific representatives in holding a christening ceremony for "Locomotive 5000," shown in the cover photograph.

The new locomotives will be used to haul passenger and freight trains in mountain territory. They embody many improvements and refinements of design and were described in detail in the February issue of the Bulletin.

The outstanding feature of the locomotive is a third cylinder placed inside the main frames, slightly above and between the two outside cylinders.

The addition of the third cylinder corresponds to the increasing of the four cylinders in an automobile to six, with consequent smoother application of power. The locomotive has a "4-10-2" wheel arrangement and is a step in advance of the "2-10-2" and "4-8-2" types which heretofore have been considered the most modern development of locomotive construction.

It is the most powerful, non-articulated or single unit locomotive yet designed.

The new locomotives are 101 feet 1 inch in length, weight 685,600 pounds and have a maximum tractive power of 96,530 pounds.

If a railroad can be judged by its motive power, then every Southern Pacific employe can realize that he is a part of one of the best and most modernly equipped transportation organiza-

This view of Locomotive No. 5000 shows location of the three cylinders, one on each side and one in the center.

tions. The placing of these locomotives in service marks another forward step in the operating efficiency of the Company.

Besides increasing power, the locomotive is designed for fuel economy.

In addition to these 16 Southern Pacific type locomotives, the Company is now adding eighteen "4-8-2" mountain type locomotives which are under construction in its Sacramento shops. These locomotives are of the same type which have been making through runs of 815 miles between Los Angeles and El Paso.

MASSACHUSETTS WOMAN IS APACHE TRAIL BOOSTER

Natural beauties of Apache Trail and the genial hospitality received at Apache Lodge were such a revelation to Mrs. Lucy J. Hersey of Wellesly Hills, Mass., that she wrote President Sproule a letter of appreciation. "Of all my western trip, the two days spent at Apache Lodge were the best," she wrote. "Apache Trail is so beautiful and so little known. I shall tell all my friends what a wonderful trip it is. Mr. Still at the Lodge is a delightful host."

Here's a squib which could well be reprinted in house organs and posted on the bulletin boards of department and other stores:

"Can you serve company?" asked the mistress.

"Yes, both ways, mum," replied the new maid.

"Both ways?"

"Yes, mum; so they'll come again or so they'll stay away."—Forbes.

After Reading Your Bulletin, Pass It Along

Sacramento Shops Have Built 143 Locomotives

By H. C. VENTER
Superintendent of Sacramento General Shops

IT WAS in October, 1872, that the first locomotive, C. P. 173, was turned out of the General Shops at Sacramento for service on the present Southern Pacific Lines. The little engine, big in its time, was a beautiful appearing machine, fairly glistening with its brass trimmings and bands holding the jacket in place, the elaborate brass builders plate between the mud guards, numbers and lettering in red shaded with green and gold, great broad stripes of gold edged with green and red around the tank and on the steam and sand dome, cab grained and varnished, and a beautiful painting of Yosemite Valley on each side of the headlight.

The building of this locomotive marked the establishing of a great industry at Sacramento which now ranks with the greatest in the West; where everything from tiny bolts to huge locomotives and working beams for steamers can be manufactured; and where employment is provided for a normal force of 3100 people.

A few days ago the most recently built locomotive, No. 4328, a 4-8-2 or mountain type, was christened at the shops. Business men and civic leaders of Sacramento joined with Southern Pacific people in celebrating this crowning achievement. The 4328 is the first of eighteen locomotives of the same type to be built at the Sacramento shops during 1925-26 at a total cost of about $1,500,000.

Comparison of the new No. 4328 and the first No. 173 gives an idea of locomotive development during 53 years.

No. 173 had a tractive power of 14,-480 pounds, weight in working order 66,000 pounds, length about 52 feet, diameter of drivers 54 inches, and cylinders 17x24 inches. No. 4328 has a tractive power of 67,660 pounds, weighs 614,200 pounds in working order, is 97 feet 6¼ inches long, diameter of drivers 73½ inches, and cylinders 28x30 inches.

Since No. 173 went into service in 1872 a total of 142 locomotives have been built at Sacramento shops, 63 having been built since 1917, including the new No. 4328.

The first ten locomotives built at the shops, all of the 4-4-0 type, were used in both freight and passenger service and gave wonderful performance. Most of them were in service until 1906. The first engine was equipped with a crosshead pump for putting water in the boiler and the steam chests were equipped with C. P. automatic self-feed oilers, but evidently were not relied upon as tallow oilers were also applied. G. A. Stoddard was in charge of design of the locomotive, W. H. Milliken was master mechanic of shops, J. J. McCormack was foreman of erecting shop, C. F. Shields was superintendent of boiler shop, A. F. Lashells was superintendent of blacksmith shop, and A. J. Stevens was general master mechanic.

In 1881 the shops were given an order for 7 engines to operate the local trains at Oakland. Mr. Stoddard designed an engine of the 2-6-2 type that operated equally well in either direction, and the first one was turned out

in December, 1881. These engines had cylinders 16 in. x 24 in.; drivers 48 in. diameter and weighed 100,000 lbs. when built. The weight was later increased to 103,800 lbs. These little engines were in service at Oakland on the local runs until the line was electrified.

In 1882 the C. P. 229 was designed and built at Sacramento Shops. This was something bigger than had ever been attempted and interest was attracted from all parts of the country. This engine was of the 4-8-0 or 12-wheel type and was designed for service over the Sierra Nevada mountains. It was equipped with steam brakes on drivers and had a steam reversing gear on account of the construction of valves, which were heavy and worked one on top of the other so that it was hard for enginemen to reverse her.

The following is the general data of this engine as originally built: Diameter of cylinders 19 in. x 30 in. stroke, diameter of drivers 54 in. later increased to 56 in. diameter, steam pressure 140 lbs., weight of engine loaded 123,000 lbs., weight on drivers 106,050 lbs., maximum pull of draw bar 22,500 lbs. on a grade of 105 feet to the mile 8 degree curves, hauling 19 loaded cars of 388½ tons.

This engine was quite successful and later on the Cooke Locomotive Works was given an order for 19 or 20 more just like her including double valves, steam brakes and steam reverse gear. This locomotive has been rebuilt and renumbered and is now 2925 and is still in service.

The 229 was such a success that Mr.

Sacramento pattern shop force and some of the patterns used in constructing locomotive No. 4328. Reading left to right: C. E. Hardy, G. R. Scott, foreman; W. B. Butler, assistant foreman; A. L. Grimes, W. H. Blaney; A. Burzlander, C. Shalag, J. B. McCain, C. H. Belknap, F. C. Smith, C. J. Lindstrom, W. B. DeCosta, J. V. Rapius, T. W. Pugh, R. R. Jensen, J. H. Nelson, J. M. Peek, A. Hauschildt, R. D. Oliphant, E. F. Halverson, H. G. Griswell and C. A. Smith.

Sacramento shops are now working on eighteen of the largest locomotives ever built at the shops, the 4-8-0 or Mountain type, one of which is now in service. Miss Marie Smith, left, of the shops store department, aided in the christening. This new powerful locomotive is shown in the lower picture with G. A. Knoblauch, general foreman of erecting shop, standing alongside. Top picture shows C. P. 55. It had the exact appearance of C. P. 173, which was the first locomotive built at the shops in October, 1872. The queer looking engine just below is of the type used in the Oakland local service before those lines were electrified. The "El Gobernador," shown next below, was one of the most famous locomotives ever built at the Sacramento Shops and attracted world wide attention when built in 1883 because of its size and power. A description of the various types of locomotives built at Sacramento since 1872 is contained in the accompanying article. Sacramento shopmen shown working on various stages of building the latest mountain type locomotives are, reading top to bottom: Machinist S. J. Dennis, better known as "Stew," turning up a side rod collar; Boilermaker Mike Brushia and Helper Joseph Lewis driving 1¼ inch rivets with a 90-lb. air hammer on one of the boilers of the locomotives. Strain in driving such large rivets has been relieved by using a spring yoke attached to the hammer, a device gotten out by General Foreman Boilermaker F. J. Hickey. Machinist Charlie Hills, a veteran in such work, is drilling one of the cylinders.

Stevens was tempted to build a still larger engine, and in 1883 built the now famous C. P. 237 and named her "El Gobernador," which is Spanish for "The Governor." This locomotive had a type of valve gear that was the beginning of the Stevens Valve gear. The valves were rotary balanced, with double admission and were driven by Stevens Valve gear employing one eccentric on each side and reversing from the rocker, which was connected to cross head with a union arm. Wide interest was created in this new valve gear and this, coupled with the fact that it was such a big engine, caused the 237 to be spoken of and pictures to be published in journals all over the world.

The cylinders of this engine were 21 in. diameter x 36 in. stroke, driving wheels were 57 in. diameter, diameter of boiler 56 in., No. of flues 178—2¼ in. diameter 16 feet—2½ in. long, steam pressure 140 lbs., total heating surface 1,839.2 sq. ft., weight of engine loaded 146,000 lbs.

This engine was considered the last word in locomotive construction. The brakes were steam operated and on each side of wheel similar to the clasp brakes on tenders and cars of today. Howard Stillman was draftsman in charge of design and Jim Ray had charge of construction. After the engine was built it was kept at Sacramento for nearly a year as the bridges were not deemed strong enough to support it. While it was at Sacramento, whenever a large tourist train was expected the 237 was steamed up and when the time was right would pull a long string of cars past the passenger station to the wonder and amazement of the passengers. She was finally taken to Sumner, now Bakersfield, and was used as a pusher in heavy trains, but was considered too big to be turned on the table. Also for a long time all trains had orders to take the sidings when meeting C. P. 237. Her days were finally ended when in July, 1894, she was broken up at Sacramento. Her boiler did service for years afterwards as a stationary boiler.

In 1885 the first of the Stevens valve gear engines was built at Sacramento and from the records at hand No. 19, later 1762, and still later 2208, was the first of these to be built. No. 19 was a ten-wheeler, 4-6-0 type, cylinder 18 in. x 30 in., drivers 56 in. diameter and weighed 112,200 lbs. in working order. In 1887-1888 there were a number more built of this same type. It was placed in service on the Oregon run using wood for fuel. Coal was used over the mountain and down the valley, but wood was used on the Siskiyou mountains.

In 1886 the 122, 123 and 125 were turned out of Sacramento Shops. These were 8 wheelers, or 4-4-0 type and were equipped with A. J. Stevens valve gear. They were passenger engines designed for speed and had cylinders 17 in. x 26 in.; drivers 68 in. diameter and weighed 88,500 lbs. in working order. They were found to be too heavy

Harry Ingram, machinist at Sacramento shops, shown here operating a double headed shaper, probably has a record for length of service at one particular machine. The shaper was installed in 1903 and Mr. Ingram has been running it ever since, and doing a fine job of it too, according to Superintendent H. C. Venter. Mr. Ingram is 66 years old, the oldest employe in the shops, and first went to work in June, 1879. He is hale and hearty and expects to easily stay on the job the four more years before he retires on pension.

on the truck and the truck journals ran hot, so a pair of equalizers had to be put on outside of the truck wheels and the axles lengthened so that the truck axles had four bearings.

During the years 1886, 1887 and 1888 there were a large number of engines built at Sacramento, mostly 4-6-0 type and 4-4-0 type. These were all equipped with Stevens valve gear In addition to these there were built 13 consolidation locos or 2-8-0 type. These were always called "Stevens' Monkey hogs" by the men. They were equipped with Stevens valve gear or Stevens "Monkey Motion" as it was frequently called. They got this name from the fact that when in motion the eccentric rod, or "galloping rod," had the appearance of a monkey hopping along.

The Stevens valve gear had much the appearance of the present day Walschaert valve gear, but was different in one respect. There were two valves one for each end of the cylinder and one rod worked the forward valve and was worked off of the crosshead; the rear valve was hollow to allow the forward rod to pass through. The engines were reversed through the medium of a rocker.

Of the engines built at Sacramento during the early days there remain in service today 2925, originally No. 229, 2187, which was originally 213, 2197, which was originally 241, 1421, which was originally 266 and 2 or 3 of the old local engines. These latter are doing service as shop switchers.

During the World War, Southern Pacific, finding itself in need of engines and not being able to obtain them from their builders decided on a large program of construction.

The result was the building of 14 consolidations or 2-8-0 type at Sacramento Shops the first one being No. 2839 turned out in August, 1917. These were followed by orders for 2 Pacifics

or 4-6-2 type, 5 Moguls or 2-6-0 type, 11 ten wheelers or 4-6-0 type, and 34 switchers or 0-6-0 type. All of these have given excellent service and have been in the shop but a small amount of time. The two Pacifics have given an excellent account of themselves.

Sacramento shop employes and supervisory forces are now taking great interest and pride in building of the 4-8-2 or Mountain type locomotives, and are giving their best efforts toward turning out a master piece of locomotive construction.

STATION MASTER'S KINDNESS BRINGS FINE PRAISE

The human side of practical railroading, the element of service contributed by Southern Pacific men and women toward making travelers comfortable and happy, was brought to the attention of Eliot M. Epsteen, general counsel for San Francisco Advertising Club, recently at Sacramento. Geo. F. Rupert, stationmaster at Sacramento, was the employe whose actions so impressed Mr. Epsteen, who wrote the Bulletin, in part as follows:

"Second section of the Overland Limited majestically rolled out of the station, and just as the rear end of the train disappeared from sight a hatless young man, out of breath, came running onto the station platform and asked an austere and official looking gentleman. who wore a cap designated 'Station Master', 'Is that my train?' The affirmative reply he received caused him to reel and almost fall in a faint. The look on the station master's face changed. He quickly suggested that the young man take a taxi, go to Roseville and there catch his train. In a twinkling of an eye the youth was gone, with a porter, who apparently sprang from nowhere, helping him off.

"I asked the station master, 'Do you think the young man will retake his train?' He smiled and replied, 'Certainly he will.' 'How do you know' I asked. He seemed surprised at the question and replied, 'It will take about 25 minutes for his train to get to Roseville. There engines must be changed, and the taxi can make it in time. Anyway, I wired to hold the train for him.' 'Do you mean to say,' I inquired, 'that you would hold up a couple of million dollars of equipment and all those passengers for just one young man who missed his train?' 'Well,' was the reply, 'that fellow probably spent a couple of hundred dollars for his trip. This is the last day the excursion rates are in effect East. All his baggage, hat, etc. are on that train. The delay won't be over five minutes at the most and that can easily be made up. It's due to him, do you not think so?'

"So I received a new slant on railroading. Certainly, there was a good-will building employe. I have discovered that under the surface there is a real human, living, sympathetic force and I am now an enthusiastic booster for Southern Pacific."

September, 1924   SOUTHERN PACIFIC BULLETIN

Giant "2-10-2" From Tiny "Hayburner" Grew

A Glance Back Over the Sixty Years Since the "Governor Stanford" Took the Rails Shows How Southern Pacific Locomotives Have Developed

IN 1829 a crude little engine hauled the first railroad passenger train. It was Stephenson's locomotive, the "Rocket," and it was considered a marvel because it could jolt along with two loaded coaches weighing together about ten tons.

The "Rocket" is a strange contrast to the finest present day locomotive in passenger service, Southern Pacific Company's powerful "4-8-2" type, which handles with easy smoothness *heavy transcontinental trains on one* of the longest regular locomotive runs in the world.

A quaint little relic the "Rocket" seems, too, when compared with Southern Pacific's tremendous "2-10-2" freight locomotive that can easily haul 96 modern freight cars up grade.

But this story about locomotives does not go back quite as far as Stephenson's day. It begins about 60 years ago at the time the Central Pacific Railroad Company, the parent organization of the Southern Pacific, started the epoch-making construction of the western part of the first transcontinental railroad.

While our story indicates in passing what the coming of the railroad and the constant improvement in its facilities have meant in developing the West, it is mainly about locomotives.

The First Locomotive

It was on August 19, 1863, that the freighter, "Herald of the Morning," slowly made its way through the Golden Gate and docked at San Francisco after the long voyage "around the Horn" from New York and other eastern ports.

With difficulty there was unloaded from the hold the first steam locomotive to be shipped to this coast for use in constructing the historical overland railroad. It was christened "Governor Stanford" in honor of the then Governor of California, who with C. P. Huntington, Charles Crocker and Mark Hopkins, pioneers, was responsible for the daring

plan to build a railroad over the Sierra Nevada Mountains to connect with the Union Pacific, thus forming the first transcontinental railroad. The first rails for the railroad also were unloaded from the ship. The "Governor Stanford" may be seen today in the museum at Stanford University, California.

The locomotive and rails were transferred to the schooner "Anna R. Forbes" for shipment to Sacramento. It was nearly two months later when the cargo was docked there on October 7, 1863. The locomotive almost was lost in the river in unloading. However, it finally was set up and its trial trip was held with ceremonies on November 11 of that year. Sacramento papers heralded the fact that it went as far as Seventeenth Street with the Governor, chief engineer and business men as "volunteer" passengers.

The locomotive "C. P. Huntington," or Central Pacific No. 3, (it was renumbered later Southern Pacific No. 1) was placed in service in 1864. It could haul four cars weighing 22 tons each at 35 miles an hour up a grade of 26 feet to the mile. It still is in serviceable condition, but is used only for exhibition purposes.

A short time later the first typically freight locomotive was received by the railroad. It had six drive wheels and was considered a wonder of the day because it could haul 18 light freight cars.

An article in the Sacramento "Union" of March 17, 1865, telling of the trial trip on the previous day of this freight engine, known as the "Conness," named after one of California's United States Senators at that time, said:

"The 'Conness' is by far the heaviest

and most powerful locomotive on this coast, exceeding the Atlantic in propelling power by about fifty per cent. She is designed for drawing freight up the heaviest grade on the road, which is 105 feet to the mile."

The early locomotives were picturesque in appearance. They had the large diamond shape smoke stacks. Brass fittings and gay paint made them exceedingly ornate. While it was the firemen's job to keep the brass polished, it is said that some engineers were so proud of their engines that they worked on Sundays or days off to help shine up the locomotives.

Instead of merely having serial numbers as at present the pioneers of the rails bore names, some of them of historical significance and others indicating the emotions evoked in the minds of those who first beheld these novel steel monsters.

Among the first engines were the "Pacific," "T. D. Judah," "Sargent," "Nevada," "Humboldt," "Arctic," "Hercules," "Piute," "Amazon," "Tamaroo," "Industry," "Gold Run," "Antelope," "Achilles," "El Dorado," "Colossus," "Tip Top," "Red Deer," "Grizzly," "Jupiter," "Storm," "Whirlwind," "Vesuvius," "Terrible," "Growler," "Apollo," "U. S. Grant," "Clipper," "Rattler," "Rambler," "Hawk," "Golden Eagle," "Blue Bird" and "Sunbeam."

Wood Was The Fuel

Wood was the fuel for the early locomotives, familiarly known as "hayburners." Large quantities of wood were piled along the tracks for use of passing trains and the fireman of those days had a real job.

Marking completion of the country's first transcontinental rail highway, the famous Golden Spike was driven at Promontory, Utah, on May 10, 1869. At that time Oregon largely was a wilderness of forests. California's valleys were great expanses of grazing and grain lands. Arizona, Nevada and neighboring states were thought of only as desert regions in which men risked their lives in search of rich ores.

With the completion of the overland railroad the Southern Pacific and Central Pacific began the great task of extending the rail-

This quaint looking locomotive was No. 1 of the Central Pacific and was the first engine to be placed in construction service on the western end of the transcontinental railroad. The maiden trip was made at Sacramento November 11, 1863, after having arrived from the east via Cape Horn. The locomotive was named in honor of Leland Stanford, then California's executive, and one of the "Big Four" builders of the Central Pacific. The historic old engine is now on exhibition at Stanford University, California. Cylinder was 15 inches in diameter and drive wheels 4 feet 6 inches in diameter.

After Reading Your Bulletin, Pass It Along

road tracks to various sections of the West.

Settlers followed the extension of the rails. Small communities that were built chiefly as railroad towns grew into vigorous, substantial cities. The prairie schooners in which goods were freighted slowly disappeared, along with the romantic stage coach and pony express.

While the communities of the West were expanding in size and in business importance, the people grew to realize more and more how much their prosperity depended on the ability of the railroad to meet their transportation needs.

As western pioneers in the business of transportation, Southern Pacific railroad officers and employes worked to produce better and more powerful locomotives so that large train units might be handled. At the same time the size of cars was increased. Heavier rolling stock made heavier rails necessary and the strength and smoothness of the supporting roadbed had to be increased. That process of improvement, making for efficiency and economy of operation, is being carried forward steadily by Southern Pacific today.

Famous "El Gobernador"

One of the most interesting of the early locomotives built by the Central Pacific at the Sacramento Shops was the "El Gobernador," which was in actual service from 1884 to 1893. This "iron monster," then said to be the heaviest and most powerful locomotive in the world, was the most talked of engine of that time.

This locomotive had a total weight, loaded, of 146,000 pounds. The latest 1924 type "2-10-2" locomotive weighs 623,200 pounds. The "El Gobernador" was kept at Sacramento for some time while trestles over the Tehachapi mountains were strengthened. During this time the big engine was often on exhibition and when a large tourist train went through Sacramento it was steamed up and pulled a long string of cars past the depot to the wonder and amazement of the people. It was considered too big to turn on a turntable for fear it would tip over and all trains took sidings when meeting the big engine so there would be no danger of it leaving the rails in swinging.

The "El Gobernador" was built at Sacramento in 1883 under the supervision of Master Mechanic A. J. Stevens. Jim Ray, now a pensioner living at Sacramento, was in charge of construction. It went into service in March, 1884 as a helper from Sumner (now Bakersfield) to Mojave. Mr. Ray accompanied the locomotive to Bakersfield and remained there for about a year attending to repairs and keeping the giant in condition. The engine was shipped to Bakersfield knocked down and when it was assembled ready for the first trial trip Mr. Ray says there was considerable apprehension among the enginemen. As the engine approached the recently strengthened trestles the engineer

---

Why Engine Should Be Called a "She"

The following amusing letter from "Japanese Lady" appeared in a recent issue of Shipping Register, San Francisco.

SOMETIME ago you publish in your paper voluble article on female shipping steamer. I have thought to write you about female engine on train. You know why? Yes, they call she for many becauses.

They wear jacket with yokas, pins, hangers, straps, shields, stays. They have apron also lap. They have not only shoes but have pumps. Also hose and drag train (psgr. and freight) behind. They attract men with puffs and mufflers and when draft too strong petticiot goes up. This also attract. Sometime they foam and refuse work when at such time they should be switched. They need guiding—it always require man manager. They require man to feed them. When abuse are given they quickly make scrap.

They are steadier when coupled up but my cousin say they hell of expense. Is not enough reason?

---

and fireman were at the edge of the gangway ready to jump in case of a crash through the timbers and Mr. Ray says he rode on the pilot for the same reason. However, these fears were unnecessary.

The big engine was the object of much criticism and in 1893 it was sent back to Sacramento. On March 14, 1894, just ten years after it went into service, it was broken up. The boiler was used until 1901 to furnish steam for the stationary engine in the machine shops at Sacramento.

The boiler had a copper fire box which was standard at that time. The steam valves were double, one set working on the other. One set worked constantly at full stroke, while the other set was used to regulate the cut off or hook up. There were three eccentrics on each side to work these valves and there were two reverse levers. When first built this locomotive had steam brakes which were later replaced by air cylinders.

The following data on this locomotive is of interest:

Length of engine and tender, 65 ft. 5 in.; driving wheel base, 19 ft. 7 in.; cylinders, 21 in. x 36 in.; weight of locomotive loaded 73 tons; rigid wheel base, 14 ft. 7 in.; diameter of drivers, 57 in.; weight on drivers, 64 tons; boiler pressure allowed, 140 lbs. per sq. inch.

During 1886 and 1887 twenty-two locomotives were built at the Sacramento shops, of which locomotive No. 177 was a typical design. Several of these engines are in service at the

present time. These locomotives were equipped with the A. J. Stevens valve gear, which worked as follows:

Each port of the cylinder had its own valve. This made necessary two valve rods, one hollow and one solid. The solid rod worked through the hollow rod and worked the forward valve. The hollow rod worked the back valve. The valves were operated by a combination motion from cross head and crank secured to main crank pin extended, the latter taking place of the old eccentric. The two valve stems were connected to upper end of combination lever which was operated by cross head similar to present Walschaert valve gear.

When first built these locomotives had the sand boxes mounted under the belly of the boiler instead of on top. They were equipped with the standard C. P. oilers on the steam chest and with steam brakes between the second and third pair of drivers. This was later changed to air, but was left in the same position and the brakes were set by the aid of cams.

Also during 1886 and 1887 a number of 4-4-0 type locomotives were built at Sacramento. In order to relieve the overheating of the brasses and journals due to the locomotives being so heavy in front, Master Mechanic Stevens put in longer axles which gave a bearing on each side of the wheel.

These locomotives were odd in appearance with only one dome on top. Sand box was under the belly of the boiler. The boiler was straight top with short front end and diamond stack. The head light bracket extended beyond the smoke box front. Counter weights on the drivers, instead of being next to the rim of the wheel, were placed in the wheel center between axle and rim equal distance from center of main pin and rim.

One of these locomotives, altered in many ways, is now in service at Sacramento as No. 1421.

Old "Monkeyhogs"

Thirteen locomotives of the 2-8-0 type, familiarly known by enginemen and trainmen as "monkeyhogs," were built at Sacramento during 1887 and 1888. They were equipped with A. J. Stevens valve gear and originally had steam brakes. They also had the sand boxes under the belly of the boiler. They had 19 in. x 30 in. cylinders set at an angle, 51 in. drive wheels, and weighed 114,850 lbs. loaded. Three or four of them are now being used by outside concerns.

During the thirty years after the first locomotives were operated out of Sacramento in building the trans-continental line there was much progress in the development of western transportation. By 1894 thirty-ton cars were being used. The locomotives were heavier and had more than twice the power of the earlier ones.

The "4-6-0" type of passenger locomotive and the "4-8-0" type for freight service were the pride of the rails at that time. The type classification indicates the grouping of the wheels.

The "El Gobernador" is one of the famous of the early Southern Pacific engines. It was built at Sacramento in 1883 and was the world's heaviest and most powerful locomotive at that time. It was not a complete success and after nine years in service over Tehachapi mountains it was returned to Sacramento and scrapped.

The passenger engines had four front truck wheels and six driving wheels, while the freight engines had four front truck wheels and eight driving wheels. Neither had trailing truck wheels.

Passenger locomotives of the "4-6-0" type were 57 feet, 1 inch in length, weighed 239,680 pounds and were capable of hauling six coaches weighing 30 tons each at a speed of 50 miles an hour up a grade of 26 feet a mile. These engines handled the first Sunset Limited trains of the Southern Pacific.

Freight engines of the "4-8-0" type of 1894 were 60 feet, 6 inches in length, weighed 272,300 pounds and could haul 65 cars weighing 30 tons each at a speed of 10 miles an hour up a grade of 26 feet a mile.

Greater Locomotives Needed

But the demand continued for greater and greater service to supply the increasing transportation needs of the West. Western fruits and products generally were gaining more attention in the East and each year still greater quantities were handled. The West also needed larger supplies each year from eastern manufacturing centers.

During the last 30 years railroad progress kept pace with the development of the country. In fact, locomotive construction made its greatest strides during this period, so that locomotive development came to be a milepost of progress.

The Southern Pacific locomotives of today truly are leviathans of the rails.

The latest and most powerful passenger locomotives are known as the "4-8-2," or mountain type. They are 97 feet 6 inches in length, weigh 593,300 pounds and are 20 times more powerful than Southern Pacific No. 1. Each can pull 14 modern passenger cars, weighing 70 tons each, at 50 miles an hour up a grade of 26 feet a mile. They were designed by Southern Pacific experts.

Their great power enables these locomotives to handle heavy trains with ease. New appliances and refinements in design assure smooth starting and stopping, thus adding to the comfort of passengers.

A new record for a regularly main-tained locomotive run was established by the "4-8-2" engines the latter part of 1923. On this run the 815 miles between Los Angeles and El Paso is made without change of engines. Heretofore a relay of four locomotives had been considered necessary in speeding trans-continental passenger trains across the mountain and desert territory between these widely separated cities.

The newest "4-8-2" type passenger locomotives are a companion type to the tremendous "2-10-2" freight locomotives that Southern Pacific has placed in heavy freight service during the last two years.

These powerful freight locomotives are 97 feet, 9 inches in length and weigh 623,200 pounds. They can haul 96 modern freight cars, weighing 50 tons each, at a speed of 10 miles an hour up a grade of 26 feet a mile.

Experts of Southern Pacific have given years of intensive study to the task of designing the most efficient and economical locomotives for both freight and passenger service. They designed the "4-6-2" or Pacific type of passenger engines which are making high records in handling the San Francisco Overland Limited, the Fast Mail and other trains between Ogden, Utah and Sparks, Nevada, and in various other parts of the West.

All of these late type locomotives are equipped with auxiliary booster engines, super-heaters, feed water heaters and other recently approved devices for increasing economy and power.

Running gear, frames and other parts have been designed with a view of making them as light as possible, without sacrificing strength. Piston rods, driving axles and main crank pins are of heat treated steel hollow bored. Another weight reduction was attained through use of high tensile strength steel in the connecting rods.

Booster Engine

The booster engine is a separate two-cylinder steam engine geared to the axle of the trailing truck wheels. It assists the main engine in handling trains, both in starting and on heavy grades.

Each modern locomotive is equipped with a feed water heater which per-forms the double operation of pumping water from the tender to the boiler and heating it on the way. It utilizes exhaust steam from the main cylinders to heat the water. Part of this steam is condensed and returned to the boiler for use again, thus reducing sediment and the amount of water required to operate the locomotive.

While designing the finest new locomotives, Southern Pacific is not neglecting its old ones. A number of locomotives that have been in passenger service for some years are being "rejuvenated" by the addition of the booster and the feed water heater. The results are proving most satisfactory, especially in the case of the Atlantic type engines which, after addition of the modern appliances, are much more efficient machines and have practically 45 per cent more starting power.

These as well as other improvements of a highly technical nature are made with the view of giving improved service to shippers and added comfort to passengers of the Southern Pacific.

Because of the great distance from the chief markets of the country, the Pacific Coast depends more for its prosperity upon good and adequate transportation than other parts of the country.

A vast, well-built transportation machine and a highly efficient organization are required to take fruit fresh from a Pacific Coast orchard and, preserving it in all its delicate attractiveness, move it over two or three ranges of mountains, carry it across deserts and plains, through the great central valleys in almost passenger train time to the end that it may be placed on someone's breakfast table in New England as fresh and appetizing as if it had just been gathered from the orchard. Southern Pacific's swift, powerful locomotives help make possible this triumph in transportation.

The progress of any country, it has been said, is indicated by the progress of its transportation service. Surely this may be applied to the West and its pioneer railroad, the Southern Pacific.

HELPS TO SPREAD GOSPEL OF "SAFETY FIRST"

Sam A. Matram, employed by the Telling-Belle Vernon Company, of Cleveland, the largest dealers in milk, dairy products and ice cream in the Middle West, recently noted a statement in the "Bulletin" calling attention to the dangers of carelessness, so he sent the statement to W. E. Telling, president of the company, suggesting that it be called to the attention of the other employes. Mr. Telling not only did this but used the statement as the basis for a newspaper advertisement, in which credit was given to Southern Pacific Company as the source of the material. He also wrote to E. G. Cook, general agent for Southern Pacific at Cleveland, saying that he was greatly interested in "Safety First" activities and had been watching Southern Pacific's efforts along this line.

Chapter Eight
A City of Shops

The November 6, 1863 edition of the *Sacramento Union* announced that the Central Pacific Railroad had begun construction of a new building to be twenty feet by one-hundred-and-forty feet. It was to be "a work shop to be used for fitting up railroad cars, etc." That was the beginning.

Further construction required special attention to the foundation of each building. Being near two rivers and a lake, ground water created a soft and unreliable footing for any substantial shop building. This problem was handled with unconventional techniques.

An example was the roundhouse. During August of 1867, work started on the foundation of the roundhouse which became a building with no supporting pillars. Instead, a layer of cobblestones delivered from Folsom was laid down before the area of the facility was leveled with sand. On the sand were placed cut granite stones that became the foundation of the twenty-eight stall building. When the granite work was completed, brick contractors started building the walls.

The *Sacramento Union* of December 18, 1868 states:

The fine new brick Roundhouse of the Central Pacific Railroad is rapidly nearing completion. It is of substantial brick construction, laid on granite foundation. It is built on a 378-foot diameter, there are 29 openings or stalls for the iron steeds, but only 28 will be used, as the offices for the officials will occupy the south end of the buildings. It will be modern in all respects, and is

deep enough to enable a locomotive to be run in over the pits and have the doors closed to protect the workmen from the weather.

In a city of shops, there must be an efficient way to move locomotives and cars into various buildings. To solve part of this problem, the railroad used transfer tables. A transfer table is a section of track long enough to hold a long car. The table moves parallel to the buildings into which the car on the table is to be moved. In 1948, the railroad's first transfer table was roughly described in *Sacramento General Shops, Southern Pacific Company-Pacific Lines* by D.L. Joslyn:

The first transfer table at the car shop was built in 1872-73 and was 68 feet long, had a travel of 265 feet. It was constructed of metal and wood, and was moved back and forth by an endless chain passing over sheaves at each end of pit, attached to each side of table. In moving table a draft horse operated a windless which moved chain in either direction. When table was lined up with proper track, horse was used to pull cars on or off table.

In 1898, the entire Southern Pacific System depended on Sacramento for car siding, car sills, cabinet work, upholstery, car wheels and iron car parts.

Between 1863 and 1930, the Sacramento Shops grew to be a complex of at least twenty-five different functions that served the railroad's needs. There were specialized buildings just for making springs, working with copper, servicing air brakes, storing lumber, and maintaining and constructing switch parts.

Sacramento was, by far, the main facility for the railroad in the West. But, just fourteen years after the first building in Sacramento, there were maintenance and repair locations elsewhere in California: Oakland, Tulare, Los Angeles, Mojave, Rocklin, Truckee, and Red Bluff. In Nevada, crews in Wadsworth, Carlin, Winnemucca, and Wells had similar functions.

By 1922, the passenger-car department in Sacramento had nine tracks, the upholstery shop was producing nearly 80,000 items per year, the plating department was re-plating nearly 26,000 pieces of silver each year, and the car department was handling approximately 100 pairs of cast wheels per day.

This chapter includes fourteen articles from Southern Pacific's *Bulletin*. The topics cover activity from 1919 through 1936 and include an explanation of how the railroad made very large cylinders for ferry boats. Building large castings for ships was a peripheral business that was possible given the tools used to build large locomotive parts.

The Bulletin articles are:

"Sacramento Shops Make Record Casting," August 1925

"WORK: Sacramento Shops Increase Payroll to Handle New Program for Air-Conditioning and Modernizing Cars," December 1935

"COMFORT: Man-Made Weather in Air-Conditioned Cars is Boon to Train Travel; Three Systems Used," October 1935

"Electroplating Department at Sacramento a Modern Plant," April 1919

"Reducing the Amount of Stock on Hand," January 1920

"SP's City of Shops a Story of Perfection," August 1920

"From Bolts to Walking Beams," February 1922

"Immense Stores Plant at Sacramento," February 1922

"Casting Cylinder Real Feat," July 1922

"Boiler Shop Now Great Factory," January 1923

"Facilities Added at Sacramento Shops," April 1924

"S.P. Paint Shop Fights H.C.L. With Own Products," January 1920

"Car Shop 9 Builds New Cars," October 1936

"Chairs," September 1936

❧ ❧ ❧

Development of the First Steel Passenger Coach

At the beginning of railroading, passenger cars were made of wood. Wood was light, cheap, and able to serve the purpose of the period, but there were also drawbacks to constructing passenger cars out of it. It was weak, prone to splintering, and burned easily. While other materials had been used in the frame and wheel assembly, the evolution to other materials was slow despite the problems with wood. The earliest record of non-wood passenger cars is from about 1850.

The 1851 designs of Thomas E. Warren used sheets of iron, but his idea, apparently, did not get past the design phase. Other ideas in the 1850s were those of Bernard Joachim La Mothe who used iron bands as the main structure. His development reached a model stage. But, the claim of being fireproof could not be made, as the center sills, floors, and paneling were still wooden. He continued revisions and improvements until, somewhere around 1860, he worked with metal tube framing. His primary obstacle was finding a reliable way to connect the tubes. Despite his effort over the years, he was never able to mass produce passenger cars with iron framing.

By the 1880s, the production cost of using steel became competitive with iron. So, the interest in steel passenger cars was invigorated and a new round of ideas were investigated.

In the late 1890s, steel freight cars were introduced. With freight hauling, a steel framed car allowed for more tonnage to be carried than in a wooden car of the same volume. But, a steel framed coach car carried no more passengers than its predecessor of the same size. The marketing strategy for builders of steel coach cars was to argue that steel cars were safer. And this was a good selling point.

By 1901, New York's first subway was being constructed. George Gibbs, a well-respected electrical engineer, was asked to design the cars. His first design was for a steel-framed car with a wooden exterior covered with copper layers and a floor lined with asbestos sheets to resist fire. Then, news came from France of an accident in the Paris subway during which a car burned underground. Consequently, the New York subway leaders decided that they needed an all-steel car. Events in Paris and New York were instrumental in the development of the steel passenger car for mainline railroads.

The Pennsylvania Railroad also played a major role in the advancements of steel car design. In 1904, the railroad's William F. Kiesel Jr. and Charles E. Barba, teamed with George Gibbs on a steel design. Two years later, a sample car made in Altoona, PA was tested between Philadelphia and Lancaster. This car was nearly seventy feet long—the main support was a large box girder made from two twenty-four-inch wide steel plates and two eighteen-inch channels. The floor was concrete. There was some wood: about 1,540 pounds worth. It was used for interior paneling and added to the aesthetic appeal enjoyed by the seventy-two passengers.

By early 1907, the Pennsylvania Railroad had completed plans for an eighty-foot, eighty-eight-passenger car that became part of the famous P-70 class of cars. Eight wheels supported the fifty-six-and-three-quarter ton car that included no wood except for seat armrests

and some incidental pieces. The interior was painted sheet steel with aluminum moldings. In June 1907, an order arrived for 200 of these cars. This large commitment likely spawned the era of steel passenger cars across the industry.

The following article from *The Bulletin* was written in December 1921 and describes the development of the steel car not only for coach service but also for use in cars to transport mail. It begins with the idea that the origin of the steel car was in 1902. The "origin" is related to the definition of a steel car. Do steel bands in the frame constitute a steel car? The important point is that the disappearance of the all-wooden car was a slow process, and as with most railroad innovations, the evolution from wood to steel involved the Sacramento Shops.

❧ ❧ ❧

In 1876, the Central Pacific Railroad car department employees posed with a Barney and Smith-built sleeper (1876). The Barney and Smith Company was founded in 1849 in Dayton, Ohio. Beginning in 1859, George Pullman ordered sleeping cars from Barney and Smith. Pullman continued to order cars until 1881 when Pullman was able to build enough of their own cars to satisfy their needs.

Building of the first car shop started in 1867. It was only ninety feet wide by one hundred thirty feet long. The two-story building was on fifteen-hundred cedar timbers one foot square by thirty feet long. Each timber was driven down to bedrock with the top of each just below the water line. On top of this grid of timber was put granite from Rocklin, a town about thirteen miles north. After about six thousand cubic yards of granite was laid, dirt was filled in and the construction of the brick walls was started.

Near the end of the 1890s, steel was beginning to be considered as a construction material for cars rather than wood. By 1906, Pullman had transitioned most of their manufacturing to steel. The advent of steel construction brought new challenges. But, opportunities to improve safety and durability led to a new era. The Sacramento Shops were, as usual, a focal point for innovation in steel car construction.
– Courtesy California State Railroad Museum

The Foundries

Before the Central Pacific Railroad (CPRR) operated its own iron-producing facilities, castings used by the railroad were purchased from the Goss and Lambard Machine Shop and Foundry in Sacramento, the Sacramento Valley Railroad Foundry in Folsom, and the Union Iron Works also in Sacramento.

In 1865, the CPRR purchased some of the small local railroads, including the Sacramento Valley Railroad, thereby gaining control of the foundry at Folsom, which had produced cooking utensils, cooking stoves, heaters, mine machinery, boilers, steam engines, and heavy forgings. After the purchase of the Folsom facility, activity continued at the Folsom location until sometime in the 1890s when the machinery was shipped to Sacramento. The story of the Folsom shops finally reached an end when the buildings were dismantled. But, they became part of a new building when the bricks were used to build the Folsom Powerhouse, which eventually generated power for Sacramento and the main Southern Pacific shops in Sacramento (Southern Pacific bought the CPRR in 1885).

In 1868, the CPRR bought the well-established Goss and Lambard, whose products included cast iron and brass parts, axles, and large forgings, and gave the railroad control of the one-hundred-by-thirty-six foot machine shop, a thirty-six horsepower Corliss engine, all tools, a forty-by-eighty foot blacksmith shop, and the foundry. But, despite the added production capacity, the demand for wheels was still too great for the railroad's own capacity. To supplement the items produced in Sacramento, iron wheels for cars and locomotive driving wheels were eventually bought from eastern factories.

The uncomfortable reliance on parts coming from facilities 2,000 miles away was an impetus to make plans for a new foundry in Sacramento. These plans, which were in the sketch stage in 1871, were acted upon in 1872 with the building of a shop on the shores of Lake Sutter near the southern boundary of the railroad's property. By the middle of October 1872, the output of wheels was forty per day and the daily count soon reached one hundred.

During the next four years, the variety of items needed from the foundry increased as the Shops became more involved in building large parts for ferry boats. It was not until 1883 however, that money became available for another expansion. This one was built on a fill in Lake Sutter and was one-hundred-by-four-hundred feet, of wood frame with a corrugated roof. The wheel production per day? Three hundred. The cost to make a thirty-three-inch wheel? About fifteen dollars. This was cheaper than the cost per wheel in the old foundry and about 15 percent cheaper than purchasing new wheels from outside the company. The shop's wheel production had finally caught up with demand.

Conditions stayed about the same until 1915 when the need arose, again, to purchase wheels from an outside source. This time, wheels came from Portland, Los Angeles and Salt Lake City. Expansion moved ahead in Sacramento with the addition of more floor space adjacent to the existing foundry. The result was a capacity of four hundred wheels per day. Further improvements included a sandblaster for wheel cleaning which was a chore previously done by hand using steel brushes and scrapers. By 1948, with more expansion, and more importantly, improved technology, the cost of producing a pound of a wheel was two dollars and eighty three cents, which was only fourteen cents more than the cost paid to the Ensign Manufacturing Company sixty-five years earlier.

In summary, the company kept up with demand for foundry work by outsourcing the work, then buying local railroads and their facilities, including the purchase of Goss and Lambard. Thereafter, on-site facilities were built and enlarged. With expansion and remodeling of existing buildings, the technology of casting improved along with efficiency.

❧ ❧ ❧

Sacramento Shops Make Record Casting

A four-thousand-ton train, traveling at 40 mph, with a length of one mile, passes by a point in one and a half minutes. The supporting mechanism, which needs to perform flawlessly in varying temperatures and moisture conditions, is the ballast. There is a phrase in the Maintenance of Way business: "no track, no train." Maybe an improvement could be "no ballast, no track."

Proceedings of the thirtieth annual convention of the American Railway Engineering Association held in Chicago from March 19 through 21, 1912, describe the ballast as "that portion of the permanent way of a railway which forms the firm and dry foundation for the ties and rails, which they support."

In addition to providing a firm bearing for the ties, thereby distributing the pressure of the train equally, the ballast should:

1. provide drainage to avoid standing water around the ties and to minimize frost heave

2. provide material between the ties that minimizes the movement of a tie relative to the adjacent tie

3. reduce the growth of grass around the rails to allow mechanical raising of the track as part of track maintenance

4. not degrade due to weather

Ballast is sometimes described as being the sub-ballast, ballast proper, and the top ballast. Sub-ballast is that portion used as a mat between the sub-grade (dirt in most cases) and the ballast proper. Ballast proper, or lower ballast, is the material that is from the sub-ballast to the bottom of the tie. The top ballast lies from the bottom to top of the tie. The ballast must be hard enough to not break apart under wear and tear.

The preferred depth of ballast varies depending on factors such as climate and train weight. A general guideline calls for the sub-ballast to be ten to eighteen inches, the lower ballast to be nine to eighteen inches and the top ballast to be six to eight inches.

The most common material used as ballast is broken stone. But, not any type of stone suffices as good ballast. The stone, when broken, must break in angular fractures and not into thin, flat pieces. According to the proceedings of the convention in Chicago, the size of the broken stone should be between three-quarter inch and two inches. Consequently, crushing the stone was more than a random rock-breaking process.

Using the general guideline for ballast depth, the volume of ballast needed for one mile of track is enough to fill a box fifty feet deep, by fifty feet wide, by fifty feet long.

The climate of the Sierra's summit is such that snow may be on the track for six months. Gradual melt produces water that must be allowed to percolate through the ballast. If this does not happen, water pools around the tie and causes weakening of the tie. There can be no standing water.

In 1925, a proud group of Sacramento Shop employees were asked to make a huge casting to be part of a rock crusher and ballast producer in Santa Ana. The Shop staff accepted the job. They made the design, produced the pattern for the mold and provided the finished product for their colleagues in Santa Ana.

❧ ❧ ❧

Sacramento Shops Make Record Casting

Left to right, back row: Moulders O. Lohman, J. McCarty, W. Williamson, Patternmaker J. Rajnus. Front row: Pattern Shop Foreman G. R. Scott, Core Room Foreman A. McCarty, Moulder C. Guttenberg, Patternmaker R. Oliphant, Assistant Pattern Shop Foreman W. B. Butler, Foundry Foreman J. Gageby, Chipping Foreman P. Early, General Foreman Foundries J. Geiger, and Assistant Foundry Foreman F. Miller.

By R. P. PEEK,
Supervisor of Apprentices

FEW people realize that in running a railroad there is more to the road than just the shining black locomotives, the comfortable coaches, the Pullmans, and the red box cars. How many of them stop to think that in order for them to make their comfortable trips throughout the country millions of dollars are spent annually to keep this fine equipment in a safe and workable condition, the road beds smooth and well appearing to the eye as you glide along over the two ribbons of steel, so finely ballasted that a glass of water on the dining car table is as safe as in one's home?

The photograph above shows one of the parts of the unseen expenditure that is necessary to crush the rock that goes on the road beds and in fills to make our road a successful one. It is the top bearing for one of the largest rock crushers in the West, in operation at Santa Ana, California.

It was found some time ago that the design of this crusher could be improved and the output of this plant increased, but to do this meant making a larger casting than ever before at our foundries, so when the proposition was put up to the officials of the Sacramento Shops considerable study was given it to see if the shops were thoroughly equipped to handle such a tremendous undertaking.

The drawings were made and the pattern shop started the pattern for this foundry masterpiece. Now, owing to the peculiar shape of the arms, this pattern caused considerable hard work and skillful planning, for it was necessary to get it very accurate, and the cores, that lightened the arms, so placed that they would not weaken the casting. After eight days of energetic application by Mr. Scott, the pattern shop foreman, and his crew, the pattern was turned over to the general foundry.

In building this pattern over 1,400 feet of sugar pine lumber was used. Owing to the size and enormous weight of this casting it was necessary to excavate a place in the foundry floor 18 feet long by 12 feet wide by 6 feet 10 inches deep. The pattern being made for one-half of the hub and one arm, had to be mounted on a spindle and one-half of the mould made at a time. This required great caution and accuracy on the part of the foundrymen to see that the two arms were directly opposite one another, and absolutely on the proper levels. After approximately fifteen days of work on this mould it was ready to receive the metal.

Then came the problem of getting twenty-seven tons of molten metal together so the casting could be successfully poured. In order to do this it was necessary to build a large reservoir to hold the bulk of this metal, for it was impossible, with the pouring ladles used in the foundry and the limited crane facilities, to handle so great an amount of molten metal at one time.

The reservoir was built at one side of the mould, clear of the arms, and filled with sixteen tons of metal. Two large ladles, one of five tons and the other of six tons capacity, were filled and handled on the two cranes over the job.

At a given signal pouring was started from all three containers at the same time, and in two minutes after the signal was given this monstrous casting was successfully poured, much to the relief of all concerned, for the least little error would have meant the total loss of the casting.

After the casting had been left in the mould for five days, to give it a chance to cool gradually, it was taken from the pit, the pouring gates and risers were cut off, and the casting cleaned. Upon inspection it was found to be absolutely free from flaws—a perfect casting, weighing twenty-three tons, the largest ever made at the Sacramento Foundry.

Much praise is due the men who were responsible for this casting. Such work only shows that the railroads have a class of work that requires the best mechanics available and from the work on the castings mentioned it seems that we have a large number of such men in our employ. Many of the men who worked on this responsible piece of work were men who served their apprenticeship with this company and are now holding the responsible positions.

Many members of El Paso Traffic Club saw railroad shop work for the first time when they visited Southern Pacific shops in that city recently under guidance of J. J. Finney, chief of tariffs, who is president of the organization. Shown in the picture, left to right, are: J. B. Sauer, soliciting freight agent, Missouri Pacific; E. D. Ranck, rate clerk, Santa Fe; L. V. Gardiner, city freight and passenger agent, Santa Fe; F. W. Grueling, traffic manager, Western Purchasing Co.; H. D. McGregor, assistant general passenger agent; F. D. Norwood, claim investigator, and R. N. Davis, city ticket agent, of S. P.; C. D. Johnson, general agent, Texas & Pacific; J. J. Finney, chief of tariffs, S. P.; William Bleick, master mechanic; C. W. Waterman, chief clerk, and K. E. Gibson, freight agent of G H. & S. A. Railway; W. T. Harrell, traffic manager, International Warehouse Co.; W. C. Long, general agent, American Railway Express; F. H. Evers, traffic manager, Mine & Smelter Supply Co.; E. McClannahan, assistant general freight and passenger agent, S. P.; F. C. Tockle, assistant traffic manager, El Paso Freight Bureau; H. W. McDermid, assistant to vice president, American Railway Express, San Francisco.

Man-made weather - Air conditioning added to cars in the 1930s

After the golden spike was driven on May 10, 1869, in Promontory, Utah to commemorate the completion of the railroad, passengers who had previously become experts on the efficiency and comfort of a stage ride across the West had a new experience that surely generated conversation about the pleasantries of train service compared to the carriage. The stage coach certainly did not travel fast enough to create its own breeze, as did the train traveling at fifteen miles per hour. But neither was the stage coach passenger ever frustrated with embers, smoke and steam.

Before 1869, there were some attempts to alter the path of air entering the car to minimize the amount of dust and soot. In the 1850s, Henry Jones Ruttan developed a method to clean the air in the train that relied on a tank under the floor with water three inches deep and a water surface area of two hundred square feet. Air entered the car near the roof, was directed down to the water bath where the dust settled then was allowed to enter the passengers' area. In 1868, Ruttan stated that three hundred cars used his method, which was a much larger number than any competitor could provide. The process worked fine on trips less than about fifty miles, after which the tank became clogged with cinders. Maintenance issues such as leaks that rotted floor sills led to the demise of Ruttan's device.

The efforts to provide creature comforts continued. In 1855, sleeping car operator, W. D. Mann developed a process that started with outside air being drawn in through a low hood that reached across the car's roof. Then, the air was channeled into a cabinet where it passed through a filter made of wet mattress packing. In the summer, the air was then cooled by passing it over three hundred and fifty to six hundred pounds of ice. Winter warmth resulted from the filtered air being blown over steam-powered pipes. Mann professed that summer temperatures could be lowered 12 to 20 degrees. Less biased sources more accurately estimated the cooling to be around 10 degrees. The healthfulness of the air, with its path over damp mattress material, was questionable as was the air-flow rate to the passenger.

At the turn of the nineteenth century, Santa Fe had the marketing dilemma of convincing customers headed west to traverse the southwestern desert. Summer temperatures often reach 100 degrees through this area and with alternative routes available, the company continued the search for devices to aid customer comfort. In 1911, Santa Fe inaugurated a train named the De Luxe. The dining car was fit with an ice-cooled ventilating system where the air was filtered by moving it through a water mist before it was sent into the dining area. An important improvement to conquer the poor flow volume, a challenge for fifty years, was the addition of a large fan that circulated air over the passengers. As an air cooler, the method was not good enough to last past the mid-1920s because having ice ready at various locations along the route was an ongoing challenge. But, the idea of forced air being part of the cooling process had been introduced.

In 1929, Willis Carrier tested another idea on a Baltimore and Ohio dining car. His process involved a 110-volt motor powered by an ammonia refrigeration compressor. The unit cooled water that was pumped into pipes in the roof. A fan then circulated the cool air over the pipes into the dining area. The kitchen, which was probably the hottest part of the car and occupied by the paid, rather than those paying, was not cooled, resulting in hot and humid working conditions.

Southern Pacific, in 1935, stated that the railroad added no extra charges to travel in an air-conditioned car.

Nearly four thousand cars were ice cooled by 1940. The lesson learned about cooling with ice was that, especially in the South, the quantity needed and the labor to handle it was more of an expense than the ice itself.

WORK: *Sacramento Shops Increase Payroll to Handle New Program for Air-conditioning and Modernizing Cars*

"Say, this new air-conditioning program's a swell idea. I've been off work five months, and now I'm back on the job again. Sure, I think it's great. Course, it'll make train travel more comfortable, and that's important too. Then there'll probably be more business, and when there's more business there's a better chance for jobs to be steady."

John Korich, car builder in Car Shop No. 3 at Sacramento General Shops, offered those comments to a *Bulletin* representative who was taking pictures at the shops recently. He well expresses the general sentiments of all employes now working full five-day weeks at the shops, and particularly the 150 men recently called back to their jobs.

In the complete air-conditioning and car modernizing program for 1936, the Company will spend $1,026,719 on its Pacific Lines and the lines in Texas and Louisiana. Of this amount $620,081 is for air-conditioning of the Company's own cars; $380,698 to modernize and improve passenger cars; and $25,940 for electrical facilities at terminals for air-conditioning operations.

All the work on car re-construction is to be done in the Company's shops at Sacramento and Houston. The air-conditioning will total 118 chair cars, coaches, diners, lounge, and cafe-lounge cars, 75 of which will also be modernized.

In addition to the Company's cars being air-conditioned in our own shops, the Pullman Company will also air-condition a total of 105 standard and 26 tourist cars for use on our lines. The completion of this program by early summer, will give Southern Pa-

PICTURES: At work on air-conditioning at Sacramento Shops. Top to Bottom— John Korich, car builder; Wm. F. Long, cutting angle iron with torch; Nicholas LaFranco, sheet metal worker, preparing an ice box; Lionel Walsh, sheet metal worker, forming and fitting air-conditioning ducts; A. J. Shugar and L. Wachenfield, inside finishers.

cific a grand total of 588 air-conditioned cars in operation on its lines between the Pacific Coast cities, Ogden, New Orleans, and Guadalajara, Mexico. Of this number 256 will be our own cars, and 332 belong to Pullman Company.

With this additional equipment available, many Pacific Lines trains now only partially air-conditioned, will be completely equipped. These trains will include: Apache, West Coast, Daylight, Owl, Klamath, Shasta and Lark; to supplement the Overland, Golden State, Sunset, Pacific Limited, Cascade and the San Joaquin. Also there will be more cars for other trains.

Work included in the air-conditioning and modernizing of chair cars, coaches, club cars and calls for the equipping of 31 coaches with the same type of de luxe reclining and rotating seats, deeply cushioned and with backs adjustable to four positions, which type of cars have already proven so popular in overnight travel. These cars will also have commodious wash room and smoking room for men, and larger washroom and lounge for women.

In all the modernized cars the interiors will be painted in light colors, rubber tiling will be placed on the floors, except in the carpeted lounge cars, and white porcelain for washrooms which will have hot and cold water.

Outstanding in the modernizing program is the conversion of 9 club cars into air-conditioned lounge cars. Design of the new cars will differ in some respects from the lounge cars now in use. The lounge room, seating 15, will have modern parlor furniture upholstered in frieze plush. Four lower open standard sections, to be used only for lounging, will have the same upholstering, as will furniture in the smoking room. At one end of the smoking room will be an open bar of modern design and fixtures, with buffet facilities in connection. There will also be a barbershop and shower bath.

COMFORT: *"Man-Made Weather" in Air-Conditioned Cars Is Boon to Train Travel; Three Cooling Systems Used*

When Mark Twain said, "Everybody talks about the weather but nobody does anything about it!" he was right, for his remark was addressed to a public that still depended upon palmleaf fans and heating stoves to overcome heat and cold.

Today the quip is as pointless as it was pertinent in Mark Twain's time. Something has been done about the weather, notably by Southern Pacific and other railroads. That something is the air-conditioning of passenger train equipment.

By means of this modern method of "climate" control, railroads are now able to shield their passengers from the vagaries of weather—summer or winter. But air-conditioning means more than a conquest of heat and cold. It also furnishes a constant supply of fresh air, keeps car interiors free of dirt and dust, and shuts out noise.

Cost Three Millions

Spending approximately $3,000,000 on the work, the company now has 339 air-conditioned cars, 274 of which were placed in service this year. Southern Pacific owns 138 of these cars, which were turned out of the shops at Sacramento and Houston. Pullman owns the other 201 cars. Taking into consideration the joint service of the Union Pacific and the Northwestern on the transcontinental Overland Route, and

These double, reclining and rotating seats are used in the company's newest 72-foot air-conditioned chair car. Upholstered in deep pile mohair plush of pleasing stripe design and color, the double-deck spring cushioned seats offer maximum riding comfort. Seat cushions are 8½ inches deep and the back cushions 5½ inches thick at shoulder height. Also the arm rests are heavily padded. The back cushion can be adjusted to four sloping positions between 20 and 40 degrees.

the Rock Island on the Golden State Route, 441 cars of this type are now in operation over Southern Pacific Lines.

Fourteen dining cars were air-conditioned to start the work in 1932, but the program was soon broadened to include Pullman sleepers and room cars, observation, club and lounge cars, tourist sleepers, coaches and chair cars. The five leading transcontinental trains—Overland, Golden State, Sunset, Pacific Limited and Cascade—are completely air-conditioned, as is also the San Joaquin traveling the Valley Route between San Francisco and Los Angeles. Many air-conditioned cars are also used on the Owl, West Coast, Daylight, Apache, and on trains of the Texas and Louisiana Lines and the Southern Pacific of Mexico.

In connection with this expensive improvement of passenger equipment, it is of first importance to point out that Southern Pacific, unlike many other roads, charges no extra fare for transportation in air-conditioned cars.

Public response to the introduction of air-conditioning in rail travel has been most encouraging, and Southern Pacific plans to extend its program as rapidly as possible. The company's advertising department and news bureau have worked hand in hand to acquaint travelers with features of the improved service. Low rail and Pullman fares, plus the new train comfort, have convinced thousands of motorists that it is economy and good judgment to leave the family automobile in the garage and use the train for journeys of any distance.

Dust Storm Test

Air-conditioned cars, put to the severest possible test during the devastating dust storms this year in the middle west, proved their worth when the interior air of cars remained fresh and clean, while outside air was so filled with dust that widespread suffering and many deaths resulted.

In all the new type cars, fresh air is filtered through bronze wool coated with odorless oil. All dust, soot and other floating matter is caught in these filters, which are cleaned at frequent intervals. This filtered, fresh air—cooled in summer and warmed in winter—is forced into the cars through air ducts by means of fans at a velocity so low passengers are unaware of air movement. The slightly higher air pressure produced inside the cars over the normal air pressure outside the cars, is not noticeable but serves to prevent outside air from entering the cars through window cracks or when doors are opened. The air is warmed in winter by steam coils in the air-conditioning

unit, supplemented when necessary by the ordinary car floor heating system.

It is interesting to note that three methods of air-conditioning are used in the company's cars—"ice activated," "steam ejector" and "mechanical" systems. The following semi-technical

One corner of the spacious lounge at the rear of the air-conditioned cars especially built for the popular "Hotel Car Cruises" through Mexico. Here passengers eat, drink and visit during the two-weeks' train ride. The car is "home" for the passengers during the trip. Forward in the car are eight standard Pullman sections, and in the center is the kitchen.

description is furnished by our Motive Power Department.

The ice activated system used in a majority of the company's cars, is a development of our own Mechanical Department. The cooling medium is ice water secured by the melting of ice in a large tank under the car body. This ice water is pumped through pipes to air cooling coils in the roof of the car. Filtered fresh air to be circulated through the car is cooled to a comfortable degree by being forced over the outer surface of the air cooling coils. The process is continuous, water passing from the air cooling coils back to the ice tanks where its temperature is again reduced for re-circulation.

Steam Ejector System

In the steam ejector system, the cooling process is also obtained through circulation of cold water in the air cooling coils, but the cold water is obtained differently from the ice activated system. The water, warmed after passing through the air cooling coils, is forced in a fine spray into an evaporation tank in which a partial vacuum is maintained. Some of the water entering the tank immediately vaporizes and the remaining portion, cooled in the vacuum to about 40 degrees above zero, drops to the bottom of the tank and is re-circulated through the air cooling coils.

A vacuum in the evaporation tank is maintained, and at the same time the water vapor drawn off, by intense suction created by a jet of dry steam passed at high velocity through a pipe in the top of the tank. This steam and vapor enters a condenser over which a spray of water is pumped and a strong current of air is forced by a motor driven fan. Water passing through the condenser is fed back to the cold water system to replace water lost by evaporation, entering a storage tank which is filled from time to time while the car is en route. The spray water, which is separate from the water used in the enclosed cooling system, is carried in a make-up tank on the car, and the supply is also replenished en route in order to maintain proper volume for the spray nozzles.

Mechanical System

The mechanical system differs from the ice and steam systems principally in the cooling medium. Instead of using water, the coils are cooled by a chemical refrigerant, commonly called Freon, a gas which liquefies under high compression. Circulated through the cooling coils, this gas is reduced in volume and converted into liquid by being forced through a power driven compressor. The process adds a certain amount of heat to the liquid which is in turn cooled by passing through a condenser over which a strong current of air is blown, in much the same manner as water is cooled in an automobile engine cooling system. The condenser corresponds to the automobile radiator. The cooled liquid refrigerant, still under pressure, is then passed through expansion valves into the air cooling coils where it immediately expands and again becomes a gas at a temperature of about 22 degrees below zero. The Freon gas, after cooling the air for circulation in the cars, returns to the compressor and the above cycle is continued without loss in efficiency or removal of the refrigerant.

MOVED: *City Ticket Offices In Attractive New Quarters*

Two of the company's city ticket offices have recently moved into attractive new locations. The Portland office is now at Yamhill and Sixth Streets, while the Oakland city ticket office, as well as headquarters of Assistant General Passenger Agent Garnett King, are back at the familiar corner of 13th and Broadway in a newly constructed building. For some months the offices were in temporary quarters.

RETIRES: *Robert Adams Ends Long Career in the Auditor's Office; Others Promoted*

After 54 years in railroad service, Robert Adams retired on pension as assistant general auditor of the Southern Pacific Company September 1.

Adams began his career in 1881 as a messenger for the Wabash Railway in Peoria, Illinois, and was continuously in railroad work up to the time of his retirement. He served Southern Pacific 33 years, starting as auditor of disbursements in 1902.

Robert Adams

In 1918, Adams was appointed assistant federal auditor, later serving as assistant auditor and associate auditor in his climb to the position of assistant general auditor.

Adams was succeeded by H. C. McCleer, whose service with Southern Pacific dates from 1906 when he started as a clerk. He had been auditor of miscellaneous accounts since 1929.

McCleer's former duties were taken over by P. J. Kendall, promoted from assistant to the general auditor, and Kendall was succeeded by J. A. Quinn, who moved up from chief clerk.

MOURNED: *Death Claims Two Veteran Officers of S. P.*

Harry W. Wistner, assistant superintendent of Salt Lake Division, died at his home in Ogden, September 8, following a lingering illness. Wistner was 60 years old and a native of Ohio. His service with Southern Pacific dated from 1903 as train dispatcher at El Paso. During 1910 he became chief dispatcher at Ogden and was trainmaster at Imlay for some time before advancing to assistant superintendent in 1919.

M. M. Moffitt, supervisor of sales and salvage in the Purchasing Department, died at the general hospital in San Francisco Sept. 14. Funeral services were held in Oakland, Cal., where he was born 53 years ago. He began his service with the Company as an office boy at the general office in San Francisco in 1898. He advanced through the Purchasing Department and in 1922 went to Portland as assistant purchasing agent, where he remained until his recent return to San Francisco.

NEW JOB: *D. V. Cowden Now Tax and Realty Com'sr*

D. V. Cowden, for many years tax attorney for the company, with headquarters in San Francisco, has been appointed to the newly created position of real estate and tax commissioner. All tax matters and the supervision of real estate not required in operation of the railroad will be under his jurisdiction.

PRAISED: *Oakland Committee Wins Railroad Week Award*

While thousands of railroad workers in California shared the success of "Railroad Week," staged June 10 to 18, this year, it remained for the committee handling Oakland and other eastbay cities under direction of Garnett King, assistant general passenger agent, to win the Certificate of Award for distinguished service in that state.

PELICANS: *Saline Crusted Birds Perish in Salt Lake*

Waters of Great Salt Lake have dropped two inches below the all-time recorded low, readings of the gauge at Midlake show, and are so heavy with salt that birds alighting on the surface cannot rise again, H. E. Watts, Ogden, reports.

The lake level is 18 feet below the all-time high recorded in 1868 and 42 inches below the "zero" mark on the gauge.

Watts said pelicans were dying of hunger and thirst by hundreds because salt encrustations made them too heavy to fly. A single feather from one bird weighed two and one-half ounces.

SELLS: *Shopman's Team Wins; S. P. Gets Business*

William J. "Bill" Avila, apprentice machinist at Sacramento General Shops, recently gained national recognition for himself, and at the same time helped Southern Pacific get a nice piece of passenger business.

Avila was coach of the Sacramento boys' baseball team which won the American Legion Western States Championship recently by defeating Chicago in a play-off at Stockton. Then the team went to Gastonia, N. C., to play for the national championship. The North Carolina boys won the title, but the Sacramento youngsters had a grand time and played before crowds of 15,000, Avila reports.

"Bill" Avila

Through Bill's efforts, together with the help of Ray J. Thomas, surplus clerk in the Sacramento Stores, Southern Pacific got the team's transportation of seventeen roundtrips.

railroads under Federal control involving police regulations of the several States, other than those affecting the transportation of troops, war materials or Government supplies, or the issue of stocks or bonds or rates, fares and charges, under which the State commissions will have jurisdiction in such matters as spur tracks, railroad crossings, safety appliances, track connections, train service, the establishment, maintenance and sanitation of station facilities, the investigation of accidents and other matters of local service, safety and equipment. It will be the policy of the Director General to see that the orders of the State commissions in these matters are carried out.

During the conference the Director General called attention to the fact that the Federal control act gives power to the President to initiate rates, fares, charges, classifications, regulations and practices by filing them with the Interstate Commerce Commission, and empowers that body to review the justness and reasonableness of them. The members of the State commissions took the position that the intrastate rates are nevertheless subject to their jurisdiction. In order to adjudicate these matters the Director General announced it to be his policy to expedite in every way a final decision by the appropriate tribunal of the questions at issue between the Railroad Administration and the State bodies.

The directors of traffic and public service of the United States Railroad Administration were directed, before authorizing advances of any importance in rates, fares or charges, either interstate or State, to submit them to the State commissions in the States affected for their advice and suggestion.

Educators and Police Chief Assist in Safety First Movement.

By C. A. VEALE
(Los Angeles Division.)

H. W. Watkins, master car builder, Los Angeles Division, has recently equipped a steel frame flat car with double deck and placed this car in service between Los Angeles, Colton, Indio and Yuma to transport wheels. With this arrangement it is now possible to load this car to 100 per cent capacity, and in one month this car has saved 4536 car miles, or 18 cars. In addition to this having its influence on the fuel oil savings and train load it will also serve as a splendid example to those whom we have at all times urged to load all cars to capacity.

Superintendent W. H. Whalen has for several years urged the teaching of Safety First principles in all schools, and has on many occasions personally visited and addressed the students. In this he has received valuable assistance from Mr. Connors from River Station. Recently Chief of Police Butler, recognizing the splendid work that had been accomplished, invited Superintendent

(Continued on page 15)

Coast Division Giving Attention to Matters of Service.

By H. R. HICKS
(Coast Division.)

For the benefit of conductors and the railroad as well, every effort should be made by employees concerned to see that passengers at way stations purchase tickets before boarding the train. Instructions on this subject have been issued from time to time, and on the Coast Division we are making a special effort in this direction.

The pamphlet entitled "Instructions and Information Relative to the Handling of U. S. Mail and Railroad Business Mail," effective September 1, 1910, issued by Mail Traffic Manager H. P. Thrall, and which forms a part of station records, shows in detail the correct manner of handling pouches of mail. At this time we would advise agents to read over this pamphlet and refresh their memories on the subject so as to avoid failures which are a cause of annoyance and complaint, and which mean the necessity of checking up to ascertain individual responsibility.

At this writing Form 1290, Report of Cars Loaded for the Month of February, has just reached us and the figures are not so good as we had hoped, compared with last year. Let us hope that next month will place us at the head of the divisions. It is a matter of better car loading.

On March 1st we held our regular fuel oil meeting, at which the new committee was installed. This committee consists of the following:

Engineers, J. F. McCarthy, F. Champlain, A. R. Bullard, L. J. Ficken, J. L. Farley. Firemen, F. Mason, E. Deffebach, L. P. Wilmeth, P. C. Bahnson, G. Reinhardt. Trainmen, H. S. Fritch, S. Arana.

This new committee brought up a great many items of importance that were discussed, and the chairman, Superintendent T. Ahern, is glad at all times to receive any suggestions from enginemen or others who have to do with the operation of trains and the burning of fuel oil, whether in stationary boilers or in locomotive boilers.

The Treasury Department has completed as far as possible the delivery of Third Liberty Bonds and such of the Fourth Liberty Bonds as were paid for, either by cash or through deduction. Those who have not received their bonds should write superintendent's office, giving facts so that we may arrange delivery if possible. It should be understood, however, that delivery of Liberty Bonds is made by the acting Federal treasurer.

Effective March 17, 1919, W. A. Drake was appointed general yardmaster, San Francisco, vice S. Bronstone, assigned to other district.

W. H. Brennan, trainmaster, District San Bruno to Watsonville Junction, is still in the company hospital and is on the mend. We hope to see him on the active list soon.

Electro-Plating Department at Sacramento a Modern Plant.

By J. HALL

One of the departments, small, as departments go, but important and very essential to the railroad service and traveling public, is the electro-plating department of our large repair shops.

The general public is probably less informed as to how the maintenance of tableware used in our diners is accomplished and the various plating department methods involved than any other class of railroad work. To give detailed explanation regarding this work would prove tedious and only partially intelligible to the average layman, but in general this department repairs, replates where necessary and refinishes all silverware used in dining cars, business cars, etc., as well as refinishing trimmings and fixtures, such as sash and door locks, lamp frames, hatracks, etc., for passenger cars in general, operating principally over the so-called Northern District.

On account of the acids and chemicals used in the various processes special attention is given to ventilation. The room where the plating tanks are situated is perfect in that respect, having large ventilators in the ceiling, besides having the upper half of the windows fitted with screen wire, which is left open winter and summer alike, insuring an abundance of fresh air at all times.

The tanks and cement floor are washed off daily, washing being preferable to sweeping, as the nature of the dust to be found on the floor of this room is not very pleasant to inhale.

To conserve the health of the men in the polishing room it is equipped with a system whereby the dust created by the act of polishing is drawn from the room and deposited on the outside of the building.

In addition to daily sweeping of the polishing room, in order to reduce fire risks to a minimum, it is subjected once a week to a special cleaning, accumulated dust being removed from the walls and ceiling, compressed air being used for the purpose.

Until a few years ago we were able to get along, though at great inconvenience, with a fifteen-gallon jar of acid for dipping purposes. Now we have a tank large enough to hold three men without displacing the cover. More capacity has been the cry of this department for many years, until at present, through gradual additions and betterments, we have at the Sacramento General Shops probably the best-equipped plating department west of Chicago.

Mr. Sanford H. E. Freund has been appointed Assistant General Counsel for the United States Railroad Administration, according to an announcement by Mr. John Barton Payne, General Counsel.

REDUCING AMOUNT OF STOCK ON HAND

Stores Department Confronted With Big Task of Conserving Material

By GEO. R. HIMMELMAN
(West Oakland General Stores)

Recently we were confronted with the task of reducing stock of material on hand to practically the same amount that existed as of December 31, 1917. At the time the Southern Pacific passed under Government control value of material on hand in store stock approximated $14,000,000.

When the Railroad Administration ordered the reduction several months ago the value of material on hand had risen to the neighborhood of $23,000,000, due chiefly to increased cost of material and rise in prices. With these figures in hand, some idea of the enormous task which confronted us may be realized, taking into consideration the present value of material as compared to a like period of two years ago. The Railroad Administration contemplated returning the railroads to private ownership under these conditions, and to attain this end stock had to be reduced item for item until the figure, January 1, 1918, was reached.

The greatest responsibility, of course, rested with the Stores Department. Orders had been received to reduce stock and stock had to be reduced to above basis. This was required of every railroad and every railroad store in the country under Government control. Value of stock at West Oakland General Stores for a period of four months was reduced $550,000, but this was only a beginning. By concerted effort this figure was greatly enlarged from month to month, and the closing days of the year found us within reaching distance of the goal set for us by the management. At the time of going to press the final figures for December, 1919, were not available, but indications were that the value of stock on hand would be within $550,000 and $600,000 of the December, 1917, figure.

How Methods Are Applied.

Some idea of the methods applied to conserve stock and to prevent purchases may be of interest to all who are not familiar with what has been done. The procedure is still in effect, and in all probability will continue until conditions resulting from the change from Government to corporate control are perfectly adjusted.

Stock carried on hand is figured on the basis of an average monthly consumption, and only such articles that will actually be used are carried in stock. Nothing new will be purchased unless absolutely necessary, and no special material will be procured from any source when anything on hand can be used or substituted.

Before submitting a requisition to the Purchasing Department for a new article the store must be positive there is nothing like it anywhere on the road that could be used, and to make sure of this the system is thoroughly canvassed. For example, a requisition is received for a five-inch foot valve and the store has none in stock. A wire or mailgram is sent to every superintendent and storekeeper and within a day or so it is known whether a valve is available or not. Many a warehouse, dredger or shop may have stored away for an "emergency" just the valve wanted. If nothing can be procured in this way and the valve is absolutely necessary, a purchase requisition is prepared.

There is still one step more to go before buying a new article. Every month each general store of every railroad submits to the Regional Purchasing Committee at Chicago a statement showing all surplus material on hand and material not required for immediate consumption. Every purchase requisition is sent to Chicago and is checked against these statements, and if the material is finally purchased we can be pretty sure there is nothing like it available on any railroad in the United States.

This system has resulted in saving hundreds of thousands of dollars in material, and when the railroads are returned to their owners it is safe to say there will be very little if any surplus material anywhere over that on hand January 1, 1918.

Where Waste Results.

Surplus material represents an investment of capital tied up and unapplied. When an unused article is allowed to depreciate in storage it is nothing more or less than a sinful waste of money. Many a foreman with perfectly good intentions has ordered an article and stored it away in his shop for an "emergency." It may not be used for some time—perhaps never. More often it becomes covered up and forgotten until the day a general house cleaning is ordered, and a deluge of obsolete material is unloaded on the storekeeper. It is up to him to dispose of it the best way he can. It is now of little or no value, while if it had been stored where it belonged it would have long before been used to good advantage or disposed of at a profit.

Orders for material should always be restricted to immediate wants, as material can be more effectively protected in the store than anywhere else. Only such material necessary in the actual progress of the work should be ordered. Material drawn for a particular purpose and not used should be returned at once to the store. Supplies or material should

SUPERINTENDENT MERCIER ON JOB AGAIN.

By T. J. WHELAN
(Brooklyn Shops)

The many friends of Superintendent Mercier are glad to learn that he is at his post again after an illness of about six weeks. Mr. Mercier has been recuperating in Arizona.

December 9 was ushered in with the season's worst snow storm. It came unexpectedly, and not having the snow-bucking facilities at hand, there were a few minor train delays. Master Mechanic McLauchlan and Asst. Hammond stood on the job night and day keeping engines from freezing and clearing snow and ice from the tracks, so that engines could be moved in and out of roundhouse and about the yards.

In addition to the regular business of the Brooklyn shops, since November 1, 258 refrigerator cars have been fitted up with false floors for apple shipments from Rogue River Valley in the vicinity of Medford, Ore. It is expected that this shipment will continue for at least 30 to 60 days longer. This is the heaviest movement of apples from this district for years.

The Brooklyn shop employees are organizing a brass band under the directorship of Draftsman T. C. Lewis and Chief Clerk A. Kropp. This will be a cheerful addition to our surroundings. Let us encourage the boys in the good cause.

The banner which has been awarded to the Portland Division (lines north of Ashland) for a clear record in the "No Accident Drive" was presented to the officials and employees at a meeting at the Union Station December 17. It was accepted with appropriate remarks by Federal Manager O'Brien, Superintendent Mercier and J. F. Grodski of the Safety Department. The banner will be exhibited at various points on the division.

never be ordered or kept on hand in the shop to avoid making requisitions or to meet contingencies that may never arise.

The matter of conserving material is of vital importance to all branches of the railroad. In questions of economy in use of material, the management's first orders are always to the store, yet without the hearty co-operation of all departments along this line nothing can be accomplished. Regardless of the good result of our efforts to reduce stock during the past year everyone must continue to look forward with the same idea in mind. Under any circumstances we can be assured that economy will be one of the main requirements, and the storekeeper must of necessity pass on all requisitions submitted to him and curtail them if expenditures are to be restricted within reasonable limits. The responsibility rests with the man who uses the material as well as with the man who buys it, and is of vital importance to everyone who has anything to do with the purchase, storage and use of company property.

During World War I, women volunteered for non-traditional jobs in support of the war effort. This photo, taken in 1917, shows men and women before a pile of scrap in the Southern Pacific yard in Sacramento. Beginning in about 1910, the Southern Pacific operated supply trains across the system. In the mid-1940s, the home for these trains was either Sacramento or Los Angeles. The process started when orders for materials in Sacramento came in from around the system. The needed items were assembled in an order to facilitate efficient loading of the supply train when it arrived from its last run. Loading time was about a week, followed by about three weeks on the line making deliveries. The crew on each train, usually about six workers, also had the opportunity to pick up scrap and unusable items from each station. So, the train returned back to Sacramento with a load of discarded items. Some of the scrap could be re-used with no restoration while other items such as switch frogs, lamps, and tin might need attention before being put into service again. Some metal parts could be melted and re-cast. As the photo shows, the piles of scrap were not small. – Courtesy California State Railroad Museum

S. P.'s "City of Shops" a Story of Perfection

Newspaper woman finds at Sacramento much of human interest in tour of our facilities at that point

PERFECTION is made of trifles, but perfection is no trifle," says an old adage, and a trip through the local Southern Pacific shops, the largest shop industry west of the Mississippi, and on a par with the largest industries of the East, would suggest to a visitor's mind that the railroad company works its biggest shops with this moral in view. And there is "some moral" contained in a story of the inner workings of that huge industrial center.

Although 45 per cent of all the equipment on the Pacific Division of the Southern Pacific Company is made in Sacramento, as well as most of the steamship and ferry equipment and machinery for the Bay district, not to speak of all the repairs and dozens of new engines turned out of the shops; not a lead pencil is used that has not been tested for efficiency first.

Oil for use in lanterns must be able to burn so many hours, and woe unto the firm that sells the Southern Pacific as much as a pencil that does not come up to specifications. Everything is bought by specification. The laboratory nestled in among the buildings in the yards is one of the least known but most important cogs in the working of the great detailed system. No imperfection can remain in the dark after the searching, trained eyes in the laboratory have ferreted out causes.

Trace Cause of Wrecks.

"Why was that wreck?" "How did that engine break down?" The laboratory can tell you that metals and chemicals were not properly fused in the beginning, or whether it was the work of a faulty shopman. Through its system of identification, the railroad company can narrow down to the person basically responsible practically every time. But more about the laboratory later.

The writer was privileged recently to view the multitude of shops, foundries, and specializing departments under the guidance of H. C. Venter, general foreman of shops, at the invitation of A. D. Williams, General Superintendent of Motive Power for this division of the railroad. She returned footsore, weary, and ready for the tub, several hours after she started, but with a whirling mind full of appreciation. Details, details, details; every one at work on something small, but what a smooth and perfect whole it made.

$400,000 Monthly Payroll.

Sacramentans rise to the early blasts from the shop whistles, or we casually glance at the hundreds (nearly thirty-five hundred to be exact) of blackened men who dot the

Miss Jacques Wilson, a member of the Sacramento Bee's reportorial staff, recently visited the Southern Pacific shops in that city. In a special article, which is reproduced for readers of THE BULLETIN, *Miss Wilson gives her impressions of the activities and methods used by the 3500 employees to obtain the greatest efficiency. Seeing ourselves as others see us is refreshing, especially when the viewpoint is as comprehensive as that taken by Miss Wilson.*—THE EDITOR.

streets nightly on their way home swinging their dinner pails, and vaguely remark to our visitors "they come from the shops"; but how few know anything real about that great thundering place that pours into the coffers of Sacramento nearly $400,000 from its monthly payroll, wages which amounted to $3,720,000 in 1919 while still under Government control. With the return of private control the payroll is larger.

Shopworn though the phrase is, the Southern Pacific shops really are one big city to themselves. They even have their own money system for exchange in the yards. This is necessary because the shops are divided into two separate great departments, the mechanical and store departments.

Each worker is given ten "checks" when he enters the employ of the company and he buys his working tools each day with them, and is accountable for these checks and what they bought, when he finally leaves the employ of the railroad.

Then the foreman has his "4218," or paper draft, with which he "buys" all the parts he needs for whatever he is assembling. Though 75 per cent of these parts have been manufactured in the shops, they are turned over to the store department and belong there as definitely as if these two departments were separate concerns. This rule holds even to a bolt or a screw.

Watch Most Minute Expenses.

In this manner, the company is able to keep track of the cost of production, a regular system of bookkeeping and pricing is kept according to the fluctuations of the regular markets, and if a locomotive costs too much, officials can account for the actual cost of each part down to the last rivet.

The big storehouse where the rolling stock is kept, looks much like any huge hardware store would look. Parts all boxed in high rows are in avenues down the center and properly marked. Messengers with "4218's"

continually are on the run back and forth with their loads for the foreman.

The Laboratory Testings.

Speaking of the laboratory and its importance in the general scheme of work it might be mentioned that practically all the men employed there are graduate chemists. To facilitate improvement in building perfect parts, broken pieces of engine wheels, broken springs or vital parts of engines that have caused wrecks or trouble somewhere along the Southern Pacific road, are all piled together near the laboratory to be tested and the trouble traced.

Analysis is not made only after damage has been done, however. By a new process in the newly constructed steel foundry, excellent steel is turned out from second-hand material through a giant six-ton electric furnace, and by the aid of certain chemicals and minerals. This work is done under the supervision of a metallurgist, L. J. Barton.

Each day, to make sure that "the mixture" is right, a sample of the metal is taken from the furnace to the laboratory, where it is analyzed by an expert. This care insures less future trouble of parts being faulty.

A great deal of manufacturing is done within the walls of the laboratory, also, and from the Sacramento shops come most of the liquid soap used on the trains, the ink, mucilage, car cleaner, and a dozen more articles used upon the Pacific System of the Southern Pacific.

Elaborate Salvage System.

But the salvage system of the Southern Pacific is one of the real wonders of modern business. There is a huge refuse pile, but there is no waste. Even the ashes are conserved, placed on freight cars, and sold for the chemicals that are in them.

In that refuse heap is every conceivable article, or piece of article, that might be used on a railway. Iron and metals are scraped, bundled in the "cannery," and then taken to the foundry where they feed the great molten mass of fiery lava in the steel furnaces. By a treatment of certain chemicals, a good grade of steel is the result, and waste becomes very fine and economical material.

Getting Rid of Rust.

Then there are a variety of "laundries." Some metal parts still are serviceable if it were not for rust and corroding. There is a huge lye boiler, called the "rattler," that eats off the outside coatings and turns out a supply of perfectly good metal again. This idea of cleaning is carried down even to the "packing" waste that is used in engines. Barrels of gritty and dirty packing arrive weekly at the shops, to be turned over to its individual laundry, steamed, reoiled and made ready for reuse.

Speaking of the "rattler," Foreman Venter reminded me, as we stood next to the heavy machine hardly able to think, that the shopmen were so accustomed to hearing the conglomerated noises that the slightest variation of sound was noticed by them, even if it came from a distant shop.

Feed Old Wheels to Furnace.

In line with the salvage work is the reuse of old train wheels. Piles of discarded wheels lie in the vicinity of the general foundry, and are broken up by a huge steel pounder and then fed to the furnace. In this manner 85 per cent of the old wheels are reused, and an average of 277 wheels a day, or 94 wheels an hour, are turned out by these men, who begin work about 3 o'clock in the morning to escape the additional heat of the day.

Then men in those shops work hard and steadily. No piecework is done any more, but there is no loafing on the job anywhere apparent. It would be impossible to go into detail about the work of tearing down and building up in all individual shops, the list including the machine shop, boiler shop, blacksmith shop, spring shop, general foundry, airbrake shop, babbitting shops, tin shop, car shop, the cabinet makers, hammer shops, rolling mill, car machine shop, steel foundry, paint shops, brass factory, frog shops, and a multitude more.

The recruits for this vast army of skilled workers are drawn from the apprentice school, which is operated there under practical instructors. Young men between the ages of 18 and 22 serve four years' apprenticeship, going to school four hours a day and drawing wages for the time they work and study. During these four years the boys are given an all-round mechanical training, and at the end of their training records show that 77 per cent of the boys remain there as journeymen, perfectly satisfied with their treatment and training. There are 215 apprentices learning trades at the local shops at this time.

Some Women in Shops.

Working shoulder to shoulder with the men, wearing the same sort of overalls, and operating the same machines, are the remaining remnants of the women's army that worked for the company during war times. There are only thirteen left, and Foreman Venter says they are leaving one by one, usually marrying shopmen. He says they do their work well and are found to be very expert upon certain lines of fine work, particularly in the pattern department.

There is no doubt but that the Southern Pacific Company is co-operating with its employees in warding off danger of accident whenever possible. Over every machine is a warn-

ing, and broken glasses showing how accidents have actually happened on that particular machine. Each man is provided with goggles and safeguards of all kinds. One of the most familiar signs about the walls is, "You can see through glasses, but not through glass eyes." According to statistics, 95 per cent of the injuries to shopmen are eye injuries.

Have Own Fire Chief.

To further safeguard the lives of the men a fire chief is employed at the yards, who does nothing but investigate fire conditions. There are two fire drills a month, and every one in the place turns out, not knowing whether something is actually wrong or only a drill.

Many highly specialized machines are monthly being installed by the company to improve the grade of work they turn out. For instance, there are three magnet cranes at work on the grounds picking up scrap, each operated by a couple of men, accomplishing work that originally took from fifteen to twenty men to do. Two huge air compressors, costing $50,000, have just been installed to compress air to be piped to all departments. There is a friction saw with no teeth that cuts through metal; there is any variety of wonderful machinery to enthuse the heart of a mechanic, but all too technical to be described in this article.

The already big shops are branching out methodically to the north on the filled-in sand lot near the river, and officials claim their dream of a time when 8000 men will be employed there is not a too far distant prospect to be a reality in the next few years.

INSTRUCTING NEW TICKET SELLERS.

Before placing a new ticket seller on duty, he should be fully instructed by agent or experienced ticket seller as to the location, form, and use of different kinds of tickets, tariffs, and current circulars; the arrangement of instructions, fares, and other details in tariffs; the method of issuing tickets, including cutting of simplex tickets; operation of ticket dater; writing on interline and local tickets, including any route endorsements required and as to handling and care of office equipment.

The new ticket seller should take pains to familiarize himself with these details and see that he understands them in order that he may be able to readily ascertain the correct fares, and issue tickets in the proper manner.— The North Western.

DON'T DELAY REPLIES TO CORRESPONDENCE.

Get the habit of answering all correspondence promptly. Every hour you delay reply to a telegram or a letter. you are delaying some one else, probably several people.

S. P. WAGE INCREASE IS ESTIMATED AT $17,500,000.

The railroads on the coast have not yet had time to determine how much of the $625,000,000 annual increase in railroad wages granted by the Labor Board will have to be borne by them, but it is believed that all of the Southern Pacific lines combined will have an added bill of about $17,500,000. On the Pacific System line there are over 50,000 employees. The dividends paid the stockholders last year were $17,-478,460. So this one increase equals the amount that the stockholders got altogether.

The annual report to the stockholders shows that the increase in wages and cost of materials used in operation in 1919 showed increase compared with 1918 of $18,833,500. This increase is approximately $1,400,000 more than the stockholders received.

The stockholders of the railroads have not had any share in the increased revenues arising from any source, whether increased volume of business or from higher freight and passenger rates. The dividend rate on no large system has been increased for many years. The Southern Pacific stockholders got 6 per cent in 1919, just as they did in 1914; but this actually means out of every dollar earned by the Southern Pacific the share that goes to the stockholders is 40 per cent less in 1919 than it was in 1914.

The progress of taxation in railroad affairs is also graphically represented by comparing 1911 with 1919. In 1911 for every dollar paid the stockholders of the Southern Pacific about 33 cents went for taxes; but in 1919 for every dollar the stockholders received the company paid out 73 cents in taxes. Taxes increased from $5,461,570 in 1911 to $12,842,270 in 1919.

The new increases in freight and passenger rates are due directly to the increased cost of operation: First, wages; second, higher cost of materials used in operations, and, third, taxes. The fourth item is the necessity for providing revenue by taking care of the interest charges on money borrowed to furnish more locomotives, cars, additional tracks, terminals and other facilities to give a better service.

The gross earnings from all sources exceeded $260,550,000 in 1919, out of which the stockholders received $17,-478,460. In 1914 gross revenues were $152,623,950, out of which the stockholders received $16,361,090. The increase in dividends was not due to an increase in rate, but to an increase in capital stock, the bondholders having exchanged some of their 4 and 5 per cent bonds for capital stock under their rights as bondholders.

From Bolts to Walking-Beams at Sacramento

Wide Range of Activities Found at Our General Shops. Came Into Existence with Pioneer Road and Grew with Southern Pacific Lines

By A. D. WILLIAMS
Superintendent Motive Power Northern Division

WHENEVER one pauses to consider the history of the organization and growth of the Southern Pacific Company, the Sacramento General Shops should be remembered, for they sprang into existence with the parent organization, the Central Pacific Company, in 1863 and have kept pace with the growth of the Southern Pacific Lines.

In 1863 these shops employed about 15 men in the Car Department and about the same number in the Locomotive Department. The first permanent buildings were erected in the years 1868 and 1869 and formed the nucleus of the present plant. The shops now cover an area of approximately 145 acres and employ a normal force of 3100 men, this growth spreading over a period of over half a century.

Since the beginning repair work on locomotives and cars has been taken care of and the production of new equipment followed up at the same time. The first locomotives to be built at these shops were 10 freight locomotives and two passenger locomotives in the years 1872 and 1873. From then to 1888 a large number were built, the most of them being "Monkey Motions." These did away with the "Stephenson" or link motion, and were quite similar to the "Walschaert" motion of today. During the world war the difficulty of obtaining rolling stock and the high price of the same made it imperative that we help ourselves and from 1917 to the present time 53 locomotives have been constructed in these shops.

First "Twelve Wheeler"

In 1882 a "Twelve Wheeler" was built which was at that time the heaviest locomotive owned by this company. It weighed 69 tons and had a tractive power of 29,140 lbs. Today we have single engines weighing 193 tons with a tractive power of 75,150 lbs. and Mallet or articulated engines weighing 219 tons with a tractive power of 94880 lbs.

The first passenger cars operated by this company were shipped around Cape Horn in sections as were the locomotives and put together at Sacramento. These cars were 42 ft. long, little larger than a box car, with flat roofs and wooden seats of the plainest character. Since then constant improvements have been made. In 1847 twenty first class passenger coaches were built which were acknowledged at that time for design and workmanship to be the best and neatest in America. Many cars have

A. D. WILLIAMS

been built since but none have given better service or longer life. Several of these cars are still in existence in our Maintainance of Way service.

The first private car was built about this time for Supt. A. N. Towne. In 1876 a number of short street cars were built for use in San Francisco as there were no car builders on the coast at that time. In 1883 the car "Stanford" was built for the sole use of the late Governor Stanford. This car when received at Washington was acknowledged by all eastern car builders to be the finest and best built car in America. Every convenience which could be thought of at that time was installed in this car. The interior was finished with choice woods. The frame work was of white oak and the workmanship of the best. In 1906 this company built its first all steel passenger car. This car is No. 1806 and is still in service. The next year we built an all steel postal car, C. P. 4097 besides having an all steel body and underframe the inside lining was of fireproof construction and an electric lighting system was also applied. This car is still in service and no material changes have been made in it.

A very large number of freight and non-revenue cars have been constructed in the last 40 years. In the last four years about 3000 have been constructed.

Steamer Machinery

While a good many of the steamers which this company uses in the

Bay and River service were purchased, a very considerable number of them have been built at the Company's ship yards in Oakland and the most of the metal work furnished from Sacramento shops. The boilers and all metal work for the steamer Solano with the exception of the engines proper and the wheel shafts were gotten out here. At the time this boat was built in 1879 it was the largest car transfer boat in existence. About the same time all the machinery, boilers and metal work for the river boats Modoc and Apache was gotten out at these shops. In later years the machinery and boilers for the Seminole, Navajo, Thoroughfare, Alameda, Santa Clara and Contra Costa were also built here.

In 1863 the shops proper consisted of a small repair shop for locomotives and another for cars. The machinery of the shop of Goss and Lambert on I street in Sacramento was utilized. Few years later this shop was taken over by the company and the machinery moved to the present site. An iron foundry was built which was put into operation in 1868. Since that time all the general iron casting used for cars, locomotives and steamer work have been made at Sacramento. The present foundry is divided into two sections, one specializing in wheels for freight cars and the other doing miscellaneous work.

The Wheel Foundry

When under full operation 300 standard wheels are made in the wheel foundry every day, requiring about 125 tons of molten iron. This is made from pig iron in a large furnace known as a cupola. This consists of a long steel shell, 90 in. in diameter, with a high stack. A lining of fire brick runs from the bottom to the top of the stack. Coke is thrown into this cupola, followed by a layer of pig iron, this method being followed until the furnace is full. A fire is then started at the bottom, and as soon as the lower layer of coke becomes ignited the air blast is turned on which strikes this hot coke. A very high temperature is thus obtained which is sufficient to melt the pig iron which drips into the furnace bottom. As soon as enough is collected the cupola is "tapped" and this metal allowed to run into a large pot. From this smaller amounts are taken for the pouring of the wheels. As the coke burns out and the iron melts more coke and iron are thrown in. This process continues through the day until the required number of castings have been made, when the

THE BULLETIN

What the Camera Found at Sacramento Shops

An apprentice-pattern-maker is shown above paying close attention to a machinist who is showing him some fine points.

Difference in size of the boilers of Locomotive 1611 in service in 1883, and of Locomotive 4210, in service in 1911, is indicated above. Below, the pouring of steel from electric furnace into a ladle.

Pieces of machinery of enormous size are handled at the Sacramento shops as shown by the picture above, the turning of crank pins for the steamer "Alameda."

Another big job is the finishing of the walking beam for the steamer "Newark," pictured at the right. The first steel car built by the Company is shown below. It was a product of the Sacramento shops.

air is turned off and the furnace cleaned, patched and made ready for the following day.

These wheels are made in sand molds, with a metal chill for the tread. When the metal is poured in this mold, the chill quickly cools it on the tread, making it "chilled" or very hard iron, the hub remaining soft and machineable. As soon as solidified these wheels are taken to the annealing pits for gradual cooling which eliminates any danger of cracking or breakage from internal strains, set up by the quick cooling of the outer layer. After four days they are taken from the pits and inspected and are given various tests to determine their strength, etc.

The General Foundry

The general foundry makes all of the general castings, of which there are nearly a half million different patterns in use by this Company. Everything is made here from brake shoes to the largest steamer casting, weighing many tons. One of the castings for which the Sacramento foundry is noted is the locomotive cylinder. This casting is very intricate, besides being of large size, and requires most careful work.

Iron for the general foundry is melted in the same manner as in the wheel end, but here the quality of iron is very soft as most of the casting must be machined. Patterns for the jobs are placed in a box, known as a flask and sand is rammed in all around. This pattern is then carefully withdrawn and the mold closed. After filling with metal it is allowed to stand until solid, when the casting is taken from the flask, the sand shaken free, and it is then sent to the cleaning room. Here the final sand is either chipped or sandblasted free from the casting and the work is then ready for the machine shop. About 55 tons of metal a day are made into castings here during the rush season.

Brass Foundry

Brass as well as iron is very extensively used in all of our work. The earliest brass foundry work was carried on in one corner of the iron foundry. It was later moved to a building by itself as the work increased and is now housed in a new building erected during the last year which has about 20,000 feet of floor space. Of the brass castings used for locomotives, cars and steamer work, about 95 per cent are made in this shop, the daily output of which has at times been as high as 15 tons. Castings are made here which weigh as

little as one half an ounce while others weigh hundreds of pounds. Journal brasses are cast, machined and lined with babbitt in this shop and as they go out are turned over to the store department for delivery to all points on our lines. Different varieties of metal are made to suit the different demands from a strong "gun metal" to meet the demands of the steam

Success of Great Plant Reflects Spirit of the Workers

In connection with the story of the Sacramento General Shops, A. D. Williams, Superintendent of Motive Power, Northern District, writes:

BE it said after all, these shops and the vast machinery activities of which we write would be as naught, were it not for the royal, loyal men and women employed therein.

The spirit of faithfulness, interest and earnestness manifested by these men and women in earning a livelihood and the joy gleaned from doing a worthy piece of work, is the measure of the success obtained and related.

So it is at Sacramento Shops, and, so it is in any industry, the real factor of success is the manifestation of the brain power of its employes.

To these men and women then, of the rank, from the laborer up through the scale to the topmost supervision, mostly belongs the credit and honor for the accomplishments recorded herein.

boat inspection service to a "soft brass" which is to be manufactured into trimmings.

Steel Foundry

Steel, the master metal, has been the deciding factor in the world's development. From the days of the ancients, when steel was made in crude air-blown retorts, to the present time, numerous changes in the manufacture of steel have been made, some to increase tonnage, some to improve quality. The latest of these developments is the electric furnace and the largest installation west of Chicago is in the Sacramento shops.

This furnace is composed of a heavy cylindrical, steel shell, lined with brick and having a brick roof, similar to the top of a coffee pot. The bottom is made of a material known as magnesite, burned into a monolithic mass, shaped like a saucer. In addition to the furnace itself, there are innumerable auxiliary appliances for tilting the furnace, regulating the electric current, etc. At the present time this one furnace is making about 750 tons of steel a month, pouring 7 to 8 tons at a time. Of this, about 30 tons a month go into steel castings, the rest into slabs, known as ingots, for use in the rolling mill.

Another one of our industries as

distinguished from the repair shops proper, is the rolling mill. The late Charles Crocker, one of the builders of the Central Pacific Railroad, said, "I do not see why we cannot make anything that we need at our Sacramento shops." Working on this assumption a 12-inch rolling mill was built in 1878. At that time practically all rolled iron came "around the "Horn" a long, slow journey. There was plenty of scrap material all around us and when the original mill was built we were able to help ourselves in the matter of bar iron and did not have to wait for deliveries. This industry has now developed until we have two 12-inch and two 18-inch mills served by 8 large reverberatory furnaces and with these we roll practically all the bar iron and steel that is used by this Company between the sizes of ½-inch round or square to 3-inch round, or 1½-inch by 6-inch in flat section. The manufacture of tie plates the upkeep of its thousands of miles of track is also the work of the steel foundry and rolling mill. The ingots are cast in proper forms and delivered in carload lots to the rolling mills where they are rolled into long bars. These bars are then conveyed on a series of rolls to a gang punch that is equipped with special attachments which punches, loads and records the number of plates each car contains when loaded.

These mills roll 235 different sizes and shapes of merchant bars and the tonnage output is approximately 5,-600,000 pounds per month.

Forge Shop

Another of our industries which is closely allied to the rolling mill is the forge shop.

A peep in the forge shop at the Sacramento General Shops will show you three mammoth steam hammers, four lighter steam hammers, two belt hammers, five air bending machines, four forgings machines, one heavy bulldozer, and several blacksmith forges, which are continuously in operation, turning out the forgings that are used on the new locomotives and new freight equipment, that the Southern Pacific Equipment Co. is now building.

Besides the forgings for the new equipment, this shop is turning out forgings of all descriptions for the repairs to the locomotives and cars of the Pacific System. Some of the different articles manufactured in this department are locomotive main and side rods, spring hangers, brake beams, rudder stocks, equalizers, draw

THE BULLETIN

bars, draw bar yokes, brake hangers, arch bars, brake rods, brake chains, brake shoe, keys, pipe clamps, roof clips, truss rods, break shafts, steps, sill steps, pinch bars, radial stay bolts and other forgings too numerous to mention.

The heavier forgings which are made for the floating equipment and the company's largest locomotives are forged from heavy ingots, under the three mammoth steam hammers. These forgings are then taken to the machine shop and finished.

Many thousands of bolts are made daily in this shop for use in the manufacture and repair of cars and locomotives in the different structures scattered over our lines, and in the track itself upon which our freight and passengers are transported.

Frog Shop

About three years ago a building was erected for the making and repairs of frogs, crossings and other track material. This work was done previously in the machine shop and blacksmith shop. This track work, from the repairing of picks to the building of the most complicated crossings is now centered in one spot. This relieved the machine shop and blacksmith shop to a very considerable extent and made it possible to greatly increase the productive capacity of these shops.

Spring Shop

Sacramento Shops can boast of a spring shop as well equipped as that of any railroad in the United States. Originally this work was done in one corner of the blacksmith shop. About thirty years ago it was moved to a building of its own, and this year sees it installed in a new and commodious building, where there is room for considerable future expansion. When in its original quarters it employed 11 men and had a capacity of about 150 tons of springs made and repaired in a year. Now 40 men are employed in this shop and the capacity has been increased ten times. All kinds and characters of springs, from one weighing but a fraction of an ounce to a heavy locomotive spring weighing 635 pounds are made here.

Pattern Shop

The patternmaker is one of the most important men in the shops. The iron foundry, brass foundry and the steel foundry have to depend upon him to furnish the pattern from which their product is made. He must take the drawing or the sample and from that build up his pattern. He not only must be a close and accurate worker but must thoroughly understand drawing and in addition understand the work in these different shops so that when his pattern is completed it can be used. Upon him the machine shop depends, for after the casting is made it must be finished in the machine shop and if his work has not been close and accurate it may be necessary to throw the casting away. In our pattern shop between 30 and 40 men are continuously employed and their work is the foundation for

the activities in the foundries and machine shops.

Saw Mill and Cabinet Shop

Another industry which is very important is the cabinet shop and saw mill. These shops manufacture the wooden parts for cars, both for freight cars and the finest passenger and private cars. During the past year 4,620,000 feet of lumber were used in getting out certain parts for new freight cars. In addition to this 7,800,000 feet were handled for other work consisting of parts for passenger and freight cars. In order to form an idea of what these shops do, it should be known that in the last year it has completed 200 office desks, made 340,000 feet of various kinds of moldings, 45,000 signal flag staffs, 19,000 freight car end and side posts, and other articles in like proportions.

General Machine Shop

This shop was first built in 1869. The size was doubled in 1872, and it was extended to its present length in 1888. In 1904 its capacity was almost doubled by adding a section the whole length of the shop for use as an erecting shop. The old erecting shop was turned into a bay for large tools and an electric crane was installed to serve them. In the new erecting shop a 240,000 pound crane was installed, which lifts the heaviest locomotives off and on their wheels. In addition a smaller crane is also used for taking care of the lighter parts. This shop is equipped with machinery for handling all classes of work needed on our lines from turning the smallest pin to the finishing of the largest shaft or cylinder on our steamers, as well as the making and repairing of all parts on locomotives.

Here you will find a lathe which will take a shaft 30 feet long or turn a piece 76 inches in diameter. Adjoining this lathe is found our last installation, a double head slotter on which we can finish four locomotive frames at one setting. Next to this is a reversing motor planer which will take a job six feet wide and 24 feet long. On this planer all frames are planed before they go on the double head slotter just mentioned. The cylinders for locomotives as well as many large jobs are handled on two large Morton draw cut shap-

Interior of the wheel foundry at the Sacramento General Shops. This is the only place on the Pacific System where car wheels are made. The wheels made here are of unusually fine quality.

ers which plane all sides, bore out the valve chambers and after the cylinders are bolted together plane a radius on the top to fit the smoke arch, by use of a machine feed which was designed here.

In addition to the above machines, this shop is equipped with many lathes, planers, shapers, vertical and horizontal boring mills, milling machines and drill presses of various sizes and capacities.

Historic Machine

The most interesting piece of machinery in the shop is a Brown & Sharpe milling machine which was purchased by the Central Pacific Railroad Co. in 1869 and has been continually in service up to the present time. This machine has been operated by the following old employes:

J. H. Andrews, 1870 to 1873.
R. A. Renwick, 1873 to 1876.
E. B. Hussey, 1876 to 1879.
H. Ingham, 1879 to 1886.
C. J. Little, 1886 to 1899.
E. S. Bechtold 1899 to 1911.

Since that time different men too numerous to mention have been turning out work on the little machine. There is only one other machine of this type known to be in existence and is now in a glass case in the factory of the manufacturers at Providence, R. I.

Another very necessary part of the machine shop is the tool room. The general idea of a tool room is that it is a place where tools are passed out to the workmen. This is correct but only reflects a very small portion of the activities. The tool room force not only passes out tools but must keep them in first class working order. They also manufacture new tools, not only reamers, taps and other small hand tools, but they make jigs for the expediting of duplicate work, punches and dies for our punches, presses, etc. to the extent of many thousands of dollars per year. While a large amount of money is spent on these tools it is money well spent as their use cuts down the bill for labor and cheapens the parts produced.

The Boiler Shop

The boiler shop is one of the most important shops at Sacramento. The boiler is the heart of the locomotive. Our boiler shop does not confine itself exclusively to the manufacture of locomotive boilers for boilers of any and all purposes are made and repaired here, and we have the reputation of making as fine a boiler as is made anywhere in the country. In addition to boilers, tenders, structural work, bridge work and a large variety of work where plate metal is used, I beams, channels and other shapes are fitted up in this shop. Rolls, punches, shears and drills here handle plates up to 1¼ inches in thickness. The plates and other pieces which make up the boilers are sheared, punched, rolled, flanged and otherwise worked until they are in proper shape to fit their respective locations. Close and accurate work is demanded

No Increase in R. R. Expenses from Sale of Oil Lands

Money Received Used to Increase Railroad Facilities, Thus Increasing Opportunity for Employment

IN a newspaper item appearing recently it was said that some employes believed the sale of Southern Pacific oil lands to the Pacific Oil Company has resulted in an increase in operating expenses of the railroad. The following are the facts:

The operating expenses of the Southern Pacific Company have not been increased one dollar through the sale of its oil lands; instead they probably have been decreased through some employes being paid by the oil company who were formerly paid by the railroad. Away back in 1903 the rule was established that the market price of oil used for fuel should always be charged in the operating expenses of the Company. The Company has not at any time paid more than the market price and it is at that price that it is now purchasing oil and charges are being made.

The practice of the Company was adopted by the United States Railroad Administration. Three-fourths of the oil produced was and is light oil not suitable for fuel and it had to be sold for refining purposes at market prices and fuel oil purchased or exchanged for it on the basis of market prices.

The amount of money received by the Southern Pacific Company for these oil lands was $43,750,000. It has had this much more money to invest to provide additional railroad facilities, which means offering more opportunity for employment. In other words, this amount of capital has been released from oil lands where it was tied up and put into active railway service. It was not given to the bondholders or stockholders or anybody else, but has been put to work. The rate of dividend to stockholders of the Southern Pacific remains as it has been for many years, 6 per cent. Interest charges remain unchanged by this transaction, except that as the Company has had this liquid capital on hand its need for borrowing money has been lessened.

The lands that were sold to the oil company were not a part of the Southern Pacific's property devoted to railway operation. The value of the land itself was not included in the valuation of the property on which a return is to be earned under the Railway Transportation Act, and its revenues and expenses were kept independent thereof. The Company's fuel oil supply has been protected in connection with the sales whereby the Company pays only the ruling market prices for its fuel oil.

---

both for steamers and locomotives as the United States sends inspectors to look after this work and see that it is performed in accordance with the rules and specifications laid down. The State of California also has its rules for the manufacture of boilers used in stationery service as well as for all tanks and drums to which pressure is applied. For these reasons if for no other it is necessary that the finest work be done.

Blacksmith Shop

While the rolling mill produces the bar iron and the hammer shop works out rough forgings with powerful tools it is still necessary to put the individual touch on nearly all forgings before they go to the machine shops for finishing. This calls for the expert work of the blacksmith and his helper and upon them depends in a large measure the cost of the work in the machine shop and also its safety when put into use on locomotives, cars and steamers.

Copper Shop

While the sheet metal shop is known as the copper shop the working in copper is but a small part of its activities. This shop handles all steel below ⅛-inch in thickness. To obtain an idea of the extent of work done we have only to look at the fact that during the last year over 300,000 pounds of galvanized iron were used. The large part of the metal used in this shop goes into tinware, as it is called, for use on all parts of the system. In making up this work, riveting has to a large extent been done away with. The sheets are fastened together by an electric spot welder which welds the two pieces together wherever they are touched. This also tends to cheapen the cost so that we are able to get out more and better work. Work which formerly required 5 hours to do by hand is now done on this welder in one hour.

Pipe Shop

Forty years ago the amount of pipe on a locomotive could be carried by a man on his shoulder. These pipes consisted of a feed pipe to carry the water from the tender to the pump, a blower pipe, and a heater pipe to warm up the water in the tender if necessary. Today it is doubtful if fifty men could carry the pipe in one of our large locomotives.

---

28 THE BULLETIN

Sacramento Shops

(Continued from Page 8)

During the past year we have used on new locomotives alone 12,128 valves and pipe fittings and 24,800 feet of iron pipe. In the shops and yards there is over 65 miles of pipe ranging from 12-inch down in diameter. This pipe is used for the distribution of water, steam, gas and compressed air.

Electric Development

Electric energy was first made use of at these shops in May, 1889, when one 35-light arc dynamo and one 500-light incandescent alternating dynamo were installed. These machines were run by a high speed steam engine and furnished light for seven years. When the first hydro electric power plant was installed at Folsom 22 miles away and electric energy brought by a long distance transmission line to Sacramento it was decided to shut down this steam driven plant and purchase power. On August 10, 1896, the first electric motor was installed in the Sacramento Shops in our spring shop. This was a 10 horsepower motor and is still in service on a machine in the rolling mill, having given 25 years of continuous service. During the same year four more motors were installed, all of which are still in service. Since that time more and more electric power has been installed until we now have a total horse power of 6059.5 These motors range from ⅛-horsepower to 400 horsepower and there are 409 of them. We also have four electric welders in constant use. The amount of work which we have for these welders to perform makes it necessary to keep them in operation during the whole 24 hours, and still more of them are needed. There is also an extensive fire alarm system protecting the shops, with 16 general firm alarm boxes and 23 still alarm boxes of the Gamewell type.

Passenger Car Department

The Sacramento passenger car shops today have nine tracks accommodating our largest cars with ample room to work in and around them. On these tracks the heaviest of our work is done. Our truck shop is one of the best equipped shops in the country for passenger car trucks and there are four tracks in this shop with overhead geared pneumatic hoists running on tracks for the purpose of lifting the trucks from their wheels and thus saving much time and work as without them the trucks would have to be handled by jacks.

Upholstery Shop

Our upholstering shop is a complete factory in itself. It would be a long task to enumerate the many articles manufactured in this department. Canvas hose for filling locomotive tenders, engine seat cushions, mattresses and furniture, and an estimate shows that 78,420 articles are manufactured here for shipment in a year.

When a car is shopped for general repairs all upholstering must be first removed—seats, cushions, rubber matting, curtains, etc. These are all thoroughly cleaned and renovated so that they are as good as new. All hair is removed from cushions, repicked and sterilized. Feathers are also treated by passing through a steam renovator.

One of the new industries taken up by this department is the vulcanizing of rubber. This is making wonderful progress in reclaiming old rubber matting, etc., which in the past, when badly worn, had to be cast aside.

Plating Department

The electro-plating department performs all classes of plating work, so essential on all modern railroads. Everything possible is done in this department to perform this work economically and thoroughly and it may be said that this is one of the most efficient plating rooms of any large railroad shop.

Seven hundred locomotive headlight reflectors have been resilvered here during the past year. There are from 500 to 750 pieces of silverware on each dining car. Approximately 26,000 pieces of silver are replated during the year. This does not cover the tableware used in depot restaurants or on ferry boats.

The cleaning and polishing of all interior trimmings from passenger coaches is also taken care of in this department.

Car Machine Shop

This shop is ordinarily known as the wheel shop as it fits up all the cast iron and rolled steel car wheels used under cars, locomotives and tenders. It handles approximately 100 pair of cast wheels per day in addition to the rolled steel wheels and a large amount of machine work which pertains strictly to cars. All car wheels come to this shop, the wheels are taken from the axle, and if either wheel or axle is in condition to be used again are fitted up to be put under other cars. A few days ago a pair of cast iron wheels came to this shop for dismounting. On looking at them it was discovered that the wheels were cast in the year 1868.

Car Paint Shop

The car paint shop has track room for 23 cars. Here the cars are thoroughly renovated, cleaned vanished, and made ready for service. On wood coaches, when paint has become worn and cracked, it has to be removed by the burning process. When steel cars become corroded and the paint worn out they have to be sand-blasted under very careful conditions. This necessitates the same painting procedure as on the wood car.

The Sacramento paint shop is not only a paint shop for cars but is a supply house for the whole system, store orders from all parts of the system being taken care of in this department.

Furniture glass, mirrors and signs for all purposes are made here. We mix paints of all kinds and for all railroad purposes. During the past year ten thousand gallons of assorted paints have been mixed for the Maintenance of Way Department for division use.

Over 500 mirrors of all kinds and sizes are manufactured here during the year as well as art glass, filling orders from all parts of the system. A large saving is made by substituting glass made from old gothic sash. This glass is first sand blasted and then dyed green, bringing it near the color of the rough green glass, and is much more durable as it is plate glass.

In addition to the car paint shop where all passenger cars are handled there is a small paint shop known as the locomotive paint shop which takes care of the painting of locomotives and tenders. In old times locomotives, particularly those on passenger runs, were made very brilliant by the art of

The little "engine" with the big smokestack at the left is C.P. 173, the first locomotive constructed at the Sacramento General Shops, in 1872. Compare it with the Pacific type passenger locomotive below which was built at the same shops in 1918. The latter weighs 234,000 lbs. loaded and the tender weights 141,000 lbs. loaded.

the painter. Red, green and other bright colors were used and the sides of the tank and the headlight were decorated with truly artistic pictures. This is now all done away with and a plain coat of black paint, with varnish to protect the surface, is all that is desired.

Freight Car Repair Work

Freight car repair work has developed along with the increased demand until today we have about 1200 freight cars passing through these shops for repairs each month. This shop as a rule does not appeal to visitors and others but it is the backbone of the plant. The freight car earns the money paid the men on our rolls. Freight equipment has changed very greatly in the last 25 years. Previous to that time all cars were practically of 15 to 25 tons capacity and of all wooden construction, and were used in 20 to 30 car trains. Now nothing is built less than 40-ton capacity and all new cars are of steel or at least steel underframe construction. This is necessary for the reason that they are handled in many cases in trains of 100 cars.

Store Department

All raw material for use in the industries and for use in manufacturing material on store orders in the various shops, all repair material and all material required in the construction of new cars and locomotives is handled by what is known as the "Store Delivery Service" under the jurisdiction of the store department.

All finished material manufactured in the industries and all finished material manufactured on store orders is taken from the shops by this same service.

To efficiently handle this material, the Store Department has seven automobiles and seventy trailers of various capacities, and sixty-five men. The equipment handles the heavy material, the lighter material is delivered by men on foot with trays and push carts made for the purpose.

The Laboratory

In any large company or concern that purchases or manufactures a large amount and variety of material, a prime requisite for safety economy and convenience, is that the material furnished be the best suited for the purpose used. To insure this many tests are required, physical, chemical and mechanical, apparatus for making same and a corps of men trained in this work. Such is the equipment and personnel of the laboratory.

The work of this department has to do with all branches of the railroad, operating, maintenance of way, operating, mechanical and stores, and requires of the laboratory force familiarity with the vicisitudes and requirements of all these departments.

The laboratory analyzes and tests the paints that are used on buildings, bridges and rolling stock, both passenger and freight. In the rolling mills, steel furnaces and foundries

Distinguished Service Order
Is Your Name Here?

LOS ANGELES DIVISION

H. A. Sullivan, conductor; C. E. Girard, Frank Soutar, R. C. Stiewel, firemen, Downey, Cal., for removing car of shooks off spur track upon noticing that some bales of burning hay had been thrown on to spur from warehouse which had caught on fire, action taken no doubt saving car and contents.

H. W. Mitchell, conductor, and B. R. Crowley, brakeman, for detecting by close observance peculiar sound coming from wheels under car in their train and instructing engineer to not exceed speed of 18 miles an hour to terminal, at which point inspection developed that the wheels were seamy, necessitating removal.

A. H. Ross, conductor; W. H. Donoho, C. A. Wood, brakeman, for quickly responding to short call for extra work, resulting in train leaving promptly for scene of trouble due to washout.

M. G. Green, brakeman, for discovering broken yoke on refrigerator car in his train due to close inspection, which action permitted car being handled behind caboose into terminal without further hazard.

J. S. King, passenger brakeman, for discovering brake beam down on tourist car, his close inspection and action taken no doubt preventing derailment.

C. Hendershot, engine foreman, and H. C. Patterson, yardman, for interest displayed in locating properly open switch to private industry until same could be attended to by section foreman, thereby eliminating serious hazard.

F. S. Watkins, engineer; D. L. Sullivan, fireman; W. S. Dressler, engine foreman; M. Rees and N. H. de Bourck, yardmen, for valuable assistance rendered in saving company property endangered by fire which destroyed gin belonging to Globe Mills.

F. S. Watkins, engineer; E. M. Fink, fireman, and L. P. Woodruff, yardman, for responding promptly to fire alarm by engine whistle, proceeding to nearby station and extinguishing blaze in car of cotton at gin.

P. R. Cook, signal maintainer, for action taken in clearing several telegraph wires which were in trouble, thereby minimizing delay to telegraph business.

R. Armstrong, telegrapher, for detecting oil on rail at his station after departure of light engine and notifying all concerned promptly in order to correct conditions.

L. J. Pletke, utility clerk, and W. L. Somerby, warehouseman, Santa Ana, for action taken recently during heavy wind storms, re-

sulting in telegraph pole being broken down, by clearing wires and bracing poles in such manner that telegraph service was uninterrupted.

C. P. Taylor, clerk, River Station, for accomplishing diversions on car of oranges and live poultry, although diversion orders carried wrong numbers, but by display of proper interest mishandling and delay was avoided.

J. Huntsman, section foreman; J. L. Fifer, pumper, and R. H. Freeman, telegrapher, Bridge 685B, for interest displayed in protecting company property by extinguishing fire which had been started at bridge.

H. Sullivan, conductor, for valuable assistance in correcting defect on engine which permitted locomotive to deposit oil on rails resulting in delay account slipping.

E. J. McGinty, engine foreman, and C. Massingill, yardman, River Station, for interest manifested and prompt action taken after discovering pilfered car on private spur by reloading the stolen articles.

J. McEnerney, fireman, Iris, for discovering broken rail in siding and taking necessary action to prevent possible accident.

TUCSON DIVISION

Yardman A. C. Williamson for energetic action taken at time of fire to protect company property.

Brakeman A. F. Corkran for his discovery of cracked intermediate iron on freight train.

Brakeman H. F. Brown for his report of broken rail in stock track at station.

Engineer C. P. Kramer and Fireman J. F. Newman for assistance rendered crew of passenger train to re-brass hot journal on baggage car.

Section Foreman D. B. Perkins for discovering blazing hot box on car in passing train and signalling train crew that corrective measures might be taken.

Fireman T. J. Delehanty for prompt action taken in providing another fireman for stationery boilers when regular fireman was incapacitated.

Engineer C. A. Thumm and Fireman P. E. Stickler for assisting train crew to repair steam hose on car in train, thereby assisting train to keep their uniforms clean, and expediting the movement of their train.

Car Foreman Chas. Leon for the assistance he rendered train crew to re-brass car in passenger train. This man was not on duty at the time and offered his service in the interest of the Company.

Car repairer D. McIntyre for offering services and helping re-brass car in passenger

---

it is required that materials used, coke, coal quartz, feldspar, magnesite, tin, lead, copper, babbitts, etc. be of required and desired quality. It has to do with every item of the approximately million dollars' worth of material purchased each month by the store department at Sacramento which is the distributing point for the system.

Office Force and Drawing Room

The activities of the shops would be very much hampered without the assistance of our office force to do the bookkeeping and prepare the pay rolls and the drawing room to prepare the drawings and blue prints necessary for the prosecution of many lines of work.

There are some fifty employes in these two departments.

A few items covering some of the activities of the year 1921 are given in closing:

Total Pay Roll.........$4,500,000.00
Iron rolled, lbs.......... 60,000,000
Iron Castings, lbs......... 16,000,000
Brass Castings, lbs....... 3,600,000
Steel Castings, lbs........ 525,000
Steel Ingots, lbs........ 12,400,000
Locomotives repaired 400
Passenger Cars repaired.. 624
Freight Cars repaired.... 15,000

During the past four years the following new equipment has been built here at an approximate cost of $5,000,000.

Locomotives 53
Freight Cars 2,750
Cabooses 50

First Unit of Immense Stores Plant Occupied

The new warehouse of the Stores Department at Sacramento is shown above and the accompanying interior views show the carefulness of detail that characterize this plant. Center inset, E. Harty, District Storekeeper, Sacramento.

Below, office employes, Store No. 1, left to right: Harriet Desmond, Curt Grolla, Fanny Schliss, Lillian Parker, C. W. Kripp, A. C. Palm and Edna Palm.

By E. HARTY
District Storekeeper, Sacramento

THE Stores Department at Sacramento has just occupied two new buildings which comprise the first unit of what is to be one of the largest facilities devoted to storage and distribution of railroad materials in the country.

These consist of a main store building of two floors, 500 feet long, with platforms making it 700 feet long in all, and a fireproof oil house equipped with all modern handling facilities.

The first floor of the main building serves as a warehouse where materials are stored in original package lots ready for shipment to other stores, thus saving the cost of packing. The second floor contains material stored in racks for delivery to local shops, for equipping the three supply trains operating from this store and for shipment to smaller stores. In all there are 82 racks, or nearly five miles of shelving.

Like Department Store

Materials are divided into sections, or departments, similar to the arrangement of modern department stores. These sections are in charge of sec-smaller quantities, while power pumps, tion stockmen, each one attending to

ordering, receiving, storing and issuing the materials coming under his jurisdiction.

Every item of material in the store must be counted once each month. This count must be accurate for it is upon this that the purchases for that month are made. Considering that these purchases may run as high as a million dollars monthly, the necessity of accurate count can be easily recognized.

To facilitate this monthly inventory, various methods are followed to secure a positively accurate inventory in the least possible time. Large items of materials are piled in unit piles; sheet iron in piles has sheets numbered successively from bottom to top so that the top sheet always shows the number on hand; small materials on shelves are arranged in trays, a certain number in each tray and these stacked one upon another. Each tray shows the number of units in it and all others below it.

The oil house is a fireproof structure of concrete and steel. In the basement are the supply tanks of oil of various thousand gallons of oil of various kinds. Self-measuring and register-

ing hand pumps handle the issues in which automatically measure and deliver amounts ranging from a barrel to several thousand gallons, handle the oils that are distributed to the line in supply trains.

Aids Other Departments

The vast manufacturing facilities of the Company are located here, producing bar iron and steel; brass, iron and steel castings; cast iron car wheels; tie plates, and many other materials, the total value of which amounts to over four million dollars yearly.

To supply these industries, the Stores Department operates the largest scrap assembling and sorting dock on the system. Here the accumulation of scrap from practically the entire railroad is sorted. Materials which can be placed in condition for further service are repaired and returned to the Store. No item of material that can be repaired is overlooked; even old oil cans and used track spikes are repaired and returned to active service. That which is scrap is sorted to suit the necessities of the various manufacturing plants, or for sale.

The activities of the Department, in

Having read your Bulletin—Pass it along.

112

16 THE BULLETIN

a large measure, are devoted to service to other branches of the railroad. In order to keep closely in touch with the needs of these operating branches, the Stores, wherever possible, deliver materials and supplies direct to the workmen.

At the shops with motor trucks, cranes and cars, they deliver direct to the mechanics all materials ranging from huge locomotive cylinders down to ice for their drinking water.

Monthly Supply Trains

These activities extend beyond the immediate shops, as means are provided for direct delivery to the user on the line of road through medium of the Supply Trains.

From Sacramento each month three large supply trains, carrying materials of every kind, including oils, lumber, etc., move east to Ogden, north to Portland and south as far as Santa Barbara, delivering these direct to the consumer, at the same time affording means of inspection and a personal knowledge of requirements by Division Officials. This feature of personal contact of Supply Department and consumer makes for economy and efficiency.

In addition to the delivery of materials, these trains gather the worn tools and other materials to be repaired and collect the scrap necessary to the operation of the manufacturing industries.

The outstanding feature of the delivery system is that it provides for regulation of quantities of materials stored on the line, reducing this to what is actually required to carry on the work, resulting in important economies, and making it possible to concentrate materials at central depots.

The proper care and economical handling of materials at the central depots, or stores, together with the distribution facilities mentioned, establish control over the material assets of the Company. To many the importance of this item in the expenses of a railroad is not apparent.

Handled Like Cash

The illustrations indicate the attention that is given the care of materials. Stored in open racks in full view; marked plainly with the proper description of every item; the quantity on hand showing at all times. This means that the section stockmen are in touch with each item of the many thousand at all times, and that the million dollars stored in this building is as carefully taken care of as if it were in cash.

While the main building is only a part of the storage facilities within the plant at Sacramento, they give an idea of the extent and nature of the Stores operation. To these buildings it is proposed to add a fireproof magazine for storing inflammable and explosive materials, which on account of their nature, constitute a fire hazard to the shop buildings and the equipment in

them; also a fireproof paint store and mixing room which will contain all the modern equipment to be found in a plant of its kind.

Around these central buildings will be grouped the heavier materials, such as locomotive tires, wheels, castings, lumber, etc., forming a store handling a business ranging from twelve to fifteen million dollars yearly, with a stock averaging four million dollars in value, and where four thousand carloads of material are received and a like number shipped each year.

STORM EMPHASIZES SUBSIDY GIVEN JITNEY LINES

That the recent rainstorm has given a graphic illustration of one of the direct subsidies granted to bus operators using the public highway is the contention in a statement issued by D. W. Pontius, Vice President and General Manager of the Pacific Electric, says the Los Angeles Herald.

"I have just returned from a sixty mile trip over our San Bernardino line, which has been kept in service despite the tremendous handicap of the record rainfall. On this one line alone, which a fair standard for the comparison of the balance of our lines, we have had more than one hundred men employed and thousands of dollars worth of equipment used in checking flood waters and keeping them under control. At other points along our line where the State and County highways parallel it, I was able to look out the car window and see large forces of men, upon the payroll of the State or County, engaged in the work of repairing highways, while the bus men sat idly by awaiting their re-opening."

ROBINSON NEW ENGINEER OF EAST BAY DIVISION

Announcement of the appointment of P. T. Robinson as Division Engineer of the Southern Pacific Company's East Bay Electric Division, effective January 1, 1922, has been by Superintendent J. C. McPherson.

Mr. Robinson began his railroad career in 1900 as rodman-instrumentman in the engineering department of the Union Pacific Railroad. In 1908 he entered the service of the Southern Pacific Company as assistant Engineer, Sacramento Division. In September 1919 he was promoted to Assistant Division Engineer of the Western Division.

J. P. Edwards, formerly Assistant Division Engineer of the Salt Lake Division, succeeds Mr. Robinson as Assistant Division Engineer of the Western Division.

The Record

Howell: "Last night was the hottest night in the year."

Powell: "Not for me; the hottest night for me was when my wife discovered that my pay had been raised and I hadn't told her of it."—New York Sun.

HEADS EAST BAY LINES

Captain J. C. McPherson

J. C. McPHERSON, superintendent of the Southern Pacific Electric lines in Oakland, Alameda and Berkeley from 1913 to 1918 and during the war a captain of engineers overseas, returned January 1, 1922, to his former position after several years with the Pacific Electric at Los Angeles.

In resuming direction of the Company's electric lines in Oakland, Alameda and Berkeley, Captain McPherson took the title of Superintendent of the East Bay Electric Division. He was transferred to this position from Los Angeles where he had been assistant general superintendent for the Pacific Electric Railway, since he returned from the war in 1919.

Captain McPherson was born in Ireland, received a grammer and high school education and came to this country when 16 years old. He entered railroad service with the Santa Fe at Raton, N. M. serving in the machine shops and later as fireman, locomotive engineer and roundhouse foreman.

He came to California in 1895 and started as motorman on the Los Angeles and Pasadena Railway, the first link in the great Pacific Electric System of today. He advanced as that system developed, being Superintendent of the Northern Division, Pasadena, in 1913, when he was made Superintendent of the Southern Pacific electric lines in Oakland, Alameda and Berkeley.

Captain McPherson is well known to business men and residents of the Bay district and to railroad men throughout the West.

When the transcontinental railroad was completed in 1869, the facility in Sacramento was hundreds of miles from any others that could provide assistance with supplying tools, equipment, track parts and locomotive maintenance. The crews in Sacramento needed to be self-sufficient and have quick access to material used to make items. The lost time and expense of having supplies shipped from the East was prohibitive. The photo above shows the large materials yard northwest of the roundhouse. The date is unknown. Even though the Southern Pacific network eventually spread to many other cities in the West during the first fifty years, Sacramento continued to be a major hub. – Courtesy California State Railroad Museum

CASTING CYLINDER REAL FEAT

Huge cylinder constructed at the Southern Pacific Shops at Sacramento for the Ferry boat "Piedmont." 50,000 pounds of cast iron were poured in making the casting for this monster cylinder.

BY R. P. PEEK
Supervisor of Apprentices, Sacramento Shops

NO, not a souvenir of our recent Arms Conference at Washington or a German mortar being scrapped to satisfy the Allied Countries; but just a peaceable cylinder for one of the Southern Pacific Ferry Boats that ply back and forth across the San Francisco Bay, the Ferry "Piedmont."

Little did you realize that while you were standing upon the decks of this spotless and comfortable boat that below such an enormous piece of machinery was necessary to make your trip possible; but if you were permitted in the engine room of this boat you would find this monster doing its daily work of driving the graceful hull of the Piedmont through the beautiful green waters of San Francisco Bay.

This cylinder was recently completed at the Southern Pacific General Shops at Sacramento, where the patterns for the sweeps and ports were made. The cylinder was cast at the General Foundry and the Machine Shop gave it the finishing touches before it was shipped to the ship-yard to be installed.

The pattern work for this casting was very little compared to the work necessary to build the mould in which

this monster was cast and to give an idea of the magnitude of the undertaking, I will outline a list of the materials used in the manufacture of the mould.

Immense Casting

The weight of the cast iron that was required for this casting to be poured was 50,000 pounds. This being too heavy to be handled on the cranes in the foundry, and to save the special manufacture of ladles or containers that would hold this amount of metal, made it necessary to build a large reservoir into which the molten metal was poured until a sufficient amount was on hand to pour the casting.

The mould itself was the undertaking that required the best moulding talent in the country and which the Foundry here can readily boast of, having probably more men capable of handling this class of work than any other foundry west of the Mississippi River. In constructing the mould it was necessary to use 7,092 bricks. It was built up in a manner that a large stack is built, which is the secret of the success of this casting. To make the mould strong enough to with-stand the enormous strain placed upon it

when the molten metal was poured in and yet be elastic enough to give when the casting set and the shrinking started was the problem. The cylinder was 18 feet in length and 56 inches in diameter and the thickness of the metal in the barrel of the cylinder was 3 inches. To get the mould into position so the metal could be poured it was necessary to excavate the floor of the foundry to the depth of 22 feet. After the mould was set in this large hole and properly clamped together the earth was solidly rammed around it to help the walls of the mould withstand the enormous pressures.

Great Care Necessary

It was an exciting afternoon when everything was in readiness to let the many tons of moulten metal into the mould, for the slightest oversight of any one of the workmen or foremen at this time would mean the loss of thousands of dollars to the Company. At the word from the foreman the gates were opened and the molten metal rushed into the moulds and in just 1½ minutes the mould had been filled and another great monument for the moulders who made it had been completed. The hard work did not end here for as soon as the metal had set it was absolutely necessary to remove several bricks from each row the full length of the cylinder to insure themselves that the casting would not be lost in cooling.

The casting was left in the pit for several days, until all strains of shrinkage had taken place before 'it was taken from the Foundry to the chippers to be cleaned so that it could go to the Machine Shop for its finishing touches.

A special boring bar had to be manufactured to finish out the inside of this monster and the best of mechanics was placed on the finishing job. A slight error in machining would have meant the loss of many days of foundry work and thousands of dollars in delay of holding up the ferryboat at the shipyard.

Much credit is due to Foundry Foreman J. Geiger and his assistants J. S. Gageby, G. Derman and F. Miller, and the many moulders who participated in building this working monument; also to Machine Foreman G. A. Holmes and his assistants and mechanics for the creditable manner in which the casting was finished, ready for installation in the ferryboat at West Oakland Shipyard.

Slight Miscalculation

A wealthy American's aunt had died abroad and he cabled for the remains to be sent to him in this country. Great was his astonishment when the undertaker whispered to him that the body received was that of a man in the uniform of a high ranking officer. After the passage of additional cables, he received this one from some G. H. Q.:

"Keep the General. Your aunt has been buried with full military honors."

THE BULLETIN

Sacramento Boiler Shop Now Great Factory

In 1867 Little Shed Housed Its Force of Four Men; Now Great Plant Turns Out All Classes of Plate and Structural Work

BY F. J. HICKEY
General Foreman of Boiler Shops, Sacramento

GOING back to the old days of the Sacramento General Shops, we find that the Boiler Department was established in the year 1867. The shop was a small shed, at that time situated on the present site of what is now known as Sixth Street. The force of mechanics in this department, at that time, consisted of four men, under the supervision of J. Hall, Foreman Boilermaker, whose duties were to keep up the running repairs on the small locomotives then in service. These locomotives were shipped around the "Horn" in sections and assembled at Sacramento Shops.

In 1873 a new Boiler Shop was constructed on the site of what is now used for the transfer table between Boiler Shop and Machine Shop. This department was under the supervision of Charles Shields. It was at this time that the Boiler Department began to expand and locomotive construction was started. This development eventually bringing the present great boiler shops into being.

The boiler of the Steamer "Yosemite" in 1876, followed by the ferry steamer "Solano's" boilers, eight in number, built under the immediate supervision of Charles Hooper, General Foreman Boilermaker and J. M. Dunigan, Assistant Foreman, lasted for many years, being replaced by new boilers in 1905. These eight new boilers were built in the present Boiler Department, under the immediate supervision of James Dunigan, who at that time was General Foreman Boilermaker. Following the construction of the first battery of boilers for the "Solano," the next large job was that of constructing the boiler for Engine 237, known as the "El Gubernador," which was the largest locomotive constructed at that time. Its construction was wonderful work and those of that time thought that many years would elapse before a larger boiler and engine would be constructed, but now it would be comparable with the little engine in the station park to one of the present-day engines.

Important Activities Begun

Great activities began in all the shops about 1885. Many boilers and engines were built under the direction of A. J. Stevens, who was General Master Mechanic of the Central Pacific Railroad. At one time the Boiler Department had orders totaling fifty-two boilers. This work was under

Large boiler built at Sacramento Shops for the steamer "Thoroughfare." The white circles surround the government inspector's stamp put on the different sheets at the mill where they were rolled.

the supervision of James Dunigan, assisted by M. J. Desmond, who was then acting in the capacity of Assistant Foreman and Layer-out. Mr. Desmond was later employed by the City of Sacramento as City Clerk for 26 years, having left the service of the Southern Pacific Company to accept that position. He died last December.

Beginning in the late "Nineties," the Boiler Department again began to show activity. In 1898, new boilers for the Steamers "Apache" and "Modoc" were constructed, these boilers being of the locomotive firebox type. Since this time the following floating equipment has had new boilers installed, which were constructed at Sacramento General Shops Boiler Department: Steamers "Bay City," "Garden City," "Melrose," "El Capitan," "Encinal," "Navajo," "Seminole," "Cherokee," "Contra Costa," Tug "Collis," "Newark," "Oakland," "Piedmont," and "Transit." A number of these boilers were built under the supervision of James M. Dunigan who passed away in 1906, being succeeded by Geo. Woodall, January 1, 1907 as General Foreman Boilermaker and F. J. Hickey as Foreman Boilermaker of Boiler Shop No. 1. Mr. Woodall held the position of General Foreman Boilermaker until December 31, 1919, when he was retired on pension and was succeeded by F. J. Hickey, as General Foreman Boilermaker and F. W. Strauchauer as Boiler Foreman of Boiler Shop No. 1, succeeding Mr. Hickey; T. R. Barrett, foreman of boiler work in Erecting Shop.

Real activities began the latter part of 1917, when the building program for boilers of locomotives of the larger type began together with the fabrication of a vast amount of car work for new cars which were being built at this time. At this time the work was under the supervision of George Woodall, General Foreman Boilermaker, assisted by F. J. Hickey, Foreman of Boiler Shop No. 1. This program has not been deviated from, to any great extent, and we are looking forward to greater activities in the line of construction in the way of the heaviest class of power. It should be understood that the Boiler Department is not strictly confined to boiler construction alone but is doing all classes of plate and structural work, sixty per cent of the shop's output being other than boiler work.

We have now completed the construction of ten S-12 class switch engine boilers and tender tanks, also steel work for new car construction at Sacramento and Los Angeles, still leaving four S-12 boilers and tanks to be built.

Tube and Cab Shops

Leaving the Boiler Shop, we will now go to the Tube Shop, a separate department, although under the immediate supervision of the Boiler Department. Here is where all tubes taken from the boilers are cleaned, new ends welded on and tested before being again applied to the boiler. Not only are the tubes from boilers taken care of here but also from many outside points on the line. We have a reclamation shop, where all flues that are serviceable but too short, are welded together and used over. This reclamation process, from August, 1919 to date, represents a saving of $110,-000.

From the Tube Shop we will go into the Steel Cab and Pilot Shop. Here is where all cabs and pilots are made, together with all other light sheet steel work, such as breechings, petticoat pipes, etc. The pilots are all made of old tubes which in the past were scrapped but are now utilized for this work. This shop is also used for general sheet metal work, from 1-16" to ¼" in thickness. heavier work being done in the Boiler Shop proper.

Modern Shops; Fine Employes

The Boiler Shop of the Sacramento General Shops is equipped with the most modern tools and machinery and is capable of doing any and all classes of boiler, plate and structural work and is considered the most up-to-date and modern shop west of the Rockies.

The names of a large majority of the old-time employes of the Boiler Shop are enscribed on wooden headboards or memorial stones. Sturdy rivet boys have become men of families. A few of the mechanics left of the group are grey-haired grandfathers but one thing could always be said of the old shop, namely: those that entered it at the time, stayed to the finish and are numbered among the stable citizenry of the City of Sacramento. Their children and their children's children have chosen this city as their home for all time and have come to be a credit to their forbears and honorable in the community where they reside.

All Come Back to Oregon!

A woman rushed up to one of the ticket clerks at the Union Station, Portland, the other night and demanded a round trip ticket. quick.

"Where to?" inquired the ticket agent.

"Back here—where do you suppose?" shrieked the woman.

Harding on Highway Problem

President Says Motor Truck Should Be Railroad Feeder Instead of Destroying Competitor

RECOGNIZING the vital necessity for increased transportation facilities in the United States and pleading for the elimination of destructive motor truck competition, President Warren G. Harding, in his message read before a joint session of Congress on Friday, December 7th, asked Congress to give the subject careful and constructive consideration.

"I know of no problem exceeding in importance this one of transportation," declared the President. "In our complex and interdependent modern life transportation is essential to our very existence. Let us pass for the moment the menace in the possible paralysis of such service as we have and note the failure, for whatever reason, to expand our transportation to meet the Nation's needs.

"The census of 1880 recorded a population of 50,000,000 In two decades more we may reasonably expect to count thrice that number. In the three decades ending in 1920 the country's freight by rail increased from 631,000,000 tons to 2,234,000,000 tons; that is to say, while our population was increasing less than 70 per cent, the freight movement increased over 250 per cent and yet find it inadequate to our present requirements. When

The Track Runs On

IN the moonlight we see the railroad track stretching away for a little distance and then disappearing in the darkness. Yet we know that it reaches on and on to far cities beyond the limits of our vision.

Why? Because we have seen it in the daylight.

Likewise we have our moments of inspiration when we can see far conclusions. These moments are followed by others of darkness and doubt.

Then there are clouds of fear and winds of uncertainty and the deceptive moonlight of illusion. We are uneasy and distressed.

It is in these dark moments that we should remember the days of vision and keep in mind that, although we cannot see, yet we know that "the track runs on."

Most of the time, indeed, we have doggedly to plod along an obscure and uncertain path, guiding ourselves by the knowledge of what we once saw in an inspired glimpse.—San Francisco Examiner.

we contemplate the inadequacy of to-day it is easy to believe that the next few decades will witness the paralysis of our transportation-using social scheme or a complete reorganization on some new basis. Mindful of the tremendous costs of betterments, extensions, and expansions, and mindful of the staggering debts of the world today, the difficulty is magnified. Here is a problem demanding wide vision and the avoidance of mere makeshifts. No matter what the errors of the past, no matter how we acclaimed construction and then condemned operations in the past, we have the transportation and the honest investment in the transportation which sped us on to what we are, and we face conditions which reflect its inadequacy today, its greater inadequacy tomorrow, and we contemplate transportation costs which much of the traffic cannot and will not continue to pay.

"Manifestly, we have need to begin on plans to co-ordinate all transportation facilities. We should more effectively connect up our rail lines with our carriers by sea. We ought to reap some benefit from the hundreds of millions expended on inland waterways, proving our capacity to utilize as well as expend. We ought to turn the motor truck into a railway feeder and distributor instead of a destroying competitor.

Motor Transportation

"It would be folly to ignore that we live in a motor age. The motor car reflects our standard of living and gauges the speed of our present-day life. It long ago ran down Simple Living, and never halted to inquire about the prostrate figure which fell as its victim. With full recognition of motor-car transportation, we must turn it to the most practical use. It can not supersede the railway lines, no matter how generously we afford it highways out of the Public Treasury. If freight traffic by motor were charged with its proper and proportionate share of highway construction, we should find much of it wasteful and more costly than like service by rail. Yet we have paralleled the railways, a most natural line of construction, and thereby taken away from the agency of expected service much of its profitable traffic, for which the taxpayers have been providing the highways, whose cost of maintenance is not yet realized.

"The Federal Government has a right to inquire into the wisdom of this policy, because the National Treasury is contributing largely to

FACILITIES ADDED AT SACRAMENTO SHOPS

New facilities added to Southern Pacific shops at Sacramento. Lower left, oxy-acetylene plant showing two carbide storage houses, generator house and oxygen manifold house and upper left, five 300 pound acetylene generators, a part of the new plant. New planing mill is shown in upper right; below it is one of the planing machines with motor drive and safety guard.

BROOKLYN STORE FORCE IS TREATED TO LUNCH

A farewell luncheon was given to the men of the office force of Brooklyn District Stores during the noon hour March 10, by the girls of the same department, the occasion being the transferring of part of the forces to the Superintendent's Office, Portland, account of recent consolidation with Portland division store.

The girls tempted the gentlemen with choice viands and, being leap year, are hopeful of results. After the luncheon, dancing was enjoyed.

On behalf of those leaving, Theodore J. Richter thanked Mr. Dobbs for his helpfulness and co-operation at all times with those under his jurisdiction.

The luncheon was in charge of the Misses Marie Borns, Sue Lewis, Monetta Monilaws, Helen Harkson, Bertha Fountaine, Lorea Cayotte, Ruth Hansen and Marie Lowes.

"The next stop is yo' station," said the porter on Number One to a passenger as they neared Cincinnati. "Shall I brush yo' off now?"

"No," replied the passenger. "When the train stops I'll step off."—Exchange.

By D. S. WATKINS
Superintendent, Sacramento Shops

A NEW oxy-actylene plant is now practically completed and a new planing mill is fast approaching completion at the Sacramento shops. These additions will greatly add to the facilities at the large shops and are in line with general policy of the Company toward keeping a step ahead in meeting transportation needs.

Acetylene gas is manufactured in the new plant and piped to all parts of the shop grounds, doing away with the portable generators or the cartage of tank gas to the job. In the acetylene generating house five 300 pound generators are being installed. The oxygen is distributed from manifolds so that the man doing cutting or welding has only to carry his torch and hose around with him, coupling on to the outlets which are placed at convenient points. The oxygen is purchased in steel tanks into which it has been compressed at a pressure of about 2000 pounds per square inch. These tanks are attached to the manifolds and the pressure reduced to 75 pounds per square inch for delivery into the pipe lines. On the acetylene line the pressure is 15 pounds per square inch.

The new planing mill is practically completed and the installation of machinery is now being made. A notable improvement will be that all machines will be operated by individual motors instead of being driven from a shaft. The exhaust system is also very complete, so that all sawdust and shavings will be taken away from the machines and the floor without having to be handled by hand. In the old planing mill, which was one of the original shop buildings erected during early construction on the Central Pacific, long timbers, such as car sills, were hard to handle, having to be switched around for different operations. In the new mill this work enters one end of the building and is not stopped to be turned, but goes straight through and is delivered ready for use at the other end of the building. The building is 126 feet wide by 360 feet long.

H. E. BARBOUR HEADS SOCIAL CLUB AT BOWIE, ARIZ.

Homer E. Barbour, manager of the Southern Pacific hotel and dining room at Bowie, Ariz., has been elected president of the "Good Times Club" of that city. The club will promote dances and other social activities which will be held at the Southern Pacific hotel.

S. P. PAINT SHOP FIGHTS H. C. L. WITH OWN PRODUCTS

SACRAMENTO PLANT OPERATES ON BIG SCALE

By JAMES HALL
(Master Car Repairer, Sacramento)

Not many employees appreciate the vast amount of material and labor that is required to keep the Sacramento general passenger car paint shop working at capacity.

This department is supervised by Foreman Painter W. B. Oldfield, and on his staff there are two assistants, C. E. Ball and J. D. Dannaher.

One hundred and thirty-five painters are employed in this department including helpers. The average output runs from 1000 to 1200 cars per year. This includes renovating, general, heavy, light and running repairs. All classes of private, business and passenger train cars are included in these figures.

In addition to this car output approximately 1100 store orders are handled per year. These orders call for the manufacture of many kinds of paints, soaps and car cleaners, and the painting of hundreds of articles manufactured by the many other departments of the general shops, such as doors, sash, desks, chairs, tables and furniture.

In the large mixing room there is mixed yearly many different kinds of paints in large amounts. The following figures of a few of the paints mixed will give an idea of the volume handled:

Special primer .... 400 gals.
Varnish, Pullman .... 156 "
No. 99 drab color .... 720 "
Station car color .... 960 "
Lead putty .... 8800 lbs.
Flat white .... 900 gals.
Flat Pullman .... 800 "
Frontene .... 6200 "
Renovene .... 2200 "
Engine renovating oil .... 600 "
Rubbing oil .... 300 "
Truck color .... 900 "
Substitute structural black 2760 "
"Saclene" car cleaner...... 10,000 "
Metallic .... 36,000 "

In the varnish room there are handled yearly over 4500 doors, 32,000 sashes, 8400 deck sashes. More than 10,000 glass of all kinds; plain, art and leaded glass and mirrors are fitted and glazed. Leaded glass is also manufactured in this department.

It might be well to mention the fact that there are also manufactured here annually over 500 mirrors of all kinds and sizes. Quantities of art glass ordered by the Store Department and needed all over the system are supplied here. Such glass is ordered by blue print in 438 different designs.

The rough green glass used in the outside gothics is a glass manufactured in the East, and at times is very hard to secure when most needed. The supervising force of this department has devised a plan whereby a large

saving is effected with substitute glass made from the old gothic glass removed from cars and glass removed from discarded steel sash. This glass is first sand-blasted and then dyed something near the color of rough green glass. It is a much more durable glass than the rough green, as it is made of plateglass. The rough green glass costs more than $2 and the substitute glass is all made from old scrap glass.

The paint mentioned as substitute structural black, of which 2760 gallons are manufactured here per year in the mixing room, is all made of old paints, oils, varnishes and the scraps of same ground up and boiled down to a very fine substitute black. This makes a large saving per year for this department.

Another noticeable thing in this department is the large varnish removing iron tank which holds when three-quarters full about 500 gallons of varnish remover. This liquid costs approximately $2.75 per gallon.

When removing old varnish and paints from sash, doors, blinds or deckscreens a whole set for one car can be placed in this tank and left over night and removed in the morning cleaned and free from all paints and varnishes, thus saving many hours of labor.

On the average dining car more than 575 parts are scrubbed, paint and varnish removed and repainted, finished and varnished and parts polished. Business and private cars have a greater number of parts than dining cars.

On coaches and chair cars there are more than 425 parts to be finished. This does not include any part of the general body of the cars.

From these very few figures given here, one can readily see why this shop is said to be the largest of its kind west of Chicago.

Foreman Painter W. B. Oldfield (center) and his two assistants, C. E. Ball (at left) and J. C. Dannaher.

ILLUSTRATED BOOKLET FEATURES NEVADA.

An illustrated booklet about Nevada, for distribution among homeseekers, has been issued by the Agricultural Section of the United States Railroad Administration. The information was compiled by Charles A. Norcross, director, Agricultural Extension Department, University of Nevada, assisted by the representatives of the railroads operating in that State.

The publication has the endorsement of Governor Emmet D. Boyle, who says:

"I say, without hesitation, that the statements of fact in this book are reliable, and that whatever there may be of prophecy in it is based upon the most conservative judgment. Nevada will welcome new population, and would regard it as an injury to herself, no less than to the prospective settler, if she should, by over enthusiastic statements, induce persons to come here without having a fair chance for prosperity and happiness."

Another article is by Chas. S. Knight, dean of the College of Agriculture, University of Nevada.

This booklet contains thirty-five pages of dependable information regarding topography, climate, schools, roads and highways, transportation, principal crops, markets, grazing lands, dairying and poultry. There are special chapters on Nevada's land settlement policy, Newlands' reclamation project, dry farming and live stock department.

OH, THESE LAWYERS!

Judge—Are you positive the prisoner is the man who stole the horse?

Witness—I was, your lordship, till that lawyer cross-examined me. Now I'm not sure that I didn't steal it myself.

Here's the Start

Wheel trucks shoved into place.

Steel underframes fitted over trucks.

Then comes welding, spotting and reaming.

Rivet tosser is a sharp-shooter.

Rat-a-ta-tat of the rivet machines.

Wm. J. "Bill" Bartle (right) general foreman, Freight Car Dep't, Sacramento Shops, and his supervisorial staff at Car Shops No. 9. To his right are: W. B. Butler, gang foreman in charge of building new stock cars; E. H. Sachs, gang foreman in charge of light repairs; Jas. G. Van Baaren, gang foreman supervising building steel sheathed box cars, stripping and dismantling cars, and steaming, cleaning and repairing tank cars; and Walter L. Williams, departmental foreman directs new car fabrication.

THINGS are humming these days at Sacramento Car Shops No. 9.

There hasn't been so much activity in 6 years as is under way now on the 60 acres that comprise the expansive and capably organized freight car construction and repair shops.

Here 325 men, most of them freight car builders of many years experience but who have had mighty little work of this kind to do for some time, have quickly caught their oldtime strides and are swinging right along with the job Sacramento Shops has to do as a part of the Company's $8,000,000 program for building 2,950 new freight cars of various types.

Jobs for 140 New Men

The increased shop work has been particularly welcomed by the more than 140 men, many of them former employes, recently given jobs in the freight car forces.

Of the 2,950 new freight cars, Sacramento Shops will build 150 stock cars, 200 steel flat cars, and 100 steel gondola cars, a job that will cost about $907,000.

Orders for construction of the remainder of the new freight cars, to include 1,750 box cars and 750 automobile cars, all steel sheathed, have been placed with various car manufacturing companies throughout the country. The whole program is being pushed right along in order to

have the equipment ready for service as soon as possible.

The job of major importance at Sacramento Car Shops No. 9 right now is the construction of the stock cars that will be particularly needed for the fall movement of cattle.

A new system for building the cars has been worked out. It's a "progressive" plan that moves the cars along to the man, instead of the man and materials being moved back and forth to the cars.

Under this plan, car ends, sides and roofs are assembled separately on tables or jigs. After these separate steel and wood units are built they are lifted into their places on the car frame by means of traveling cranes.

Speed and Safety

This method does away with handling quantities of materials over the heads of the men. Nor are the men required to climb so much on stagings or scaffolds where there is always the hazard of falling or of materials dropping on someone. There's greater efficiency and safety all around.

While the ends, sides and roofs are being constructed close to materials and where the men can work while standing on the ground, the trucks and underframes are being assembled at the upper end of the working track.

On this working track there are 11 distinct positions which when completed find the brand new, freshly painted car at the lower end of the

Car ends are fitted up separately.   *Sides quickly fitted on jigs.*   *Hammers beat tattoo nailing the roof.*   *Sides moved into position.*

track ready for final inspection and an official O. K.

The job positions are passed through somewhat in this manner:

In the first three "spots" the steel underframes are fitted over the wheel trucks; which trucks had been previously assembled at the shop's own Truck Plant.

Then there follow two "spots" where body bolsters are fitted to the under frames; where welders with electric and acetylene torches weld the center sills of the underframes; and where torches are also used to spot and start the holes for the reamers.

Rat-a-tat-tat of Riveters

Then come the fellows with the powerful pneumatic drills that bore the holes for the riveters, who in turn set up a din of racket with the rat-a-tat from a battery of power riveting machines. The fellows that heat the rivets to a red-hot glow are as accurate in their tossing as the receivers are in their catching. These shopmen are all careful, as well as efficient workmen.

With the underframe welded and riveted, this part of the car is moved along the working track to where the sides, ends and roof are swung into position by the traveling crane. Then comes more spotting by the torches; more reaming, as the car takes definite form and is moved further along the track.

In the succeeding operations the flooring is laid; all safety appliances

And when the new cars are OKed for road service here is one of the yard crews that sends them on their way from Car Shop No. 9. Left to right: F. E. Urban, fireman; T. K. Schulze, engineer; E. T. McDonald, yardman-brakeman; Wm. L. Lewis, yard foreman-conductor; and Jas. Warisch, yardman-brakeman.

put on; couplings and air line hoses fitted in; side bearing clearance adjusted; and the car completely finished for painting and stenciling.

It takes only 25 minutes to spray the whole 40-foot 6-inch car with the three-jet spray apparatus that was specially designed at the shops. Two coats of red paint are applied and then the car is adorned with a big white S.P., a number and all other customary numerical designations.

All metal parts used in construction of the cars have been manufactured at the Sacramento Shops, with exception of the castings, such as coupler striking castings, door hangers, center plates, front and back draft lugs. The timbers are all cut and shaped in the shop's big mill.

Five Cars a Day

An average of four stock cars are now being completed daily and it is expected that five will be turned out when all phases of the progressive operation begin functioning with maximum smoothness. The stock cars will cost approximately $1800 each.

In addition to this new construction and the continuance of ordinary repair work, the freight car shops are applying steel sheathing to 100 former wood boxcars, and installing Evans Auto Racks in 200 automobile cars. The latter job averages about $757.50 a car, and the steel sheathing will cost about $500 a car.

Here's the Finish

Here it is ready to give a steer a ride.

Car designated by many stencils.

One coat of paint in 25 minutes.

Sliding door fits in place nicely.

Brake wheel and coupler given attention.

End drops snugly into place.   Roof for 40½-foot car is good sized load.   Nailing the flooring is next.   Side bearing clearance is adjusted.

CHAIRS: *Skill of S. P. Craftsmen Shown In Furniture Built for New Type of Lounge Car*

"These lounge car chairs are so comfortable. Their rich finishing and upholstery would do justice to any club, hotel lobby or well furnished home."

Remarks such as this are typical of the reactions of patrons to the high quality of the furnishings that enhance the attractiveness of Southern Pacific's new de luxe lounge cars now operating on the Sunset and Lark.

And it's true! For into the construction of chairs and lounges go the finest of materials and workmanship. A brief tour of the "birthplace" of this furniture, the Sacramento Shops, bears out the statement.

First there's the cabinet shop. Here are stored hundreds of feet of birch and ash lumber used in the construction of lounge, dining, club and chair car furniture. Most of the birch, which is used for all of the exposed parts, is shipped from Michigan and Wisconsin. Ash comes from Mississippi, Tennessee, Arkansas and other Southern states. It is used in the construction of the "under cover" parts.

One section of the cabinet shop is devoted to a good-sized lumber mill where, under the expert supervision of Wallace Glasford, legs, arms, seats and backs for chairs are roughed and framed. These parts are then sent to the cabinet makers for assembling and finishing.

Skilled Artisans

The cabinet makers, 16 of them, are skilled artisans, many boasting more than a quarter of a century of experience in this work; for the history of the cabinet shop hearkens back to the days of the all-wooden cars, when it was one of the major departments of the Sacramento Shops. Among other

Picture saga of a lounge car chair: 1— Arthur Soares shaping up top rail of chair on band saw. 2—Jerome Barry chiseling a mortice on machine that bores square holes. 3—John Harrison shaping legs at the shaper. 4—Richard Kirchner finishing frame work. 5—Robert Schu and Richard Crow covering chair cushion and back. 6—Laura Renwick, one of the seamstresses. 7—Louis Gutenberger, right, foreman upholstery shop, and Steve Guzules fitting an arm covering. 8—John Andrade hauling finished chairs to the lounge car.

things, the cabinet shop turns out frameworks for tables, dressers,—in fact, any and all articles of furniture incidental to passenger car equipment.

In addition to the cabinet makers, Foreman George Scott has a force of three helpers, nine mill hands, four apprentices and two laborers. The average time necessary to complete a chair framework is twenty-five hours, after which it goes to the paint shop.

Work in the paint shop consists of the application of a walnut stain and two coats of varnish. This represents a three-day process, allowing sufficient time for drying.

Goose Feathers

When the chair reaches the upholstery shop, certain preparations are necessary before actual upholstering is commenced. The bottom of the chair must be webbed, springs set firmly in place, and a covering of canvas applied. A border of plush is tacked around the edges and the seat is covered with heavy lining on which a reversible cushion rests. The cushion is stuffed with goose feathers and covered with a plush material, attractively patterned in shades of tan, green and rust, designed exclusively for Southern Pacific. The back of the chair is also webbed, edged and covered with sheeting preparatory to donning its "dress." Then the arms are covered and the upholsterer's work is done. Chair upholstering requires from 15 to 17 hours, not including two hours necessary to make a cushion.

The upholstery shop personnel includes a foreman, four seamstresses, twenty-two upholsterers, three regular apprentices, six helpers and one laborer. Some of their many other accomplishments are the making of locomotive seat cushions and curtains, brakeman seats, back rests, arm rests, canvas water spouts, awnings, tarpaulins, vestibule curtains, caboose cushions, and carpets for club, dining, chair and business cars. The carpeting, by the way, is also designed especially for Southern Pacific's use.

Foreman Louis Gutenberger furnishes

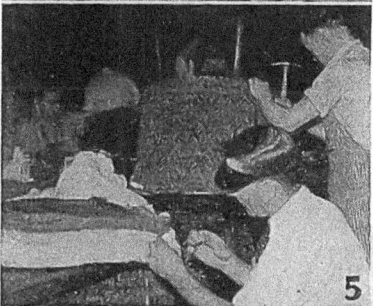
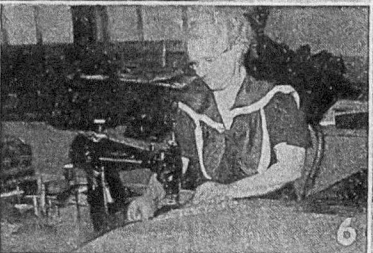

All the furniture used in the Company's newest air-conditioned lounge cars in service on the Sunset Limited and Lark, was built at the shops in Sacramento, where skilled artisans also rebuilt the interior of the cars and installed many new facilities designed to provide greater rest and recreation for passengers. A feature of the new full-length lounge cars is the refreshment stand shown in above picture.

some interesting statistics concerning the carpeting and upholstering in the nine lounge cars, just recently completed. Each car contains 19 chairs and 4 lounges, and one hundred yards of carpeting, representing a total of 1400 square yards of material for the nine cars, at an approximate cost of $5,250.

Add to this the fact that it takes about eight days to complete one chair, and you'll realize it's no small task furnishing one of these "traveling hotel lobbies."

BASEBALL: *S. P. Stores Team In Fight for No. Cal. Title*

Interest of San Francisco bay district baseball fans is centered on the semifinals of the Oakland Tribune League to determine the championship of Northern and Central California. Southern Pacific West Oakland Stores team, which in three previous years has placed among the five winners, is again in the spotlight as a favorite to repeat. With a new pitching staff and the acquisition of other fine talent, Manager Gerichten has an aggregation that will be hard to beat. Two finalists from twenty-four entries will meet for the championship at Oakland Coast League Ball Park, Sunday, Sept. 13.

SOLVED: *No Mystery Now to Visits of Woman to Station*

The well-known story of the man who stood on a street corner gazing at the sky and drew a crowd of curious passers-by, recently had its counterpart in a humorous incident at El Paso.

According to a United Press report,

a woman has been meeting the trains almost daily for 17 years. A tiny figure in ragged coat, she would arrive in the morning, before the Golden State and Sunset Limited rolled in; and would remain until late at night, peering eagerly through the gates at the stream of passengers alighting and departing.

Speculation ran rife as to the reason for her vigil. Was she waiting in vain for a son who, perhaps, was killed in the World war, or a long-absent sweetheart or husband? Investigations shed no light on the subject until recently when the aged woman ended her long silence.

"I'm a widow and have never had any children," she finally confided. "Why do I come here? Because I like to watch the trains. I know my own business. I don't bother anyone."

And so, El Paso's 17-year mystery has been solved.

EDUCATION: *Rail Institute Resumes in S. F. on Sept. 15*

Commencing September 15, sessions of the Second Railway Educational Institute will be held each Tuesday in the auditorium of the Pacific Gas & Electric Building, 245 Market Street, San Francisco, from 5:10 to 5:50 p. m.

The Institute will continue until December 1, and will be conducted along the same lines as the one held last year which was so well attended by railroaders and others interested in rail transportation.

Held under the auspices of the Pacific Railway Club and sponsored by Southern Pacific, Santa Fe, Western Pacific, Market Street Railways, and the Key System, the expense of the Institute is borne by the sponsors and there is no expense to those who attend the sessions.

At the September 15 session, Stuart Daggett, professor of transportation at the University of California, will deliver an illustrated address on the History of Transportation. Lecturers specially versed will present the succeeding subjects, which will include: The Economics of Freight Transportation, The Economics of Passenger Transportation, The Economics of the Railway Dollar, and similar topics.

Dennistoun Wood, engineer of tests of the Southern Pacific, is a member of the Directing Committee, and William S. Wollner of the Northwestern Pacific, who was Director of the First Institute, has been appointed to the same position in the Second Institute.

Those wishing to enroll in the course should communicate with Mr. Wollner at 269 Southern Pacific Building, San Francisco.

WHO'S SLOW? David Johnson, Ashland, Oregon, vouches for this one: When asked by Johnson if he knew what Sloe Gin was, a Mexican boy working on a section gang at Steinman replied: "Sure, it's the Chinese cook with the B&B gang at Steinman!" And, to quote Johnson, "he made the remark in dead earnest."

EARNINGS: *S. P. Net Railway Operating Income Shows Gain During First Seven Months*

For the seven months ended July 31, 1936, Railway Operating Revenues of Southern Pacific Lines amounted to $107,386,430 or $17,770,630 over the same period for 1935, an increase of 20 per cent.

During the same period Railway Operating Expenses were $81,930,364 or $11,867,901 over the same period for 1935, an increase of 17 per cent.

After deducting Tax accruals of $7,785,317, net rentals paid for use of equipment and joint facilities totaling $5,490,878, there was left Net Railway

Operating Income of $12,179,871, or $4,129,306 over the same period for last year, an increase of 51 per cent.

Net Railway Operating Income is the amount left after paying the expenses of operating the property as shown above. From this must be paid interest on bonds and Equipment Trust notes in the hands of the public, interest on the borrowed funds, as well as improvements to the property and dividends, if any. No dividends have been paid, however, since January 2, 1932.

Chapter Nine
Spirit of Employees

During the very early days of the Shop's history, the education of a new employee was on-the-job. Then, as the complexity of tasks increased, classroom training became more common. Sacramento was one of the class locations because it was one of the largest yards in the West where almost any kind of railroad work was completed. Some of the training programs were as long as four years and enrollment was not limited to new recruits.

There was a challenge in training employees who were often in remote towns hundreds of miles from Sacramento. The railroad's solution was an instruction car that traveled around the system. The car seated thirty and was designed to provide information about dispatching and engine work. So, the use of this mobile classroom brought instruction to the employee rather than having the employee travel to Sacramento. The lessons were also meant to update established employees on new rules and techniques.

The work environment in Sacramento also included choirs, sports teams, and bands comprised of employees that traveled to other railroad facilities to perform and compete. Singing, performing in a band and playing sports provided after work activity rather than watching television, playing computer games, and using apps on smart phones. As with any work environment, friends are made and assist each other outside of work.

Southern Pacific's *Bulletin* had many articles that depicted employees in activities not directly related to railroad work. Some of those articles follow. Also included are articles about apprentice programs in Sacramento.

"Shopmen Paint Home of Veteran," May 1924

"Veteran Shop Craftsman Proud of Skill in Foundry," May 1919

"Sacramento Employees Form Important Industrial Group," April 1919

"Musicians at Sacramento Win Recognition," February 1922

"Speedy SP Team at Sacramento," September 1923

"New Instruction Car is an Academy on Wheels," September 1919

"Southern Pacific Instructors Teach Apprentices Eleven Trades," March 1920

"Apprentice Schools Have Enrollment of 638 Young Shop Men," May 1925

Painting and Unknown Hazards

In the early 1920s, the Southern Pacific's paint shop in Sacramento had enough rail to hold twenty three cars. The cars were renovated, cleaned and varnished. On wood coaches, worn paint was removed by burning. Paint on steel cars was removed by sand-blasting. For many years, locomotives primarily used on passenger trains were made attractive with bright colors. But, in the early 1920s, a plain coat of black was the custom.

In one year, ten thousand gallons of paint were mixed for the just the Maintenance of Way Department.

But painting was not the only task. That department was also the location where furniture, glass, mirrors, and signs for all purposes were made, and the workers there had many more skills than just using a brush. In one year, over five hundred mirrors were produced for all parts of the system. In addition to the car paint shop where all passenger cars were handled, there was a small paint shop, which took care of the painting of locomotives and tenders.

Some reports indicate that the Southern Pacific began using spray-painting in the late 1880s. By the end of the ninteenth century, one coat could be applied in about twenty-five minutes. Due to the known dangers, some supervisors in the industry required painters to fasten sponges over their mouth and nose.

At a facility that was large like Southern Pacific's in Sacramento, the volume of paint used was tremendous. Railroad cars, shop vehicles, locomotives, furniture, shop buildings, and tools needed to be preserved with paint.

Imagine an employee who was on their way home from work after painting all day on an August day in 1924. On the trip home, the idea of spending the upcoming weekend painting could not have been appealing. But in 1924, a former foreman painter, Mr. Oldfield, and his wife needed their house painted and were treated to help from at least two dozen painters from the paint shops who went to work on the Oldfield's residence.

The show of friendship involved not only giving time and effort to the task but also exposing themselves to the usual known and unknown safety risks involved, such as climbing a ladder to the second floor. The friends of the Oldfields might have used lead paint which eventually was regulated strictly. When the men painted Mr. Oldfield's house, they did not know about this hazard. Regardless, on March 30, 1924, over two dozen shop employees worked at the Oldfield's. With quality as a top priority, the crew came back the following Sunday for the finishing coat. This reveals the fraternity among the workers at the Shop.

BOARD OF INQUIRY REPORTS PROVIDE OBJECT LESSONS

Collision—two engineers, a conductor and a rear brakeman, through failure to comply with rules governing rear end protection and control of train in yard limits, were charged with responsibility for passenger train colliding with rear end of a freight train. Freight train crew failed to protect rear end of their train when they had stopped on main track at water column, in spite of rule covering protection and the fact that they were on time of a superior train. Passenger train engineer failed to exercise care in observance of yard limit control rule, especially in view of fact that track ahead was somewhat obscured.

Fire—collection of charcoal dust and other refuse in charcoal bin of car was the origin of a fire which caused $1800 damage to a dining car. It is a known fact that charcoal, particularly in powdered form, is subject to spontaneous ignition. This fire emphasizes the necessity for effective cleaning of charcoal bins of cars and points to advisability of continued attention to other charcoal storage.

Collision—engineer overlooked the fact that cars were standing on siding, he having observed the cars as he moved through house track, and failed to stop on passing track at cross-over switch until switch had been lined to main track. As result engine was backed into cars on siding, completely demolishing a refrigerator car. Damage sustained $2,022.

Collision—sideswipe of two passenger train engines. Engineer of passenger train on siding, failed to bring train to stop before fouling main line, despite the fact that fireman had notified him twice that he had arrived at clearing point, with result that engine of passenger train approaching meeting point sideswiped engine fouling switch.

RECENT CHANGES MADE IN SAFETY COMMITTEEMEN

The following changes have been made in safety committeemen account expiration of term, transfer and leaving the service:

San Joaquin Division: E. G. Roche, secretary, vice O. Oswald.

Steamer Division: C. McNulty, Captain, vice L. P. Cooley; Alfred F. Kerr, fireman, vice John J. Murphy.

Portland Division: C. J. Pearce, asst. division storekeeper, vice J. F. McAuley.

Sacramento Gen. Shops & Stores: S. McDonald, blacksmith, vice J. Comber; Geo. Harris, machinist, vice Geo. E. Glick; R. Gesswein, crane operator, vice J. D. Rowe.

Stockton Division: E. S. Hodgson, route agent, vice C. C. Graves; R. Askien, brakeman, vice F. G. Talbot.

Two small boys were puzzling their brains to invent a new game. At last one of them said, eagerly: "I know, Billy, let's see who can make the ugliest face."

"Aw, go on!" was the reply. "Look what a start you've got!"—Pathfinder.

Shopmen Paint Home of Veteran

Painters from the Sacramento shops recently swooped down on the home of W. R. Oldfield, retired foreman painter, and treated it to a coat of paint. Mr. Oldfield is seated in front of the band of helpful fellow employes.

TWENTY-FIVE painters from the Southern Pacific shops at Sacramento played the part of Good Samaritans Sunday, March 30, when they brightened the appearance of the W. B. Oldfield home with a coat of paint. They returned with a second coat the following Sunday and now the old homestead of the veteran foreman painter fairly glistens.

Mr. Oldfield, who was retired on pension in September, 1922, has been in ill health for some time. One of his former shop-mates, following a visit to the home, suggested that the men who formerly worked with Mr. Oldfield contribute something toward the happiness of that family. It was then decided to paint the home, material for which was furnished by the W. P. Fuller Company at Sacramento.

"The work was done like magic," says J. Hall, master car repairer. "Before Mr. and Mrs. Oldfield realized what was going on the scaffolds were erected and in two and a half hours the first coat of paint was applied. The two old folks were overcome with emotion in attempting to express their appreciation. They declared it one of the happiest moments of their lives."

The Sacramento "Independent-Leader" in commenting on the thoughtfulness of the Southern Pacific men, said:

"It is a sample of human kindness that carried more and better lessons than were preached in all the sermons of Sacramento that Sunday."

The men who participated in the affair were: F. L. Caton, E. F. Hannon, A. Pellegrini, K. F. Knecht, J. Matranga, W. A. Higgins, H. L. Ravellette, E. M. Smith, J. Enos, J. Pugh, J. P. Stefani, L. Birchler, H. Hoskins, J. L. Barney, J. Favilla, C. Lindfeldt, O. R. Hammell, E. S. Schlegel, C. H. Riley, C. Cuilla, J. P. Nicholls, K. Bose, W. C. Mitchell, C. C. Delchiaro and F. Cusenza.

CEREMONIES MARK OPENING OF GLENDALE STATION

The new Southern Pacific station at Glendale, near Los Angeles, was opened March 27 with appropriate ceremonies attended by residents of Glendale, Los Angeles, Hollywood and neighboring cities.

T. H. Williams, assistant general manager of the Southern Pacific at Los Angeles, delivered the address presenting the station to the city of Glendale for the convenience of travelers from that community and also from Hollywood, Pasadena and Eagle Rock.

Acceptance of the station was made by Mayor Spencer Robinson of Glendale. Other speakers included F. S. McGinnis, assistant passenger traffic manager; T. H. Allen, president Pasadena Chamber of Commerce, and G. G. Greenwood, president Hollywood Chamber of Commerce.

An informal dance was held following the ceremonies.

The new Glendale station will be of great convenience to travelers in the north and west sections of Los Angeles and nearby communities. It is an attractive building of Mission design.

PAINTERS
OF THE
C. P. R. R. CAR WORKS,
SACRAMENTO, CAL., OCTOBER, 1882.

O. V. LANGE, Photo.

Posed in the photo above is the 1882 crew of the Central Pacific Railroad car works. A beautiful and freshly painted passenger car was a marketing tool, as it was the first impression a passenger or potential passenger had with the travel experience. During most of the nineteenth century, applying a long-lasting and shiny finish required a large commitment of time. In the 1860s, with the decorative work included, up to ninety days were needed for a complete paint job. Twenty years later, the chore had been reduced to about sixty days. A first-class job started with priming raw wood with linseed oil that was allowed to dry for at least one week. Then, after putting holes, the first light coat was applied and allowed to dry for twelve to twenty-four hours. A filler was applied, followed by another coat of color and a twelve to twenty-four hour drying time. The process continued with an additional three coats of color with up to a day of drying time between each layer. Striping, lettering and varnish were near the end of the routine and done in one of five wings attached to the main paint shop building. The last step was two more coats of finishing varnish. – Courtesy California State Railroad Museum

Shasta Division Studies Handling of Perishable Freight.

By F. E. SLATER
(Superintendent's Office, Shasta Division.)

A very interesting OS&D meeting was held at Dunsmuir April 16th, at which time a large number of agents and other employees were present. Among the items discussed was the prompt and careful handling of perishable freight now that the season for shipping such commodities has commenced. Attention was also directed to claims for loss by pilferage, particularly as to such articles packed in cartons and shipped in boxes. Often empty cartons are packed in the boxes to fill up space and care should be exercised to check shipments against the invoice before claim is allowed to obviate payment for articles which were not included in the original shipment.

During the past few weeks men have been busy beautifying the station grounds at Dunsmuir. All dead wood has been cut out of the trees and cobblestone curbing placed around the walks.

About May 15th Shasta Springs and Shasta Retreat stations, both summer resorts, will be open for business. These stations were not opened last year owing to the war conditions and extreme shortage of help. Now that the country is again assuming normal conditions it is anticipated that these resorts will be well patronized this year.

Eighteen Years Ago.

Eighteen years ago the first through train on the Southern Pacific Coast line from Los Angeles to San Francisco passed through Santa Barbara.

Conductor John Hartnell, who has been a guest at the Neal for the past few days, was conductor on the local train whose terminus was Santa Barbara, many years previous to that. Neal Callahan, proprietor of the Neal, says that Hartnell was conductor of the train on which he first came to this city 25 years ago. Hartnell now travels on the Lark and will soon be in line for the pension list.—Santa Barbara Press.

Help! He's Out Again.

The "Sacramento Union" has discovered the champion fool of the world. Hearing that the Southern Pacific had a fine roadbed on the Sacramento Division this man drove his automobile onto the main line track and had traveled quite a distance when he was pulled out of his car by A. A. Stoddard, a switch tender. The police took care of him. Yes, he was trifling with Rule G.

C. F. Stewart has been appointed manager of the Troop Movement Section, vice George Hodges, deceased.

Little strokes fell great oaks.

Veteran Shop Craftsmen Proud of Skill in Railroad Foundry

By H. C. VENTER
(Sacramento General Shops.)

In the safe operation of railroads, the operating department's first duty is to see that the passenger and freight equipment is dispatched with care, safety and speed. In order to fulfill this requirement, it is the duty of the mechanical department to put equipment in first-class shape in order to fulfill the expectations of the operating department, in seeing that engines and cars are gotten over the road without mechanical failures.

Of the various shop craftsmen that make up the big army of mechanical employees, there are none whose duties are more important in the work necessary to make this equipment serviceable than those who operate our foundries in the making of iron and brass castings. Whether these castings are so constructed that they will stand the necessary strain that is placed upon them in service depends as much on the manner in which they are constructed as upon the material from which they are made.

There are very few railroads throughout the country that operate their own foundries in the making of castings. The Southern Pacific, at its Sacramento shops, is one of the very few, and ranks among the leading railroads of the country in having foundries that are capable of furnishing all gray iron and brass castings used.

At the Sacramento shop the Iron Foundry is divided in two sections, one part used exclusively for the manufacture of cast iron car wheels of various sizes, and the other for castings of various shapes and dimensions for general use. The wheel department has a capacity for 258 cast iron wheels daily, and we expect to increase this to 296 daily as soon as the new addition, which is now under way, is completed. This new addition will give 100 annealing pits and allow for an increase of thirty-eight wheels daily.

There are a number of veterans who have seen a good many years of service in the wheel foundry, with records that have always been trustworthy and reliable, and whose associates have found them to be men of skill in their duties, and among them are:

F. W. Clark, 19 years molder.
J. Lyons, 16 years melter.
J. Wagner, 19 years cupola leader.
C. A. Smith, 17 years foreman.

The other section of the foundry is what is known as the General Foundry. In this department the important locomotive, roadway and floating equipment castings are made. Castings made in this department hold records that are unsurpassed by any foundry throughout the country. This success is due to the skillful molders who pride themselves on their work. All the large castings, such as the cylinders, bed plates, etc., on the Steamer Contra Costa, which is recorded as being the largest transfer ferry boat

in the world, were made at these shops.

The foundry at Sacramento is under the supervision of the following:

J. H. Geiger, foreman.
J. S. Gageby, assistant foreman.
P. M. Early, gang foreman.
G. L. Derman, leading workman.
C. A. Smith, assistant foreman, wheel foundry.
F. Valentine, leading workman, core room.

As to records of service, the writer with his eighteen years has had pride in himself, but when he looks upon the records of the following foundry employees, he feels very young:

E. Sheehan, 40 years' service.
A. H. Bening, 38 years' service.
A. Kortstein, 37 years' service.
L. Lohman, 37 years' service.
J. S. Gageby, 36 years' service.
Geo. Williams, 30 years' service.
G. L. Derman, 29 years' service.
J. H. Geiger, 28 years' service.
C. H. Shick, 26 years' service.
Ed Galvin, 25 years' service.
M. J. Kelly, 25 years' service.

In the successful operating of the foundry, there is considerable credit due to our patternmakers who are a very necessary part of a successfully operated foundry. The patternmaker's dress while at work is possibly a little more fastidious than that of a molder, but that does not worry the foundry employees, as they defeated them in a recent baseball game held at McKinley Park. The molders poured the hot metal over the patternmakers so fast that nothing remained of them but charred wood, and this was done by a tune of 4 to 1. The batteries for the victors were McCarthy, McCaffery and Burke, and for the losers, Glacken, Peek and Moose. The features of the game were the battery work of the victors, and a one-hand phenomenal catch of T. McCaffery at short stop. The iron foundry team hereby issues a proclamation challenging all of the S. P. shop teams, any place at any time.

During the past month we have had a number of our former employees return from the Army and Navy, and they have found their jobs waiting for them.

One of our heroes, J. P. Douglass, in the copper shop, who has returned from the thickest of the fighting, received a number of shrapnel wounds, but has come back to us in good physical condition and able to go on with his work.

Base Hits.

Do not throw away pencil stubs. Put them in pencil lengtheners and get 100 per cent use out of every pencil.

Steel pens are at a premium and hard to get. Do not throw away if they become a little corroded. Scrape off with a penknife and they are as good or better than a new one. Use them until worn beyond redemption.

A rolling stone gathers no moss.

Sacramento General Shops
Form Important Industrial Groups

By H. C. VENTER.

The Southern Pacific shops at Sacramento constitute one of the most important railroad plants in the United States. In variety of operations conducted within their numerous departments they are excelled by few if any similar establishments. In addition to the usual sections constituting the average big rolling stock repair plant, this group of shops includes such important units as a large foundry for gray iron and chilled iron castings, a brass foundry for composition and brass castings, a wheel foundry for chilled iron car and truck wheels, and a rolling mill for producing a variety of sections and shapes. Then, too, the company operates here important car building shops, and in the machine shops, besides the regular work of overhauling and repairing rolling stock, a number of new locomotives are being turned out annually.

The shops also build, overhaul and keep up the mechanical equipment on numerous river steamers, ferry boats, dredging apparatus and other marine units operated by this company.

As this is the largest plant of the Southern Pacific and the most important of all the groups of shops located along its lines, it is perhaps only logical that here should be centered one of the greatest of the system's industrial endeavors, the work of reclaiming old and discarded material. This is one of the principal features of interest to the visitor at Sacramento.

Familiar as one may be in a general

(Continued from Page 5)

has been made in average tons per locomotive.

High speed freight trains mean light loaded locomotives and detract from engine efficiency in light engine loading and excess consumption of fuel.

Mr. Williams cited the record obtained by the officials of the Los Angeles Division during the second week of January. On the freight district from Indio, Cal., to Yuma, Ariz., a distance of 122 miles, this division hauled an average of 2321 tons per locomotive, as against 1845 tons per engine for the same week in the year 1918, an increase on each train run of 476 tons per locomotive this year over last, giving the division 100 per cent engine efficiency this year as against 86 per cent last year.

"To illustrate further the efforts of the organization on this division, during the second week in January each engine run on the four freight districts of the division, in the direction of the volume of business, eastward, obtained 96 per cent efficiency in tons per engine, which was the best record made, or the highest percentage of efficiency of any division on the system."

way with the handling of scrap material, it is interesting to learn what becomes of the enormous quantities of such material accumulated at different points and in what form it is destined to appear again in service.

Cast iron and brass scrap are melted and recast. Wrought iron and steel parts that have found their way to the scrap heap form another problem, and the disposal of the material in extensive quantities is of particular interest.

In no line of activity is there greater accumulation of scrap of this character than in the operation of railroads. These are, roughly speaking, 70,000 locomotives and 2,500,000 cars in service over the 260,000 miles of track in our railway systems. Our scrap piles consist of a quantity of worn-out and discarded material that constitutes in the aggregate a vast volume when expressed in gross weight or in number of pieces.

The methods of the Southern Pacific at Sacramento hold much of value and interest to the student of reclamation work, not only because of the volume of business daily transacted, but also on account of the diversified shop channels through which the scrap metal of the system is conducted in the various processes under which it is rehabilitated for service. With this in mind the following data may aid toward an understanding of the method of handling scrap from the time it reaches the scrap dock till it becomes a finished product.

Scrap is received at the scrap dock daily and with the aid of a high-power magnet is unloaded from the cars onto a large platform where it is sorted out for distribution to the various reclaiming shops. Scrap intended for rolling mill is taken to piling shed, where it is put up into piles varying from 160 pounds to 380 pounds, after which it is transferred to the rolling mill storage yard ready for heating and rolling into the various products.

The output of the rolling mill for the year 1918 was as follows:

Merchant bar iron	25,679,706 lbs.
Merchant bar steel	229,385 lbs.
Track bolt iron	311,560 lbs.
Nut iron	226,840 lbs.
Angle plate iron	3,071,640 lbs.
Total	37,025,251 lbs.

In addition to the above, billets and slabs to the amount of 27,143,850 lbs. were also hammered or rolled in the mill for use in getting out the above product, a large portion of which is shipped to various points throughout the Pacific system, but the greater part is used at the Sacramento shops in the manufacture of all kinds of bolts, nuts, car, locomotive, steamboat, M. of W., and all other forgings necessary to repair and build railroad equipment.

In getting out the above material at

Sacramento instead of purchasing in the open market, this company made a saving of $571,305.97. In doing this work 2,691,668 gallons of fuel oil were used.

The supervision of the rolling mill and blacksmith shop is under James Harkin. E. M. Lancaster is in charge of the piling of iron at the scrap dock, while A. R. Haley and J. A. Hallanan are respectively night and day foremen at the rolling mills. S. J. Uren is assistant foreman under James Harkin, with O. B. McDonald and Joseph Palm in charge of the blacksmith shop and W. S. Holmes in charge of the bolt forge shop. S. J. Uren is a son of Stephen Uren, who was in charge of this work from 1876 to November, 1907, under whom the plant grew from small dimensions to nearly its present size.

In doing this work, the rolling mill operates two 12-inch and two 18-inch mills using eight 3-door and eight 2-door reverbatory furnaces. Part of these furnaces are used to heat the iron for one 5000-lb., one 4000-lb. and two 3600-lb. steam hammers which assist in preparing the iron for its ultimate use. In addition to hammers mentioned, the number of small hammers are scattered throughout the works and blacksmith shop for working up the forgings.

In addition to merchant bar iron, we get out all sizes up to 3 inches square and round, and 1⅝x9 inches and flats as wide as 9 inches. We also get out a great many patterns of angle bars and fish plates for rail on our tracks, and a large number of tie plates.

A considerable portion of the year these mills run night and day and are under the direct supervision of the head rollers, who are Michael J. Donahue, Claude F. Dodge, Richard M. Lawrence, Peter E. Mitchell, James N. Styles, James N. Pugh and Frank De Riso.

Nearly all of these gentlemen came to the Southern Pacific Company as boys and have grown up with the work. They are good loyal employees and are a credit to the company.

District Fire Inspectors Named to Supervise Protection.

With the approach of summer, bringing with it the fire hazard, the railroads are again gathering their forces for the annual battle to safeguard property and lives. Niles Searls, General Fire Inspector for the United States Railroad Administration in this territory, announces the appointment of S. B. Howatt as District Fire Inspector for the Northern District of the Southern Pacific, Western Pacific, Tidewater Southern and Deep Creek Railroads; and the appointment of M. B. Anfenger as District Fire Inspector for the same lines in the Southern District. Both officials will have their headquarters in San Francisco.

Gain may be temporary and uncertain, but expense is constant and certain.

THE BULLETIN

Musicians at Sacramento Win Recognition

Glee Club, Band and Quintet Prominent in Life of Community, Aiding Charities and Public Movements. Center of Social Affairs

ALTHOUGH organized as recently as May, 1917, the Southern Pacific Band and the Southern Pacific Glee Club at Sacramento have a record that other similar industrial organizations may look at with envy. These two organizations in a little over four years have raised over $14,-000 for charitable purposes, have built up a music library valued at over $3000, and have equipped themselves with uniforms valued at several thousand dollars, all through the initiative and resourcefulness of the members.

club gave a concert at the Exposition Auditorium. The club is now planning its annual minstrel show which is a popular affair at Sacramento.

The glee club has a quintet which is in constant demand and gives frequent concerts. The quintet includes W. E. Thomas, director, H. Harter, G. E. Lester, T. L. Smith and V. T. Hodge.

In Liberty Loan Drives

G. E. Lester, W. E. Thomas and H. E. Weida were the organizers of

the glee club which came into being at the time of the Liberty Loan "drives." Mr. Lester was asked one day to provide vocal music for a bond drive meeting to be held the next noon. He secured some talent from the General Shops and after a half hour of practice was ready for the concert. The next day Mr. Lester met Mr. Thomas and Mr. Weida and it was decided to form a permanent organization. Rehearsals were started immediately. The work of the members of the band and glee club is purely voluntary.

Shortly after the band was organized a committee composed of workmen from the several shops was formed for the purpose of raising funds to uniform the band. They gave a ball which cleared $1100 and the uniforms were then contracted for at a cost of approximately $1400. Other entertainments followed the first ball, and the balance to cover cost of uniforming the band was soon raised.

The dances and entertainments are always well patronized by

Sacramento shops band. Front row, left to right: W. Fann, Wm. Hale, D. Alonzo, C. Jensen, J. E. Weida, F. Cippa, J. Stoneking, T. Keene, A. Lascari. Second row, left to right: F. Townsend, G. Bencich, W. Wright, H. Westphal, W. Williams, G. Milner. Third row, left to right: N. D. Mapes, G. Scott, F. Fisher, T. Conners, F. Kelleher, A. Bencich, H. Thompson. Fourth row, left to right: W. Milter, J. Curry, D. Grant, F. Schrader.

The band has 38 members, including the drum major, and the glee club numbers 48 voices, both organizations being under the same musical director, J. E. Weida, Foreman of Locomotive Machine Shop No. 1 at the Sacramento General Shops. The band members all own their own instruments individually.

During the last year the band and glee club have given a number of concerts in Sacrament and in October came to San Francisco at the instance of Mrs. Minerva N. Swain, prominent in San Francisco musical circles, to take part in the Music Week program. The band gave a concert in Union Square and the glee

Glee Club, Sacramento Shops: Front row, left to right: L. James, S. Jensen, W. I. Goodwin, G. E. Lester, W. F. Davis, J. F. Weldon, A. A. Weichert, W. L. Greaves Second row: J. G. Nilan, B Manchester, P. Mondine J. J. O'Hare, J. Nardelli, H. Savage, T. L. Smith, S. B. Curry. Third row: S. W. Walton, S. C. Morris, W. E. Nuttall, J. H. Hogan, H. Harter, C. B. Stocks, O. D. Manley, M. Shelton. Fourth row: J. A. Bayless, P. Guidero, R. E. Johnson, W. J. Dowrick, J. E. Weida, B Boyd, W. A. Trevarthen, T. M. Nightwine, S. J. Dillon, W. E. Thomas.

Having read your Bulletin—Pass it along.

Sacramento Shops Quintet, left to right: V. T. Hodge, T. L. Smith, G. E. Lester. H. Harter and W. E. Thomas.

Mrs. J. E. Weida, wife of the Musical Director, and Soloist for the Club.

Southern Pacific officials and employes generally, as well as by the general public, both organizations having won quite enviable popularity considering the comparatively short time they have been before the public.

During the war period both band and glee club furnished music for many patriotic gatherings, especially during the several bond "drives," volunteering their services without charge. During one of these drives the band played 27 different engagements and the glee club sang almost an equal number.

A special train was run out of Sacramento on two different occasions, carrying the band and glee club and a corps of speakers, in the interest of the Liberty Loan, and although the territory through which these trains passed was said to have subscribed its full quota; and in some places had been oversubscribed, over a half million dollars in subscriptions was taken. On such occasions as these, the members of both the band and glee club were allowed their regular pay the same as if employed at their respective occupations in the shops. This is the only exception to the statement that services of the band and glee club are volunteered without charge.

The Company provides a rehearsal room located upstairs over one of the Car Department shops, the room being used jointly by the two organizations. The Company also provides special trains and free transportation to employees for excursions and picnics on different occasions, which has been another source of income for both organizations. Aside from this both are self-sustaining.

Uniforms Purchased

After organization of the glee club the question of suitable "uniform" or dress for public appearance naturally arose, it being finally decided to use white woolen trousers with fine pin stripe, with about one inch of black silk braid down each pants leg, a double-breasted blue serge coat, trimmed on the sleeves only, in black silk braid; hat of soft white felt, plain white shirt, soft collar and black four-

in-hand tie, white socks and white canvas shoes. This constitutes the uniform, with the exception of linen duster and small cane for street parades.

The club "uniform" was purchased with funds raised principally through a minstrel performance, or rather a number of performances given by the club, at a cost of about $1300.

It is now a set custom for the band to give an annual ball and the glee club an annual minstrel show for the purpose of raising funds for the respective organizations. The glee club also makes a moderate charge for providing musical entertainment at lodge banquets or similar events, in this way assisting in keeping up the running expenses for music, etc., but for civic benefit affairs it has always gladly contributed their services, as has the band. The annual ball and minstrel show referred to, have been most gratifyingly successful in the past, with every indication of doing even better on future occasions.

S. P. Band and Glee Club Praised by Judges

Following is a portion of a letter written by Justices E. C. Hart, Albert G. Burnett and Wm. M. Finch, of the California District Court of Appeal, Third Appellate District, to J. E. Weida, musical director for the Southern Pacific Band and Glee Club at Sacramento.

"The organization is certainly a pronounced credit to the men through whose vigorous work and indomitable persistency it was directly brought into existence; it is equally a credit to the City of Sacramento, which can justly and proudly claim it as its own; it is, or must necessarily be the pride not only of the coworkers in the shops of its promoters, but of the Southern Pacific Company under the auspices of whose employs it was inaugurated and is maintained."

The band has a musical library that could not be duplicated today for less than two thousand dollars. It contains the cream of standard musical compositions, as well as a liberal sprinkling of the lighter grade music, including the "popular" numbers, the latter being constantly added to from time to time.

The glee club is also rapidly acquiring an excellent collection of music suited to its needs, so that at present it is in a position to furnish music suitable for most any occasion.

Christmas Carols

One custom of the glee club won instant approval at Sacramento, and that is Christmas carol singing. On Christmas morning at five o'clock, forty of the club, in automobiles, furnished by various club members, make the rounds of all the hospitals, city and county jails, to the houses of various of the city officials, railroad officials, to the homes of the shop employees who had been sick and were still unable to be out at their customary tasks, and to any place where it was felt a little music would bring cheer.

Christmas (1919) was the club's second season of carol singing, and long before Christmas inquiries came from hospitals, the orphanage and other institutions as to whether or not the club was going to sing carols, and if so, if the singers would like to have hot coffee and sandwiches or doughnuts ready for them. This was an agreeable surprise to the club, for starting out at five o'clock in the morning to sing in the open air surely puts an edge on the appetite.

The band, by special invitation, played a concert at the Assembly Attaches Legislative Reception at Sacramento. March 26, 1919, and also played the grand march. The honor of playing on this occasion would

THE BULLETIN

PERSONAL SERVICE OFFERED BY S. P. ON EUROPEAN TRIP

Attention has been drawn to the excellent system of personal service rendered by the Southern Pacific Company to passengers to and from Europe, by A. J. Poston, General Agent, New York.

The traveler can purchase a ticket to Europe and return, through any Southern Pacific representative, or should a person desire to pay for the passage of a friend or relative in Europe, the cost of such transportation can be deposited right here in America with the nearest Southern Pacific representative, and the Southern Pacific lines will do the rest. An order for the transportation is purchased and sent to Europe. The relative or friend then presents the order to the steamship line and receives passage to New York.

Arriving at New York, the traveler holding Southern Pacific prepaid order is met at the pier or at Ellis Island, and arrangements are made for the journey to destination. The passenger agent will be personally responsible for seeing that the traveler is conducted to the proper train or Southern Pacific steamer and every arrangement made for a comfortable and pleasant trip west.

If the Washington-Sunset route has been selected for the trip, the traveler will be taken to the train for Washington, D. C., and upon arrival there will be met and put aboard the through tourist sleeping car, or other car chosen. This through tourist car is ideal for strangers or families, going from Washington to San Francisco without the necessity of changing cars if the traveler is destined for either Los Angeles or San Francisco, or any other station at which the Sunset Limited is scheduled to stop.

In addition to this through rail service, the big, steel Southern Pacific steamships plying between New York and New Orleans provide excellent cabin and steerage accommodations. The passenger fare includes meals and berth aboard ship, and the voyage of "100 Golden Hours at Sea" is delightful. The steamer makes connections at New Orleans with the fast through trains of the Sunset Route to points in Louisiana, Texas, New Mexico, Arizona and California.

Mr. Poston states that the Southern Pacific company representatives will be very glad to assist American travelers going to Europe, and that the Southern Pacific European agents will be ever ready to answer inquiries regarding return journey, which service will prove of great assistance to Americans abroad.

An Old Story.

Irate Visitor—Mr. Editor, I've been told that you have printed in your sheet that I am the greatest swindler the world has ever known.

Editor—No sir! Not in my paper. It contains only the latest news.—Exchange.

SHOPS MUSICIANS

(Continued from Page 10)

have been welcomed by any musical organization in the city.

The two organizations have done a wonderful work in promoting harmony and fellowship among the employees. They have been given every possible moral support of local railroad officials and by the management generally, and it is only reasonable to expect that as time passes, interest and effort will increase and inestimable benefits will be realized.

Helped Welcome Pershing

When General Pershing and his staff came to Sacramento and was given a reception in the State Armory, the band and glee club alternated in furnishing musical numbers during the evening.

In November 1920 the glee club gave a concert in the Greek Theater on the University of California campus at Berkeley. The concert was a great success although weather conditions were unfavorable.

In December 1920 the band and glee club, backed by the "Sacramento Union," a Sacramento newspaper, launched a drive for funds for the Salvation Army and in five days raised $3600, this sum being made up of small contributions received when the organizations "passed the hat" after

Early Road Opening

A reminder of the old days during the early growth of the Southern Pacific Company is contained in a clipping received by the Bulletin taken from the Sacramento Union of August 10, 1869 describing the opening of the "Western Pacific" line from Sacramento to Stockton, which later became a part of the Southern Pacific Lines. The item is as follows:

"The first passenger train to Stockton over the Western Pacific Railroad left this city about 10 o'clock yesterday morning. It consisted of three cars, quite well filled with excursionists drawn by the locomotive C. P. Huntington. The road between this city and the Mokelumne was found to be in excellent condition, completely fenced in, smooth and well ballasted, principally with gravel.

"Beyond that river the track, having just been laid, was of course rough and toward the extreme end the rate of speed was necessarily slow. The terminus of the road, which was in the suburbs of the town, close to the asylum, was reached about 1 o'clock."

concerts in hotel lobbies, theaters or other places of public assemblage.

The two organizations and the members of the shop forces at Sacramento also raised $3000 to equip a playground for the Sacramento orphanage.

Many letters of appreciation have been received by the organization. Mrs. Swain, who was chairman of the Ways and Means Committee for San Francisco's Music Week, and who as was stated before, was instrumental in having the band and glee club come to San Francisco, said that the members of the glee club had every appearance of professional artists who had been on the concert stage for many years, that they had the musical attitude, and as for the voices, she had nothing but the greatest admiration. She is the authority of a recent statement that a number of big companies in San Francisco are planning the organization of similar glee clubs and bands. Other musical critics in San Francisco who heard the glee club and band have commented most favorably on their work. The work of Mr. Weida, the musical director has been characterized as most unusual. Mrs. Weida is one of the soloists for the club.

Letter of Appreciation

Governor Stephens, this last month, wrote a letter to Mr. Weida, expressing his pleasure at having heard the two organizations, and his appreciation for their contributions to the community life of Sacramento. The City Commission of Sacramento in 1920 passed a resolution expressing the thanks of the people of Sacramento for the community activities of the glee club and band, and for their charitable activities.

An unusual tribute to the work of the band and glee club was expressed in January, this year, by E. C. Hart, A. J. Burnett and W. M. Finch, Justices of the District Court of Appeal, Third Appellate District, in a joint letter to Mr. Weida. The Justices expressed the conviction that it was the duty of all the citizens of Sacramento to join in a cordial and hearty commendation of the "splendid work" of the two organizations.

A. D. Williams, Superintendent of Motive Power, Northern District and D. S. Watkins, Superintendent of the Sacramento Shops on the first of the year sent a letter of best wishes to the two organizations. A brief portion of the letter is as follows:

"During the past year we have watched activities of your organizations with great pleasure. You have given freely of your service not only to the railroad employees but to charity and for various civic activities. These activities have lead to much favorable comment and this has been of very considerable advantage to the Southern Pacific Company. We consider your organizations a big asset to our shops."

Having read your Bulletin—Pass it along.

BOXING BOUTS FEATURE OF SHOPMEN'S MEETING

Four boxing bouts were the features of the two hours' social program that followed a recent meeting of the Shop Craft Protective League at Sparks. The program was so much enjoyed that it has been decided to stage other bouts at the next meeting. Vocal selections were rendered by H. C. Mulcahy. Following are the results of the bouts:

Nello Gianotti, boilermaker apprentice, 175 lbs. vs. Tony Poloni, blacksmith apprentice, 150 lbs., draw. Ernest Rosa, machinist helper apprentice, 129 lbs. vs. Jim Dermody, store helper, 129 lbs., Dermody winning, decision. Patsy Clark, blacksmith helper, 139 lbs. vs. Robert Logan, tinsmith helper, 125 lbs., Logan winner, K. O. Les Odem, machinist, 165 lbs. vs. Jess Holt, blacksmith helper, 175 lbs., Odem winner, decision.

TRAINMASTER CHANGES ON SHASTA DIVISION

With the transfer of E. J. Kellum to the Los Angeles Division as Trainmaster, J. J. Sullivan, formerly Terminal Trainmaster at Los Angeles, becomes Trainmaster on the Shasta Division.

Mr. Sullivan entered the service of the Southern Pacific as a Yard Clerk on April 26, 1904 on the Tucson Division, serving in various clerical capacities terminating with Assistant Chief Clerk. For two years after March 1, 1913 he was a student in railroading, and on January 1, 1916 became Clerk and Special Inspector in the office of Vice President and General Manager. From July 25, 1918, during the remaining period of the war, he was engaged in several positions with the American Railway Association and with the Railroad Administration. In March 1918 he became First Lieutenant, U. S. Army, and later Captain Engineers, serving in France from June 7, 1918 to September 15, 1919, first as Ass't. Division Superintendent and later as Division Superintendent Railway Service, A. E. F. March 1, 1920 he returned to the Southern Pacific as Supervisor of Transportation, Northern District. In July, 1920, he became Trainmaster, Stockton Division; and was soon after transferred to Trainmaster, Tucson Division, being transferred from there to Terminal Trainmaster at Los Angeles on December 1, 1922.

H. A. Sprague has been appointed Trainmaster, Tucson Division relieving G. H. Moore who has been granted a leave of absence. Mr. Sprague entered the service as a Brakeman on the Sacramento Division, August 2, 1900, being promoted to Conductor April, 1906, and was later appointed Yardmaster at Marysville. In June, 1921, he was appointed Assistant Trainmaster, Sacramento Division, and was later Trainmaster at Marysville.

Speedy S. P. Team at Sacramento

Southern Pacific team at Sacramento. Standing, left to right—Alloway, ss; Hauser, c; L. Wagner, 2b; Pitt, c; Manger, 3b; Lefty Wagner, p; Keenan, p. Sitting—"B.ll" Blackburn, mascot; S. Blackburn, manager; Ratoni, rf; Link, cf; Clark, lf. "Babe" Downs, p, is not in the picture. Photo by D. L. Joslyn, Sacramento Shops.

THE Southern Pacific baseball team at Sacramento walked away with a well played game from the San Francisco General Office team at Sacramento August 19, by a score of 7 to 0.

A large delegation accompanied the team from San Francisco and were entertained at Sacramento Saturday evening at an open air dance given by the employes of the General Shops.

While the score was rather lopsided the game itself was much better, each side playing errorless ball, timely hitting bringing the scores for the Sacramento boys.

Sacramento opened the scoring in the first inning when "Rabbit" Clark, the diminutive center fielder, landed on McGough's offering for a homer. Clark also featured in the field, accepting all kinds of chances, and it is said one of the visiting railroaders remarked, "we have to hit it a mile to get on first base with that bunch of fielders."

"Big Bill" Keenan, although an apprentice machinist by trade, proved himself a "journeyman" when it came to pitching. He allowed only five scattered hits and only once, in the first inning, was he threatened by a score.

The Sacramento boys have had a very successful season and out of seventeen games played have lost only one, and that to the speedy Grass Valley team. A return game is being planned when the railroaders expect revenge.

Manager S. D. Blackburn, care Superintendent Motive Power, Sacramento, is anxious to arrange games with other Southern Pacific teams.

The lineups were as follows:

Sacramento—Link, cf; Clark, lf; Ratoni, rf; Houser, 1b; Pitt, c; Alloway, ss; Wagner, 2b; Manager, 3b; Keenan, P.

General Office—Ahern, lf; Estes, cf; Harrington, 1b; Ahlf, 3b; Wurm, c; Lewis, ss; Mesmer, 2b; Nelson, rf; McGough, p; Fonseca, lf.

HIGGINS IS PROMOTED TO CHIEF CLERK AT S. F.

A. M. Higgins has been appointed Chief Clerk of the Freight Protection Department, General Office. He entered the service of the Company as a stenographer in the Mechanical Department, San Joaquin Division, in October, 1910, and a few months later was transferred to the Superintendent's office in the same capacity. In January, 1912, he took the position of OS&D Clerk, remaining in that position until April, 1917, when he was appointed Division Car Distributor, serving in this capacity four years. In March, 1921, he was appointed Assistant Chief Clerk to Superintendent, San Joaquin Division, and held this position until April, 1922, when he was transferred to the Freight Protection Department at San Francisco.

A. M. Higgins

THE BULLETIN

NEW INSTRUCTION CAR AN ACADEMY ON WHEELS

Ten Thousand Pupils Profit by Practical Demonstrations of Railroad Problems

Views of the new Southern Pacific Instruction Car. No. 1. Office quarters. No. 2. Class room. No. 3. Exterior view of Southern Pacific Instruction Car C. P. 111. No. 4. Interior showing a train order projected on screen from picture room in rear of same. No. 5. Interior showing automatic block signal model on top of counter. Cover opened. Also chart of absolute-permissive block signal system.

By WM. NICHOLS

(Chairman, Board of Examiners, Southern Pacific R. R.)

An interesting piece of rolling stock, recently turned out from the Southern Pacific's shops at Sacramento, embodies the latest ideas with regard to railroad instruction cars. Designed to aid in the instruction and examination of new employees as well as old, the car is equipped with every modern device and test to be found in the railroad curriculum. Its designers and builders believe that this "academy on wheels" represents an advance both in car construction and railroad educational work.

The car is 72 feet 6 inches in length over end sills, possesses steel underframe and trucks and is equipped with vapor heating system and electric and Pintsch gas lights so arranged that connections may be established with city power mains where stops are

made. Three staterooms are provided for the chief instructor and his assistants, with a commodious office 12 feet in length at the observation end. A shower bath and other conveniences are included in the accommodations.

Modern Classroom

Approximately one-half of the instruction car is devoted to a classroom with a seating capacity of thirty. This room is equipped with an automatic block signal working model, the signals being operated from a storage battery. A display board, provided in bulkhead near the center of the car, enables charts to be effectively exhibited and illuminated. There is also a time table apparatus for displaying various conditions under which trains may be placed when a new time table takes effect.

By means of a projecting machine in what is known as the picture room,

train orders may be flashed upon a screen in front of the display board where problems are worked out in connection with the miniature road.

The car is intended to serve a territory of about 9000 miles, comprising what is known as the Pacific System, and to make possible the examination and instruction of approximately 10,000 employees on each round trip. Stops are made at points where crews lay over, making it convenient for men to attend at either end of their run. Where telegraphers are employed the car is halted the required time for thorough examination and instruction.

The personnel of the examining board comprises men who are qualified by long experience in train and engine service and in dispatching. The local examiner on each division questions employees just entering the service or those promoted to new positions. But all employees are required to attend these examinations in order that the work may be kept fresh in the minds of all.

In this manner a conductor or engineer obtains the benefit of questions and answers between a dispatcher and the examiner, and brakemen and firemen learn the problems that confront conductors and engineers, thereby preparing them for the problems they will face when the time comes for promotion.

Practical Lessons Given.

As the car passes over each division examiners check all train orders used by conductors and engineers to see if they are in proper form and legible. Attention is directed to any irregularities, and the class receives instruction on the spot concerning the neces-

sity for correction. A similar check is made upon dispatcher's train order books, train sheets, train registers, checks of train registers and clearance cards.

Division officials find it to their advantage to be present at these examinations in order to familiarize themselves with instruction and thus obtain a uniform understanding of the company's practices.

As new conditions obtain and improved methods of operation are conceived, the instruction car plays an invaluable role in practical demonstration of the ideas to be adopted. Employees take keen interest in the visualization of problems thus graphically presented in the classroom and enjoy marked benefits from the instruction.

RAILROAD FRANKNESS
A MESSAGE TO TRAFFIC MEN

By Brice Clagett
Assistant to the Director-General

It has become a byword that the traffic men of the railroads are the eyes and ears of their companies, but it is becoming more and more necessary that the railroads have mouths as well, and the traffic men naturally fall heir to this function.

Whatever may be the form of control of the railroads in the future, everyone seems to admit that there will be closer public supervision than ever before, and closer public supervision means inevitably that the public must be kept better informed than ever before of the details of railroad operations and finances. Therefore the railroads, which have been under close public scrutiny for years past, must expect that this scrutiny will be intensified in the years to come.

Recognizing, as they must, the public character of their service, it is the highest form of intelligence for the railroads to recognize this situation frankly and meet the public more than halfway, not hiding from the public view but inviting pitiless publicity and welcoming the interest of the public in the railroad business.

Unquestionably many of the difficulties of the railroads in the past, both rate and wage matters, have resulted from misunderstanding, both on the part of the public and on the part of labor. Believing, as all forward-looking Americans must, in the high intelligence of the average citizen of our country, is it not the wise policy to keep the facts constantly before the public, realizing that, once the problems of the railroads are thoroughly understood, a right and fair solution will be brought about?

The most generally supported solutions for the railroad problems appear to involve an official valuation of the properties, and, once this is brought about, there should be a stopping of the

constant suspicion that existed in the past that the railroads were overcapitalized. The removal of this suspicion should go far toward making easier the solving of wage and rate questions in the future, but this will not settle the entire difficulty; there must be a day-by-day enlightenment of the public, and it is in this work that the railroad traffic men, it seems to me, will play an important part.

The experiences of the United States Railroad Administration, occupying as it has a position where it has been constantly in the public eye, have proved how deeply the public have become interested in their railroads. This interest has existed in the past, but never has it been so intense as at the present time, and it seems certain that this interest will continue for many years to come.

The experiences of the Railroad Administration, too, have brought out more clearly what has been realized by operating and traffic men in the past: that the public does not always understand the many difficulties involved in the current operation of the railroads. I fear there has been too much of a disposition to take it for granted that facts which were self-evident to railroadmen were equally self-evident to the public. This is a tendency which, it seems to me, must be corrected in the future if railroad operations are to have the necessary public support and if a fair return on the money invested in railroad properties is to be assured.

Take the question of the granting of special reduced passenger rates for example. Every traffic man has experienced the attitude of those who urge special rates, regardless of the effect on operating revenues. Every passenger traffic man realizes that travelers should not be transported at less than cost, and yet he is constantly subjected to demands

that just this thing be done. Believing in the fairness of the average American, I cannot help but think that, once the facts of the case of railroad operations are made clear to the public, these demands will be reduced to a minimum. The average American is not used to getting something for nothing, and there is no reason why he should get it any more from the railroads than from other industries of the nation.

How often are we told that we should grant a special rate to earn the good will of the members of some organization or some special class of travelers, and how little is the cost of transportation appreciated at the present time. The sentiment against granting special privileges to a few has been growing stronger and stronger in recent years, and yet the requests for special privileges so far as the railroads are concerned continue undiminished.

It has been the policy of the United States Railroad Administration to do its utmost to establish standards, both as to rates and practices, which in their general application can be continued after the return of the roads to their private owners.

Much remains to be done in this direction and manifestly it has been impossible in the brief period of Government control and operation to standardize everything or to do away with practices which all railroadmen admit are bad for the railroads, but which have grown up in part as a result of fierce competition.

Much will depend upon how earnestly the effort is made after the return of the roads to private control to restrict a return to some of these harmful practices, and the traffic men, both of the railroads and of industries, naturally will play a most important part.

But the effort to resist the return to these practices is destined to certain failure unless the public is constantly kept advised of the facts concerning railroad operations.

It has been the deliberate policy of the Director General of Railroads to use every possible means of keeping the public so advised during Federal control, and there are indications that this effort is meeting with gratifying results.

The public will be slow to demand special concessions in rates if it is generally recognized and honestly believed that the rates in existence at any particular time do no more than insure a fair return on the money invested. They will continue to demand special concessions, however, if they think they are being fooled by the railroads—if they think there is a "nigger in the woodpile."

The traffic men of the railroads are the ones who come most constantly in contact with the public, and therefore they, as no other class of railroad officers or employees, can make certain that the public is kept advised of the facts.

Therefore it would seem as though every railroad should adopt the policy of entire frankness, both with the public and with labor, and that the traffic men should be made the heralds of this policy.

THE BULLETIN

S. P. INSTRUCTORS TEACH APPRENTICES ELEVEN TRADES

Youthful Mechanics Become Skilled Workmen in Four Year School Course That Links Classroom With The Shop

View of apprentice classroom at Sacramento Shops, and (below) boilermaker apprentice studying blueprints under guidance of instructor.

The Southern Pacific Railroad offers unusual opportunities for young men between the ages of 16 and 21 to become mechanics in its various shops over the system. Ambitious youths, mechanically inclined and possessed of a grammar school education, will find that a four-year apprenticeship course with the Southern Pacific, combining classroom and workshop, will enable them to qualify as expert mechanics in any one of eleven trades. Not alone that but they are given an attractive rate of pay from the start, and this increases every six months until graduation, when the regular rate of pay of their craft is established. In some branches of the service this is 67 cents an hour, in other branches 72 cents.

The Southern Pacific established its first apprentice school at West Oakland under Master Mechanic A. C. Hinckley eight years ago, modeling its course after the ideas of Mr. George M. Basford, father of the national movement in behalf of such schools. The idea proved so successful that similar schools were established for apprentices at Sacramento, Sparks, Ogden, Brooklyn, Ore., Dunsmuir, San Francisco, Roseville, Dallas, Ore., Beaverton, Ore., Tucson, Ariz., Los Angeles and Bakersfield.

Today there are approximately 700 pupils and a faculty of eleven skilled instructors. Last year fifty-one apprentices received diplomas and the pay rating that accompanies a full-fledged artisan. Since 1912, four hundred and twenty-two young men have completed the four-year course of instruction in Southern Pacific apprentice schools, and of this number three hundred and forty-one remained in the company's service. Many graduates of the course have risen rapidly to positions of responsibility. R. P. Peek, superintendent of apprentices at Sacramento, is himself a graduate from the railroad shops at this point. So is R. E. Hanson, instructor at Tucson, and W. J. Bartle, assistant chief draughtsman at Sacramento, as well as many assistant foremen at various points on the system.

Attendance Compulsory.

Attendance at the apprentice schools is required of all young men who desire to become railroad mechanics. In fact, at most points on the Southern Pacific Lines the apprentice instructors are in charge of shop employment. Preference is given to the sons of employees, but any willing ambitious young man between 16 and 21 who can read and write and understands ordinary arithmetic will receive consideration if he makes application to any of the following instructors: P. S. Gillette, West Oakland Shops and Shipyard; R. P. Peek, Sacramento; J. W. Hanson, Roseville; A. M. Finn, Sparks; M. W. Cram, Ogden; R. W. Quigley, Dunsmuir; John Mc-

Cullough, Bayshore, San Francisco, and San Luis Obispo; M. C. Meston, Bakersfield; L. B. Verweire, Los Angeles; R. E. Hanson, Tucson, and L. C. Lewis, Brooklyn, Ore.

Apprentice schools offer training along theoretical lines which supplements the practical training received in the shops. Unlike the public school system, instruction is individual so that a student may enter at any time and progress as fast as application and aptitude permit. The largest school is maintained at Sacramento and may be cited as typical of the standard established by the Southern Pacific. Here over two hundred young men are receiving instruction in eleven different trades. Some are learning the building and repairing of locomotives, passenger and freight cars; others general machine shop practice and blacksmith's work. Still others are studying to be boilermakers, electricians, sheet metal workers, brass, iron and wheel foundrymen and cabinetmakers.

Probationary Period.

The first six months of the first year of apprenticeship is known as a probationary period, during which time a boy's fitness for his chosen work is determined. If it is found at the end of this period that the young man does not possess the proper qualifications for a good mechanic his parents or guardians are so informed and the cause stated.

Apprentice schools are operated two hundred and eight hours per year. At Sacramento, Superintendent Peek conducts three classes a day every day in the week. The periods are from 7 to 9, 10 to 12 and 1 to 3. Shop mathematics, blue-print reading and elemental geometry and trigonometry form the basis of instruction, together with such subjects as are along the line of the student's apprenticeship. During the four years' course of study the instructor keeps a careful record of the student's progress both in the classroom and in the shop and notifies the foreman when the student has served the required time on each class of work.

Attractive Pay.

The rate of pay is attractive. For the first six months the rate is 29 cents per hour with an increase of 2½ cents per hour for each six months thereafter, up to and including the first three years; 5 cents per hour increase for the first six months of the fourth year and 7½ cents per hour for the last six months of the fourth year. Students then receive the minimum rate of pay of their craft, which, in some branches of the service, is 67 cents and in other branches 72 cents. Outside of these rates, however, there are opportunities for higher rates of pay for specialists, a considerable number of which are required throughout the various shops.

As an indication of the enthusiasm and interest shown by apprentices in their work it may be stated that over one-half of the safety suggestions received by officials of the Sacramento

MR. TRAINMAN

If You Were Paying the Bills You Would—

Not loose your head and smash your lantern against side of a box car just because the engineer or fireman did not get your signal quickly enough.

You would see that the old journal bearing is returned to the caboose when you apply a new one. These would cost you six dollars each.

You would see that the caboose supplies and equipment are properly cared for, and when necessary to replace them the articles are returned in exchange so that they can be repaired and reissued.

You would willingly accept and use a good second-hand lantern and give it the same care as a new one, knowing the cost is but one-half.

You would not use fusees in inspecting a train when your lantern will answer the purpose. Fusees are expensive and you can waste several of them on the inspection of an average train. Of course, they make a better light and are handy, but how about the expense? There are sometimes oil cars in these trains and the two make a bad combination. SAFETY FIRST.

You would hesitate about breaking a switch lock with a rock or bar just because it did not open promptly. It only takes a few seconds to make out a report.

You would look around and pick up all the tools you were using in making temporary repairs to a car instead of leaving them on a tie or on the end sill of a car only to be jarred off when the train is started.

You would not tear a flag off the stick and use it for cleaning lantern globes or to shine your shoes with, or even as a handkerchief; it would be rather an expensive practice, wouldn't it? Save the stick also, a new flag can be put on it.

You would not use waste and oil in starting fires in your caboose stove.

You would take care of chair cushions and not allow stockmen and others to carry them away—$2 shot to pieces every time this happens.

BEAR IN MIND THAT THERE ARE MANY, MANY MORE MEN IN THE EMPLOY OF THE RAILROADS WHO ARE FILLING THE SAME POSITION AS YOURSELF, AND A LITTLE SAVING FROM EACH OF YOU EVERY MONTH IN THE AGGREGATE AMOUNTS TO MANY THOUSANDS OF DOLLARS.

MR. E. P. RIPLEY DIES AT SANTA BARBARA.

By J. E. SLOAN
(Agent, Santa Barbara)

Santa Barbara and the railroad world suffered a distinct loss in the death, February 4, at his home here of Mr. E. P. Ripley, chairman of the board of directors of the Atchison, Topeka and Santa Fe Railroad. Mr. Ripley's railroad career, stretching over a period of sixty years, will long remain an inspiration to American youth. Many prominent railroad men attended the simple funeral services and telegrams poured in from all portions of the United States.

shops come from apprentices, and many of these suggestions have been found worthy of adoption.

The youth is father to the man. By giving earnest ambitious young men a chance to work and learn at the same time by extending the helping hand when it is most needed, the Southern Pacific is being well repaid by the splendid results it is getting from its mechanical forces replenished each year by skilled and enthusiastic apprentice graduates.

STEAMER SANTA CLARA RE-ENTERS SERVICE.

The ferry steamer Santa Clara, which was rebuilt at West Oakland after having her superstructure destroyed by fire on January 25, 1919, made a trial trip around San Francisco Bay on February 20, 1920, with a number of company officials and other invited guests.

The trial trip proving satisfactory, the Santa Clara went into service on the Oakland pier ferry route on the following morning, relieving the steamer Alameda, which was tied up temporarily for repairs. The Santa Clara will remain permanently on the Oakland pier ferry route, and the Berkeley will eventually be transferred to the Alameda pier route where she formerly operated.

Generally speaking the Santa Clara is a sister ship to the Alameda, but there are several points of difference, such as improved steering gear, marine telegraphs for signaling between pilot houses and engine room, different design of pilot houses, and officers' quarters on the hurricane deck, allowing more seating room on the saloon deck.

working on various Divisions as lineman, Telegraph Dept. He retired with 26 years 11 months service.

Harry Highfield, electric stationary engineer, Salt Lake Division, entered the service on July 1, 1899, as caller at Carlin, Nevada. In April, 1917, he became engine dispatcher at that point where he remained until September of the same year. From the latter date to his retirement, he has been working as electric stationary engineer, retiring after 25 years 6 months service, the age of 61 years 6 months. Mr. Highfield died February 24, 1925.

Samuel G. Wonderlich, yardman, Coast Division, entered the service in April, 1900, as brakeman, Coast Division. From that date until his retirement, he has worked as brakeman and yardman at various points on that Division. He has been in the service of the Company for 24 years 10 months.

William F. Cummins, geologist, entered the service January 15, 1903, at Houston as geologist for the Rio Bravo Oil Co. He has served in that capacity with the Rio Bravo Oil Co. and the East Coast Oil Co. until his retirement April 1, 1925, after 20 years 7 months service.

HOPKINS TRAINMASTER AT WATSONVILLE JUNCTION

L. P. Hopkins has returned to the position of trainmaster at Watsonville Junction on the Coast Division, after a leave of absence during which time he was Representative of the Western Regional Operating Express Committee Association of Railway Executives. Trainmaster L. C. Gram has been appointed representative on this committee, with headquarters in San Francisco.

EASTER EGGS DISTRIBUTED TO KIDDIES ON TRAINS

Dining car stewards played "good fairies" to children traveling on Southern Pacific limited trains Easter. Baskets of easter eggs and candy rabbits were distributed to the little tots. Easter egg hunts were staged and everything possible was done to make the celebration complete.

Special Easter dishes were served on the company's dining cars.

Apprentice Schools Have Enrollment of 638 Young Shop Men

THERE were 1318 apprentices on Pacific System during February. There were 680 helper apprentices. A total of 630 regular apprentices and 8 special apprentices were in attendance at the eighteen apprentice schools at general and divisional shops on the System. The Sacramento general shops school leads with the largest enrollment, having 298. The classes cover the entire system represent machinists, boilermakers, blacksmiths, sheetmetal workers, electrical workers, freight and passenger car men, molders and spring makers.

The apprentice schools were opened in 1913 since when nearly 1000 have graduated. Of these about 70 per cent are now in the service of the Company. Every apprentice is required to attend the school. They are placed on six months probation after which time they are either dropped or permitted to continue depending on their progress in the work assigned them. Students receive pay while at class during the four hours of each week. The complete course which qualifies them for journeyman rate of pay requires a period of four years service with the company. Regular apprentices are from 17 to 22 years of age and have had a common school education. Helper apprentices are over 23 years of age and generally without any special or general education. Special apprentices are men of college or university training who wish to take the practical work in order to gain a better understanding of the shop problems and methods.

The apprentices are assigned different kind of work so they become acquainted with a variety of problems. Their work is changed once at least every three months. The class room study is assigned to cover subjects applicable to their shop work. The subjects are mechanical and geometric drawing and mathematics. School periods are held from 7 to 9 and 10 to 12 a.m. and from 1 to 3 p.m. twice a week for each class.

Graduates of the apprentice school have proven the worth of the school by the positions they hold with the Company. Many are in positions of authority and instruction. R. P. Peek, now instructor at Sacramento shops is a graduate of the school, and has been supervisor for nearly nine years. E. M. Pitt and Ray E. Hansen, instructors at the Dunsmuir and Tucson schools, are also graduates. J. C. Hanssen night round house foreman at Roseville, and Tony Munizich, valve setter at the Sacramento shops, are conspicuous among the graduates holding important positions with the Company.

CONDUCTORS HOSTS AT DANCE IN LOS ANGELES

The annual ball of the Order of Railway Conductors of Los Angeles Division, held April 16 at the Masonic Hall, was declared to have been one of the most successful gatherings of Southern Pacific employes in Los Angeles in recent months. About 600 were in attendance.

The grand march opened the festivities, after which dancing was enjoyed until midnight. C. M. Murphy, assistant superintendent, and Mrs. Murphy led the march, at the conclusion of which Mr. Murphy in a novel manner introduced each couple to the other dancers.

Among the guests of the conductors was Wm. Nichols, chief examiner of San Francisco, who was said not to have missed a single dance during the evening.

Excellent music was furnished by Hatch's Radio Orchestra. In addition to the dancing, refreshments were served during the evening.

As an unique feature, each dance was named for a well known Southern Pacific train, the program detailing the time of departure from Los Angeles and destination. For example, the first dance, a waltz, was the "Daylight Limited," followed by the "Shore Line Limited," the "Owl," the "Sunset Limited," and other such familiar names.

Many compliments for making possible a good evening of entertainment were given the committee in charge composed of Harry A. Sullivan, Shon M. Schweizer, L. E. Smitley, Ira B. Gotfredson and Chas. A. Bell.

Section 4 of the Apprentice School at Sacramento General Shops: R. P. Peek, supervisor of apprentices, is at the extreme right. The young railroad shop men are, left to right: G. Osuma, S. Roundini, V. De Riso, R. Kieweg, A. Johnson, W. Hetherington (back), E. L. Jones, A. S. Schiro, E. Jackson, C. Nicholas, Wm. Boyd, B. Patterson, A. Dal Porta, Wm. Mansir, J. Irvine, S. Lubetich, A. Addison, E. E. Peek, G. Da Shiell, M. Loyola, A. H. O'Hara, H. Bennetts and L. Bertolucci.

Chapter Ten
Safety – Always Important
in the Shop and on the Train

On July 2, 2010, the Department of Transportation and the Federal Railroad Administration published a document called *Railroad Safety Appliance Standards, Miscellaneous Revisions*. The document, available in the Federal Register, states that the expansion of the railroad system after the Civil War led to "excessive" numbers of deaths and injuries among railroad employees. The report states, "during the (8) eight years prior to the passage of the first Safety Appliance Act in 1893, the number of employees killed or injured was equal to the total number of people employed by the railroad in a single year."

Indeed, before the Federal Railroad Administration, the Occupational Safety and Health Administration, minimum wages, steel-toed boots, safety glasses, and reflective vests existed, there was a long list of hazardous situations involved in railroad operations. For example, brakemen had the unenviable job of standing on the roof of the car to turn the wheel to operate the hand brakes. This was to be done if the temperature was 100 degrees in Colfax or 10 degrees on top of an ice-covered boxcar at Donner Summit. Early coupling methods required the brakeman to stand between the cars when they were connected or uncoupled. Benefits of using a radio to coordinate any train movements didn't exist in 1900.

Between 1890 and 1892, Congress addressed the problem with seventeen bills designed to improve the safety of employees and passengers. During March of 1893, the first Safety Appliance Act was passed requiring power brakes, rather than hand-operated brakes, on all trains engaged in interstate commerce. These trains also had to have handholds and automatic couplers. In 1903, the second Safety Appliance Act broadened the rules to include any railroad equipment in interstate commerce. Seven years later, the third Safety Appliance Act led to rules governing the number, dimensions, and location of safety features on boxcars, hopper cars, gondola cars, tank cars, flat cars, cabooses and locomotives.

In 1868, the Central Pacific built a hospital, thought to be the first industrial hospital in the world, for both employees and passengers involved in accidents. The facility, which cost $64,000, was a four story wood structure located on the southwest corner of Thirteenth and C Streets and could accommodate 125 patients. It was sixty-by-thirty-five feet with two wings fifty-two-by-thirty-five feet and a kitchen twenty-four feet square. There were ten private rooms and six large wards.

Between 1868 and 1877, it served about 3,600 patients and handled 7,750 office visits. Operated on a non-profit basis, the hospital was funded by employees, the railroad, grants, and gifts. The financial support was enough to fund all staff and materials. Some evidence indicates that the monthly contribution was fifty cents per month from each employee. These payments covered most of the monthly cost of about $1,600 per month. The facility was not full-service, however, so major surgeries were performed at the railroad hospital in San Francisco.

By 1904, improvements continued with a new brick hospital and a hospital rail car was used to transport seriously injured personnel.

In 1912, yet another hospital was open for business on Second Street.

Accidents are not always the result of poorly designed equipment. Fatigue, the work environment, and poor regulations can be a large factor with any job. As the railroad industry grew, so did the number of legislative actions to improve safety.

Other major safety acts passed during this time include:

1907 - The Hours of Service Act limited the work shift of railroad employees involved in train movements to sixteen hours. Ten hour days were common to about 1916, and by the late 1930s, eight-hour shifts were common. Up to the late 1930s, most shop workers gave their effort six days per week with Saturday afternoon and Sunday off. (In 1974 and 1976, shift durations were further reduced to fourteen and twelve hours, respectively.)

1908 - The Transportation of Explosives Act issued regulations covering the marking, loading, and handling of explosives and other dangerous commodities.

1910 - The Accidents Report Act required each railroad to file a monthly report with the Interstate Commerce Commission (ICC) listing all collisions, derailments, and accidents resulting in injuries, damage to equipment, and damage to the roadbed or related structures. The Act gave authority to the ICC to investigate the accidents.

1911 - The Locomotive (Boiler) Inspection Act established safety requirements for steam locomotive boilers.

The evolution of safety on a railroad started with nearly nothing and now has a long history of improvements. Longer work days increase the risk of injury. The heavy machinery and equipment with moving parts of a railroad shop accentuated the need for safety. Safety was also a function of the weather because some work was done outside. Beginning in the 1920s, most carmen in Car Shop 9 worked outdoors and needed to bring their own rain gear. During the cold of winter, they burned scrap lumber in empty oil drums for heat. Unfortunately, escaping summer heat was more insurmountable, as they worked at jobs such as replacing paneling inside of steel boxcars.

Noise pollution posed a problem: The erecting shop was known for being extremely noisy due to cranes, riveting, and pounding. Another safety factor was air quality. The foundry was dark, dirty and poorly ventilated. Consequently, the dust produced by grinders posed a constant hazard. Also, asbestos was used in boiler jackets and passenger cars.

It appears that the Southern Pacific did not have a formal workman's compensation policy until mandated by state law in 1913. Instead, injured workers were placed in less physically demanding positions when possible.

Accidents in the Shop and on any moving equipment were recorded and investigated. The accidents that occurred were categorized by the division and department of the railroad in which the accident occurred. The September 1922 issue of *The Bulletin* featured an article by R. H. Wells who was a conductor in the Sacramento Division. One of his points of emphasis was the importance of situational awareness in the train yard. The attention to the activity *around* the employee was as important as the work itself.

Other safety-related articles in *The Bulletin* are:

"Sacramento Safety Week Big Success," December 1923

"Piecework Inspector – Mr. Rippon Holds Record for Safety Improvements," July 1919

"The Conductor – Always a Factor in Safety Work," September 1922

"Little Things That Make Safety A Success," July 1940

SACRAMENTO "SAFETY WEEK" BIG SUCCESS

Just a few of the many original posters which were painted by Sacramento Shop employes for display during the recent "No Accident Week" at the shops.

By D. S. WATKINS
Superintendent, Sacramento Shops

DURING the week October 28 to November 3, Sacramento Shop employes made an excellent safety record by going through this week without a reportable accident. This record, considering the numerous hazardous occupations at Sacramento Shops, is a wonderful achievement and shows the loyal support of all officials and employes given this humanitarian effort to reduce avoidable suffering to their fellow-employes.

Avoidable suffering, I will term it, because over 98 per cent of our accidents are caused by carelessness, and the recent "No Accident Week" shows that if every one will combine a little thought with his work, he will save himself much unnecessary suffering. This reflects upon him as a workman, for a careless workman is not only a detriment to his employer but a dangerous man to those working with him.

The foreman plays an important part in this safety movement. It is one of his responsibilities to see that his work is carried on in a safe and sane manner. Our week proved that while every foreman was on the alert we had no accidents. If it were possible to make a check on shop output we would find that quality and quantity production was higher than during weeks where special attention was not given to safety work.

Committees Named

James Hall, Master Car Repairer, was appointed Chairman of the "No Accident Week", with H. Kriesel, Geo. Wanger, J. McNevin and R. Peek to assist him. They were assured by the officials at Sacramento Shops that they stood ready to assist in putting over this work. Much credit is due to painters, Justo Martin, C. P. Hilder, K. Knecht, W. Higgins, C. Lindfelt, C. Delchario, W. Beckett, J. P. Steffani, E. F. Hannon and F. P. V. Parisi for their valuable work in preparing the many interesting pictures and signs for this week.

Monday morning, October 29th, when the shopmen entered the gate they were greeted by a large poster informing them of the intentions of making that week a campaign against accidents. Upon entering their respective shops they found them flaming with attractive safety bulletins and pictures and were greeted by their foremen who asked their co-operation in putting this movement over.

In order to prevent the campaign from becoming a tiresome affair, the bulletin board posters, large signs and pictures were changed regularly. At the completion of each days' work, a large sign informing the men that no accidents had occured that day in the shops was displayed over each exit, along with new signs with excellent safety slogans, asking the employes to take the safety work into their homes.

The interest aroused and the comment caused by pictures and posters proved that it pays to advertise "Accident Prevention" as well as other movements.

Upon completion of the week, after coming through without a reportable accident, a large poster was placed over each exit, thanking the employes for their efforts and asking continued co-operation in keeping down accidents.

The pictures and signs are still in excellent condition and could be used in other shops to carry on programs similar to the successful one just completed at Sacramento Shops. With the united efforts of all employes on the System, and an extensive advertising program to keep safety always before our eyes we would be doing our part in an effort equal to that of stopping a world war, for more men are killed and injured yearly through accident than were lost by this country in any one year of the great world war.

RAILROADS LOWER RATES AS COSTS COME DOWN

Since the summer of 1920, the railways have reduced their operating expenses about $109,000,000 a month and have passed nine-tenths of this saving, or about $97,000,000 a month, along to the public in reduced cost of transportation, according to figures received by the "Railway Age".

The Railway Age's comparisons are made for the four months period of May, June, July and August of 1920 with the same period in 1923. This period is chosen for comparison, since the wage increase of 1920 were made effective from May 1, 1920, and the increase in freight rates of

Keen Rivalry for Safety Banners

AT the end of the nine months period the Sacramento Division is in the lead for the steam divisions, the Los Angeles Shops in the lead for the General Shops and the East Bay Electric Division has a better record than any of the steam divisions, in the contest for 1923 Safety Banners.

These banners are awarded on the record made of least number of casualties to employes per combined million man-hours and locomotive and motor miles. Rivalry among the Divisions and the two General Shops is very keen. In the latter there is a difference of less than one point, and every effort is being made to obtain the coveted prize. Commencing with December 1st, Division Safety meetings will be held monthly instead of bi-monthly.

Standing at the end of the nine months period is shown in the following statement:

DIVISIONS

Name	Number of Casualties	Man-Hours and Loco., Motor and Steamer Miles	Casualties per Mil'n Man-Hours, Loco., Motor and Steamer Miles	Rank
East Bay Elec.	10	2,908,260	3.44	1
Sacramento	59	11,635,820	5.07	2
San Joaquin	69	7,715,516	8.94	3
Portland	139	14,724,307	9.44	4
Coast	171	14,470,898	11.82	5
Steamer	30	2,493,315	12.03	6
Stockton	66	5,062,171	13.04	7
Salt Lake	209	13,926,736	15.01	8
Shasta	106	6,737,202	15.73	9
Tucson	154	9,412,006	16.36	10
Western	205	11,012,126	18.61	11
Los Angeles	371	13,999,235	26.51	12

GENERAL SHOPS

Name	Number of Casualties	Man-Hours	Casualties per Mil'n Man-Hours	Rank
Los Angeles	131	3,621,208	36.02	1
Sacramento	214	5,800,872	36.88	2

that year was not made effective until the end of August.

"The wages paid in May-August, 1920", says the Railway Age, "averaged about $334,000,000 monthly, while in 1923 they averaged only about $260,000,000 monthly, a reduction of approximately $74,000,000 a month. But all this reduction in wages was not secured by reducing the average wage paid to each employe. The average number of men employed in May-August, 1923, was almost 129,500 less

than in May-August, 1920, the saving due to this reduction in the number of employes which was a result of an increase in efficiency of operation, was about $16,500,000 a month. Therefore, of the total reduction in operating expenses secured about $57,500,000 a month was due to a reduction of wages, and the remaining $51,500,000 a month was due to reductions in the amount of labor employed and to other economies that were effected.

"Rates were not advanced until the end of August, 1920, and in consequence large deficits were incurred which were covered by government guarantees to the companies that were in effect until the end of August, 1920. The total amount the public paid in rates in May-August, 1920, was $2,037,000. When the deficits the public had to pay are added, and the taxes paid during these months by the railways to the public are deducted, it is found that the total cost to the public of the transportation service rendered was $2,465,000,000 or about $616,000,000 a month.

"In May-August 1923 the amount paid by the public in rates was $2,189,000,000. Deduction of the taxes paid by the railways to the public shows that the net cost to it of the transportation service rendered in May-August, 1923, was $2,074,500,000. This was less than $519,000,000 a month, or over $97,000,000 a month less than in May-August, 1920."

Common sense is the ability to take the worst and make the best of it.—Exchange.

Officers and committeemen of the Shop Craft Protective League at Sparks. Front row, left to right—F Lanigan, boilermaker, Boiler Shop Committeeman; Peter Smith, piper, Guard and Pipe Dept. Committeeman; H. L. Hawkins, machinist, Machinist Committeeman; Frank Loveridge, coppersmith, President; Geo. Young, machinist, Trustee. Back row—Ed. Cantlon, roundhouse, Machinist Committeeman; H. L. Burt, machinist, Erecting Dept. Committeeman; E. F. Gothberg, car repairer, Car Dept. Committeeman; H. A. Lessenger, machinist, Secretary-Treasurer; H. W. Moss, coach carpenter, Trustee; I. L. Glean, blacksmith, Blacksmith Committeeman.

PIECEWORK INSPECTOR TELLS HOW HE DOES IT.

By H. C. VENTER
(Sacramento General Shops)

It has been a pleasure to the employees of the Sacramento general shops to learn that Mr. C. H. Rippon, piecework inspector of the locomotive department at this point, has for the third time carried off the honors for the year by having done most in furtherance of safety work. He has the largest number of credits for the system, as well as for the general shops. This is his record:

Period ending January 30, 1916, he had 1923 credits; period ending June 30, 1917, he had 1895 credits; period ending September 30, 1918, he had 2841 credits, or a total of 6659 credits for the three periods, a record of which he can well be proud.

Mr. Rippon is a product of the Sacramento shop. He entered the service of the company as a machinist apprentice in the year 1875, and has been in the service continuously ever since, which gives him forty-four years of faithful service. After completing his apprenticeship he worked at his trade until the year 1913, when he was appointed to the position he now holds. His duties take him over a good portion of the plant every day, and having enlisted under the banner of safety first he is ever on the alert for the things that are unsafe.

There is a history back of Mr. Rippon's activities in safety work which should be quite interesting to the readers of this paper. How came he to get started in this work? Several years back, in the course of his duties, it happened that one day he was in charge of a number of men who were transporting some locomotive castings that had to be shipped in a hurry. In order to get these castings to the store department to load them on a car it was necessary that he take a wagon pulled by a number of men to a platform for loading. It so happened that the castings were delivered safely, by taking the proper runway to the platform, but on returning he thought he would make a short cut in order to get back to the shop in quicker time, and to do so they started to lift the truck off the platform without using the runway. The result was that the wagon tipped over, injuring one of the employees slightly. A few days later Mr. Rippon received a letter from the shop superintendent reprimanding him for such acts, and from that time to now he has been a student on safety work, and if a short cut is to be made in his work it must be the safest way.

Mr. Rippon often states: "I may offer and suggest many safety measures, but if the employee is not careful and does not take advantage of the facilities furnished for his use it goes for nothing. It is a greater loss to the individual than to the company he is serving. The company may spend hundreds, yes, thousands of dollars for safety, but if the employees would give a second thought before taking a

Safety Specialist

C. H. RIPPON
of Sacramento.

chance what a wonderful improvement there would be in the accident reports."

Baseball.

On June 4th the United States Railroad Administration baseball team, playing upon the Twenty-first and C streets diamond, administered a decided defeat to "Harty's Hustlers," score 8 to 3.

Until the eighth inning it was nip and tuck, then the Administration boys opened up with some heavy stick work and shoved five markers across the plate, Hinton, Lofquist, Asmy, Downey and Hoagland being the star clouters.

The winners are now looking for new fields to conquer. Write to Wm. Regan, 1714 Twelfth street, Sacramento.

CENTRAL WESTERN REGION LED THE FIELD.

Figures which were received too late for publication in the last issue of The Bulletin show that the central western region led all others in the Fifth Victory Loan, both in per cent of employees subscribing and average amount per subscriber. In expressing to Federal Manager W. R. Scott his pleasure at the showing made Regional Director Hale Holden announced also that the central western region ranked first among all classes of employees except roadway, the eastern region outranking us .8 per cent for that class of officers and employees.

SAFETY AND LOAN MEDALS ARE DISTRIBUTED.

By J. R. HOCKETT
(Superintendent's Office, Western Division)

H. J. Nops, brakeman; F. Schnetzler, yardman; W. Parks, yardman; H. B. Cook, engineman; F. S. Dreschler, conductor, and H. L. Taylor, agent, constitute the six employees on the Western Division who, during the period July 1, 1917, to September 30, 1918, did most in the furtherance of safety work. A gold medal suitably engraved has been presented to each of these employees.

The following locomotive and car department employees have been presented with medals for their good work in the Victory Loan drive: B. F. Harris, F. Fei, A. G. Krenkel, J. J. Foley, E. L. Greenwood, W. H. Burke, C. L. Gibson, J. H. Korb, J. J. Orell, W. A. Herb, J. P. Flinn, J. R. Breedlove, J. L. Newton, S. Clifton, C. F. Goode, Wm. Lang, R. Cremer, C. K. Murray, A. J. Moylan, A. Thurlow, C. W. Russell, O. S. Prince and Mrs. Sarah Diamond.

Latest reports are that Mr. E. E. House, road foreman of engines, is improving rapidly.

Engine 2376, one of the ten-wheel type engines being built at Sacramento shops, has just been received on the Western Division and has been assigned to handling heavy passenger trains on the Portland run. Engine 2377, now under construction, will shortly be turned out and will also be assigned to the Western Division. These engines are superheated, having 69-inch driving wheels and 23x28-inch cylinders. Their weight on drivers is 174,000 pounds, the total weight being 218,500 pounds.

The following Western Division employees discharged from the Army service have re-entered the service of this company: A. Christensen, station truckman; W. J. McNally, yard clerk; Harold W. Sanford, fireman; W. H. Lucas, freight brakeman; E. C. Young, timekeeper; Arthur M. Rico, warehouseman; Alfred D. Rico, warehouseman; F. J. Wuelzer, yard clerk; H. B. Whitehouse, electrician and hostler; J. C. Saunders, yardman; A. C. Sobranes, car inspector; E. G. Nielsen, passenger brakeman; A. Flaherty, section foreman; John W. Parker, operator; C. V. Neukirch, gateman; Jas. H. Cockrill, train electrician; L. C. Asmussen, train gateman; Harry O. Walters, station truckman; E. T. Mulligan, yard clerk; A. Rank, truck driver; C. L. White, motorman; J. P. Fernandez, operator; M. Conners, yardman; V. Skaggs, signalman; Fred Holl, station truckman; H. R. Dallas, signal maintainer; J. L. Bowe, fireman; A. A. Cabral, station truckman; S. E. Vivelich, timekeeper.

L. H. Vincent, division accountant, and C. Goldberg, superintendent's secretary, are at present enjoying vacations at Yosemite Valley and Raymond.

The Conductor as a Factor in Safety Work

'Don'ts' for the Student Brakeman Are Pointed out by Conductor who Urges Close Observance of Rules as Best Safeguard

By R. H. WELLS,
Conductor Sacramento Division

RAILROADING has always been considered one of the most hazardous, as well as one of the most fascinating occupations, but thanks to the Safety movement, traveling on a railroad like the Southern Pacific, where a vigorous Safety campaign is conducted, both among the employes and the traveling public, the hazard has been removed, until today it is safer to drive and ride on a railroad passenger train under the watchful attention of train Safety employes than remaining at home and following our ordinary callings.

The last decade has often been referred to as the mechanical age, but the present can be referred to as the Safety age, or, in its last analysis, the Christian age, for in Safety we are taught not only self preservation but that we are also our brother's keeper.

From a Conductor's standpoint the first and most important duty is to know that the train is properly protected at all times. On a freight train before leaving an initial station he assures himself that the train is properly made up according to standing instructions with steel equipment on the head end and the weaker or wooden frame cars on the rear or near the caboose; that the Federal and State laws are obeyed in the makeup of the train in regard to the transportation of explosives, acids and inflammables. He sees to it that a train never leaves an initial station, or any other station or place where the train pipe has been parted, until the air brakes are thoroughly charged and in continuous working order from the engine to the caboose.

For Student Brakeman

In train service the safety and movement of a train is vested in the conductor and this gives him a great chance to be a missionary in the safety movement. From the day the student brakeman, or fireman, fresh from school or the farm, enters the railroad service until he becomes a conductor, or an engineer, the conductor keeps his eye on him, instructing him in the avoidance of all hazardous practices, such as kicking over draw bars with the foot to make a coupling; the getting on or off fast moving engines or trains; the close inspection of the running gear of trains at all stops; care in switching, especially over street or road crossings; never to make a running or flying switch when possible to do otherwise; never to cut a train in two until the air hose has been separated by hand, because if this is not

R. H. WELLS

done it strains the hose fibres and causes spongy, or leaky hose that sooner or later burst, often while the train is running, doing other damage to equipment and endangering the lives of all on the train, often pulling or straining the train pipe connections, leaks from which are often as costly or dangerous as leaky or bursted hose; see that local freight in way cars is properly broken down in cars after unloading at all stations; and that cars, especially flats or gondolas loaded with rails, lumber, rock or gravel are not unloaded only from one side while a heavy load is left on the other side of the car, because many derailments are often traced to shifted, or uneven loading.

A conductor should see that all under him concentrate their thought on their work, but not to such an extent as to put themselves in hazardous positions. Records prove that never a year goes by, but some old experienced valuable employee is killed through over concentration on work in hand, like several conductors I have known or heard of, who were walking up one track in a busy yard booking their trains and were killed by having cars shoved over them. A case of too much concentration of thought on work at hand without the accompanying thought of Safety.

Double track, while necessary to increased business, also creates new hazards, and no one should stand on or close to a second track while a train is passing, because the noise of the train prevents anyone from hearing the approach of a fast moving train or light engine on the second track.

Care With Fusees

The use of fusees in flagging, or thrown from a moving train for the purpose of spacing following trains, should never be placed or thrown in the middle of the track, or stuck in the end of ties. This often causes the burning out of cross tie, weakening the road bed and creating an unnecessary hazard. Another reason why the fusee should be placed from between the rails, or from the end of a tie is in case the following train does not get stopped until possibly the engine and several cars have passed over it, the train may be stopped with a gasoline tank or some other highly inflammable car directly over the burning fusee, and in case of a small leak from a gasoline tank the loss of life and property can only be guessed at by the explosion that might take place in a populous community.

Almost every careless and unsafe practice is a violation of some rule already in effect, so that the first and last thing we should do is to live up to all rules, for unsafe practice leads to injury and means either loss of life or limb and much suffering, as well as loss of time and money.

From a selfish standpoint a train employe carrying life insurance cannot help but appreciate the benefits of Safety in the reduced assessments on a given policy or in the increased value of a policy with a straight assessment like the policies of the Brotherhood of Railroad Trainmen, where on a Class C policy, twenty years ago the face value per death was only $1200.00, while today the same policy at the same cost is worth $1600.00 and matures at the age of seventy regardless of one's physical condition, and often sooner under the benevolent clause. What I have said about the insurance policy of the Trainmen is equally true of all fraternal and old line life insurance policies. Many life insurance companies now insuring railroad men where twenty years ago they were placed in the prohibitive class.

Let us all pull together for Safety, because it makes our calling less hazardous, our insurance policies more valuable, and the net income to ourselves and our employers greater. While much has been done in the Safety movement, more can be accomplished.

Little Things That Make Safety a Success

"IF you take care of the Little Things, the Big Things will take care of themselves," is an old axiom that has vital bearing on the success of the Big Job of minimizing the number of accidents to railroad workers.

"Carelessness, thoughtlessness, and foolhardiness, are the small beginnings of what so often cause major injuries and sometimes death," declares L. P. Hopkins, ass't to gen'l manager, who directs attention to the pictures on this page illustrating how the careful placing of tools and the proper use of hooks, clamps and cable slings can eliminate hazards to safety.

Posed in three pictures on the left are Boilermaker S. Longmire (above) and Boilermaker Apprentice A. M. Cross (below) of the SP shops at Bakersfield. (1) Shows the boilermaker's tools (circle) in a dangerous position on the platform where they might easily be shoved off onto the head of the apprentice working below. (2) The boilermaker shows the *right way* to take care of tools he is not using. He has the apprentice put them aside, out of the way, so that (3) the platform where he is working is perfectly clear.

The other four pictures were taken in the Gen'l Shops at El Paso.

(4) Machinist Jose Rodriquez at safe and efficient electric hoist used in Rod Room. One man can handle the largest type main rod with ease, the load being raised or lowered by means of a push button on end of the cable shown. This device is far superior to the old style air cylinder type hoist which was dangerous on account of its erratic action, and several men had to be used when raising a load.

(5) and (7) Special cable slings designed for use in lifting driving wheels, trailer frames, cabs, springs, driving boxes, boilers, trucks, and other parts of locomotives and heavy materials. Each cable has hooks and clamps suitable to handle all parts in a safe manner. Chain and rope slings are not permitted to be used. Many accidents are prevented and time saved. These cables are inspected at regular intervals and those with broken strands quickly replaced. Machinist Helper Daniel Macias in picture 5, and Machinist Helper Jose Acosta in picture 7.

(6) Machinist Geo. Evans showing safety clamp hooks designed primarily for handling locomotive driving boxes. Many other uses have been found for them, however, such as the handling of guides with greater safety, cylinder heads, or almost any flanged piece of irregular shape where an ordinary cable or sling is liable to slip. The hooks are self-clamped by the weight of the suspended article being lifted. Even though the load may become unbalanced, the hooks will not release.

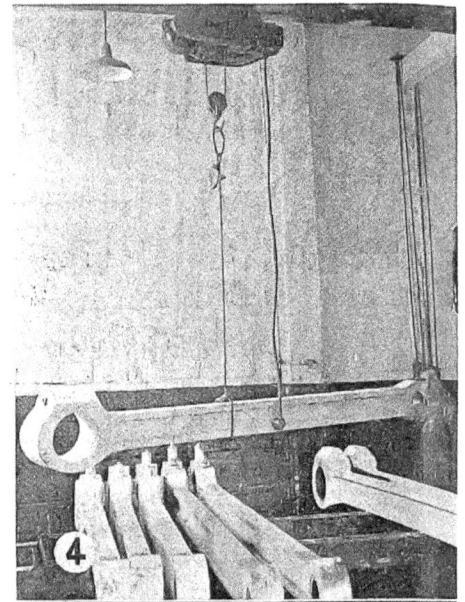

Chapter Eleven
Snow Over the Sierra – One of the Largest Problems

Even though Sacramento is in the central valley of California where snow is extremely rare, the operations of the Sacramento Locomotive Works was annually affected by snowfall in the Sierra, the summit of which is slightly over one hundred track miles east of Sacramento. The highest elevation of the original line through Summit Tunnel and Donner Pass is 7,033 feet, while the alternate track through the two-mile long tunnel just to the south is 6,900 feet. During decades of operating trains over the pass, the Central Pacific and Southern Pacific had a momentous task to provide a clear line between Sacramento and the yards in Truckee, Reno, and Sparks.

The challenges presented by snow are evident when observing the long record of snow accumulation at the Norden station at 6,890 feet. Norden was the location of a covered turntable large enough to rotate cab forwards that had been used for help up the climb from the east or west. Some of the earliest snowfall records date to 1879 when the Southern Pacific began taking measurements. Some of the snow depths were measured against a snow stake on the roof of the turntable. The Southern Pacific also monitored snowfall at Truckee, Cisco, Emigrant Gap, and Blue Canyon. The snow season can begin as early as October, but a significant accumulation in early fall is rare. Autumn storms usually produce snow above about 6,500 feet and it may melt before the next snow. The average snowfall at Norden, based on a record since 1878, is about 415 inches. This includes nineteen years during which the total snowfall was greater than 550 inches (45.8 feet.). The snow problems are aggravated by snow slides, which may occur whether or not snow

is falling. Late March or early April is usually the time of maximum accumulation. Thereafter, the melt starts and new problems for the railroad begin, including rising creeks, mudslides and erosion.

Early snow-related lessons about the difficulty in operating a railroad over the Sierra were noticed during the building of the line. John R. Gillis, a civil engineer who worked on the tunnels near the summit, told the American Society of Civil Engineers about the severity of the storms and how work continued in the snow:

> No one can face these storms when they are in earnest. Three of our party came through the pass one evening, walking in the storm – two got in safely. After waiting a while, just as we were starting to look up the third, he came in exhausted. In a short, straight path between two walls of rock, he had lost his way and thought his last hour had come. (Kraus, 146)

Gillis also referred to his image of workers on the mountain side above Donner Lake as viewed from the Dutch Flat and Donner Lake Wagon Road about five hundred feet below the workers:

> From this road the scene was strangely beautiful at night. The tall firs, though drooping under their heavy burdens, pointed to the mountains that overhung them, where the fires that lit seven tunnels shone like stars on their snowy sides. The only sound that came down to break the stillness of the winter night was the sharp ring

of hammer on steel, or the heavy reports of the blasts. (Kraus, 145-146)

According to the July 1927 issue of the Southern Pacific *Bulletin*,

> The tunnel workers stayed right in the mountains. Tunnels were dug through the snowdrifts to reach the face of the rock tunnels. Snow passages were excavated to the dumps where the rock taken from the tunnels was disposed of. In many cases the road between camp and work was through snow tunnels. Some of these tunnels were more than two hundred feet long. In addition to tunnel work, the construction of retaining walls in the canyons was carried on through the winter. A great dome was excavated in the snow, where the wall was to be built, and the wall stones were lowered through the shaft in the snow to the men working inside the dome.

The experience in keeping the line open through the winter of 1866-1867 led to the significant business decision of covering the track with snow sheds. Twenty years later, Arthur Brown, then superintendent of buildings and bridges, spoke about the need for track protection:

> Although every known appliance was used to keep the road clear from snow that winter, including the largest and best snow plows then known, it was found impossible to keep it open over half the time and that mostly by means of men and shovels, which required an army of men on hand all the time at great expense. It became evident from our experience then that the snow problem had become serious and that it was decided after various discussions on the subject by the directors of the company that the only positive means of protecting the road was by snow sheds and galleries. Although the expense of building a shed nearly forty miles in length was

almost appalling and unprecedented in railroad construction, there seemed to be no alternative." (Kraus, 159)

So, during the summer of 1868, construction began in earnest. Carpenters were paid four dollars per day, laborers from two dollars and fifty cents to three dollars. Twenty-five hundred men hit the mountain sides and by the fall of 1869, nine hundred tons of bolts and spikes had been used to build thirty-seven miles of shed at a cost of more than two-million dollars.

Despite the existence of so many sheds, they did not cover every inch of rail subjected to snow. A mechanized means of removing the snow was still needed.

Snow removal from the main line has an evolution ranging from wedge-shaped deflectors on the front of early-era steam locomotives to plows with a rotating set of blades with a diameter of about fifteen feet.

Considering the great snow-related struggles encountered during the building of the line over the Sierra, the first locomotive designs had an elementary and surprisingly simple snow-removal feature. Attached to the front of the locomotive was a set of two pieces of sheet iron. Each piece was bent to form a concave shaped sheet that deflected the snow off the track to either side. The sheet's vertical dimension was about three feet. This design was a carryover from the snow removal philosophy prevalent on eastern railroads. But, the Sierra snowpack is much different than snow east of the Mississippi and provided a valuable hydrology lesson to western railroad operating personnel. Snow is often measured by its density, which is a ratio of the snow's water content (measured in inches of water) to the depth. The density of fresh snow is frequently higher in the Sierra than in the Rockies of Colorado and in states east of the Mississippi. This means that a cubic foot of snow in the Sierra is often heavier than the same volume in eastern states. Removing snow from the tracks in the Sierra, therefore, required extra force and equipment. Because Sacramento was a major facility close to the base of the Sierra, the operations department there became involved in the maintenance and development of snow removal equipment.

One of the early innovators of snow equipment at the Sacramento Shops was Chief Mechanical Engineer George Allen Stoddard. During the 1860s, he helped develop what became known as the Bucker plow. The first plow of this type was completed in 1866. It was twenty-eight feet long and weighed twelve tons. The first contact of snow with the Bucker plow was along a flat piece of sheet iron slanted at about forty-five degrees. As the snow was forced up the slanted surface, it encountered a nearly vertical face that gradually diverted the snow off to both sides. The top of the vertical piece was about eighteen feet above the ballast. Near the top of the plow was a platform on which the plow boss would stand and, using hand signals, advise the engineer in the lead locomotive. Operators in other engines involved in the push would receive commands via whistles from the lead locomotive. Results of the first Bucker plow's field test were not good. Snow would gather and pack under the plow causing derailment. Back in Sacramento, the decision was made to add weight to the plow. The second plow built weighed an additional seven-and-a-half tons.

The snow-removal efficiency was increased as the number of locomotives used to push the plow was increased from about three to six. In some instances, up to eleven locomotives were used. The combination of added weight to the plow and more locomotive power overcame the forces wanting to lift the plow vertically and impede the forward movement. Over the next eighteen years, ten Bucker plows were built in Sacramento.

During the era of the Bucker plow, a new, completely different idea was being developed. In 1883, Orange Jull, a Canadian in Orangeville, Ontario, had the idea of a rotating fan on a railroad car to throw the snow off to one side. Mr. Jull, a mechanic, approached John S. and Edward Leslie, also of Orangeville, who operated a machine shop and were equipped to help Jull. Even-

tually, patents were issued to Jull but were assigned to the Leslies who agreed to pay royalties to Jull. Several other individuals and companies, using the trial and error method, provided improvements in the efficiency of the blades, ice cutters, and flanges (used to prevent snow from accumulating under the equipment). The principle of the rotating blade gained acceptance, and in December 1887, Southern Pacific ordered one for use over the Sierra. The winters of 1887-88 and 1888-89 were mild while the new equipment remained untested. The winter of 1889-90, however, was another story, according to Superintendent J. A. Fillmore of the Southern Pacific. That winter was a tough test for the rotary plow that had been stationed in Truckee since its purchase. The performance of the rotary was mentioned by Fillmore in a letter to a friend:

> The heavy work we have had to do has fully demonstrated that the Leslie machine's boiler should have a much larger heating surface and that the shovel of the plow should cut eight inches wider on each side. It would then be perfection. (Best, 73)

With reviews and improvements suggested, the plow was taken to Sacramento where a new sixty-one inch diameter boiler was installed to replace the fifty-two inch boiler. Some damaged blades were replaced also. But, the ability of the rotary principle had proven itself, and Southern Pacific ordered two more upgraded rotary plows. The rotary plow became an established member of the winter array of snow equipment and was the heavy-duty machine when the storms posed a challenge too great for any other method of snow removal. For over half a century, the rotary plow was powered by steam pressure. By the middle of the twentieth century, diesel-electric technology was becoming the power source for locomotives because less maintenance was needed. With the new technology, a diesel engine turned a generator. The generator produced electric power for the motors that turned the wheels and moved the train. The diesel-electric locomotive replaced the steam locomotive and created a major change in the labor force at every railroad facility including those of the Southern Pacific in Sacramento. Employees who had spent decades working with only steam power faced the challenge of learning new processes while seeing the tasks requiring their knowledge eliminated. This created a disruption to many career paths in the railroad industry. An interesting case is that of Malcom Gaddis, who was involved in adapting the diesel-electric technology to the rotary plow idea that Jull and the Leslie brothers had started about seventy-five years earlier. Mr. Gaddis told his story to the California State Railroad Museum in November 1996 as part of their oral history program.

In the early 1950s, one of the large Southern Pacific operations in Southern California was the Taylor Shops in Los Angeles. Mr. Gaddis, with a Bachelor of Science Degree in Electrical Engineering and graduate work in Industrial Engineering, held a job in the diesel facility there and later in Union Station. While at this job, Gaddis said, "I could wear a white shirt and not get dirty" while checking electric brakes. But the "not get dirty" job was abolished with the introduction of diesel-electric, and Mr. Gaddis moved to the Alhambra Roundhouse, which was the steam roundhouse, where he learned how to light fires in the locomotive's firebox and to service steam locomotives. Then, he filled a position in the coach yard where he performed air conditioning, heating, and lighting work. Yet another position held by Mr. Gaddis was a position on board the Sunset Limited to provide maintenance should there be problems en route. Later, he learned that an electrical supervisor was needed in San Luis Obispo. After filling the open position, he inspected new diesel locomotives and finished a project installing radios in cabooses. In 1956, Mr. Gaddis came to Sacramento to fill a dual role as an electrical draftsman (and was paid about $540 per month) and locomotive engineer (earning about $700 per month). Then, he was assigned to redesign a rotary steam plow to make it completely electrical. Mr. Gaddis had come back to his original educational training as an electrical engineer and had a chore to work on in Sacramento.

In November 1996, Mr. Gaddis described his work in Sacramento redesigning the rotary plow:

In the early steam plows, you had kind of a little cupola in the center of the snow plow, and up in that cupola with windows all the way around it was usually a maintenance-of-way engineer that knew what was ahead of him and knew which way to have the blade going and whether the wings were to be out. They had very poor communication between the fellow up in the cab where they flung snow all up around him and the locomotive engineer. So, the locomotive engineer just kind of went by feel.... This caused a few problems, but the engineer would look out and make sure he had a green signal and he could kind of see ahead of time how deep the snow was.... Other railroads had experimented with different types of snow plows, and it was decided that since we had diesels and we had electric power on the diesel and there were some spare traction motors that we should see what we could do about designing our own steam plow. The thought at the time was to get two controllers: one to control the speed of the plow blade and the other controller would control the motion of the entire consist—the snow plow and the car behind it or whatever else—just keep it all together. Particularly, if the maintenance-of-way engineer and the locomotive engineer were sitting side by side, they could discuss what they wanted to do, do they want to pull the snow to the left or the right, and do we want to speed along a little bit the snow is only two feet deep or is it six to ten or maybe twelve feet deep.... Then, I also discussed with General Electric and some other organizations about had they ever done this? Had they furnished motors for it? Or should we use General Motors traction motors? Well, we kind of got into a situation where GE said they would help us with it if we would use their motors. Well, we found that we did not have a surplus of their motors so we decided we'd use General Motors. We had lots of spare units on hand particularly at Roseville. So it was designed to use motors you could take out of a locomotive in the wintertime when you didn't need that particular kind of power, and put those four motors into the snow plow…. Suppose we were in a very deep cut, maybe in eight or ten feet of snow and then all of a sudden we break out. What do we do to protect the motors from just over speeding? These were some of the things that had to be designed so we had some control of the motor because a series motor will just run away with itself if you don't have control of it.

So, I started out by drawing a proposal for control, and then we tried to figure out what to do for things like: we run into a pile of rocks or we run into a tree stump or things like that. I elected to put overload relays into the system so that… if you picked a lot of rocks or it was starting to bog down the first thing that would happen is the motion stops and that allows an opportunity to clear itself. But sometimes it won't clear itself. It might have a gigantic tree stump or a ten-foot diameter rock or something in front of it. The second overload relay, which was probably at 400 or 500 amps would stop the motion of the blade. I had contacted General Electric about this and also General Motors about this…. Well, there was probably a month or two that was spent just researching [the design of the plow] and trying to decide how we're going to do it electrically. Then there was probably another month or two when we were trying to decide how to mount the traction motors and how do we tie them onto the big blade up on the front…Most of the people I

worked with were boilermakers. They were people that were going to do heavy, heavy steelwork. We were going to have to change the sides and change the cab on it and put a different type of a roof on it. It [the plow] has one controller for the maintenance-of-way engineer who controls the blade and the wings and a lot of other things on the plow…. You must have a maintenance of way engineer in there because he knows where all of the signals are along the railroad and where the switch stands are and where the bridges and entrances to tunnels. It's very important that this man have his wings pulled in when he goes by a switch stand and that he doesn't go in the tunnel throwing snow…. He's responsible for throwing the snow and picking it up from along the side. Now on the right hand side, there's another control stand and it has eight notches. That's for the locomotive engineer. It's his job to keep them going at the right speed so that the snowplow is performing as efficiently as possible. Also, he has the brakes on the snowplow for the entire train and whatever is in the consist of the units he's working with. It's very convenient to have the two of them right there together so that they can slow down or speed up or maybe even back up if he missed something. They also have a two way radio on the snow plow that he can continuously talk to the dispatcher so that if the dispatcher wants him to quit plowing and get out of the way of a passenger train or something coming up the hill they'd have immediate contact for that or any kind of an emergency….

When we started designing this, we needed a drive shaft that was about four feet long to go through the traction motor which was just about twelve feet long. On the rear end we had a large, about eight-inch diameter bearing and in between the two traction motors we had another eight-inch bearing and then forward of the truck traction motor we had a bearing and then on the end of that there was a small gear that tied in with the blade on the front, which was a large—very large diameter shaft. There were four motors in this so there were two motors on the right and two motors on the left on these long shafts. That was my original design....

So anyway, there were a number of things that were really firsts on that. And we took the plow out after it was finished and we tried it out and found that what we originally designed was something that would run over 100 rpm. Well, we found out that it could throw snow a quarter of a mile. It was way too much. Anyway, we slowed the plow down. In fact, we even put some fine-tuning in the plow so that we could get slower in between notches if we wanted…Also, on this long twelve-foot shaft, we found that it would whip in the center when we were up around 80 or 90 rpm with a plow, and this tended to wipe out the bearing in the center between the two motors. So, starting with the next plow and this little learning curve that we had gone through, I decided to put a link belt universal joint between the two motors, instead of having three bearings, have four bearings [with] one on each side of the link belt universal. This took care of the problem of whipping the bearing. The other thing that we did was we took a standard 4625 vapor steam generator, which was a small passenger locomotive steam generator, and that worked fine. It was much larger than the little tiny one we had before, which was about the size of a large chair, and it did a good job of keeping all the windows clean and the hinges clean that we wanted to keep de-iced…. They [approximately eleven plows] were converted from steam to electric right here at Sacramento General Shop.

The next thing that we found was rather than having to carry a whole lot of cable and go out and find a unit and then bring it in and start cabling that up and taking two or three days to make all your connections, we decided to take some older "B" units and make them permanent power. There were permanently coupled behind the snow plow with all the cables in place. This way, if at three o'clock in the afternoon the superintendent decided he wanted a snow plow up at Donner, you could go over and within an hour he could leave Roseville and within a couple of hours you could be up on the hill plowing snow. Otherwise, it could take two or three days to reach out and try to find a unit that would match the snowplow. So, since then we have had these designated "B" power units that are permanently assigned to a snowplow.

Mr. Gaddis mentioned a month or two of research into the strategy of the mechanical-to-electrical conversion and another couple months deciding how to mount the traction motors and connect them to the blade. Like his project and all other projects dating back to the beginning, computers were not available to test the stresses and performance of new designs and theories. The graph paper and pencils worked fine, did not need to be rebooted, and the final product met the needs very well.

The revelation by the railroad that equipment and labor, by themselves, could not keep the line open in the mountains during the winter lead to a huge commitment that eventually led to about thirty miles of track covered with sheds. The operations and planning needed to keep the line open was a subject of several issues of *The Bulletin*:

"Problems of Southern Pacific Over the Summit," June 1925

"House of 1000 Curves," April 1920

"King Snow- What's Used to Clear the Line," February 1936

June, 1925　　SOUTHERN PACIFIC BULLETIN

Problems of S. P. Line Over the High Sierra*

By T. AHERN
Superintendent, Sacramento Division

T. AHERN

SOUTHERN Pacific's transcontinental line over the summit of the Sierra, between Roseville and Reno, presents fascinating problems of operation and maintenance. By no means the least is that of combating the deep winter snows. The years of constant battling with the armies of winter have developed methods of defense which I believe are unique among railroads.

In order that there may be some order in presentation, I will divide my subject under the following heads:

(1) A general description of operating conditions on the mountain.
(2) Historical development of our snowsheds, description, types and cost.
(3) Fire protection in the snowsheds.
(4) Present and past methods of removing snow.
(5) The present day snowsheds and the necessity therefor.

Our "mountain" begins at Roseville, 107 miles from San Francisco, at elevation 162. A double track main line runs east to Emigrant Gap, 64 miles, at elevation 5218, with profile grades up to 118 feet to the mile. From the Gap we have single track main line for 29 miles to Andover. rising on grades up to 95 feet to the mile, to elevation 7017 at Summit, where we tip over the top to descend on profile grades of 95 feet to the mile to Truckee. From Andover we have 46½ miles of double track to Sparks, elevation 4425, running down the canyon of the Truckee River.

The snow territory begins at Gold Run, elevation 3226, 20 miles west of Emigrant Gap, and extends to Sparks, elevation 4425. The snow sheds are between Emigrant Gap and Andover, a distance of 29 miles, the sheds being all above the 5000 foot level. Within this 29 miles we now have 25 miles of snow sheds.

World's Longest House

I will refer later to the changes we are making in our mountain line. but I might say here that until 1921 we had 41 miles of single track extending from Blue Canon to Truckee, and in this 41 miles we had 29½ miles of sheds, nearly continuous, and aptly called "The longest house in the world."

Generally speaking, we operate two and three engine trains from Roseville to Summit, the length being governed by the capacity of the power. Fruit

*From a recent address on snowsheds and operation of traffic over the Sierra Nevada mountains given by Superintendent Ahern before the Pacific Railway Club at San Francisco. The address was illustrated by stereopticon views, presented by Loyal Himes of the Company's Duplicating Bureau. Mr Ahern's subject will be covered in two installments. Second installment will appear in an early issue.

trains usually consist of 56 loads, and between Roseville and Colfax, where the maximum grade is 1½%, are handled with one Mallet road engine and one consolidation helper cut in 13 cars ahead of the caboose. At Colfax, near where the 2 2% grade begins, a Mallet helper is added just behind the consolidation, running thus to Emigrant Gap where the consolidation is cut out and the train handled on the 1.8% grade between Emigrant Gap and Summit with two Mallets. At Summit the Mallet helper is cut out and the train is operated single Summit to Sparks. Mallet helper returns to Roseville.

The Mallets used weigh 400,000 pounds on the drivers and the consolidations 185,000 pounds on the drivers. Fruit trains make the run Roseville to Sparks in sixteen hours and thirty minutes.

Sidings are spaced approximately two miles in the single track territory, and operation is by staff system, there being staff operators at each siding. Light signals are used exclusively, which, being connected with the staff system, indicate, by various color combinations, whether an approaching train is to go through, hold the main track or head in on siding. At some points switches are electrically controlled from the staff office. Dispatcher is in constant touch with all staff stations by telephone.

Traffic density may be appreciated by visualizing a movement over this single track territory during the summer season of a train or engine every 26 minutes for a thirty day period, or on peak days every 23 minutes.

History of Snow Sheds

Instrumental surveys of the Central Pacific were made in 1861 under the direction of Theodore Judah, chief engineer, and although construction was

not to reach the Summit for several years, it is evident from his report of that year that he looked forward to troubles to come. He says in part as follows:

"Thirteen ft. limit of snow at Summit. Due to accumulation of successive storms during the winter....It is only necessary then to start an engine with snow plows from the Summit each way at the commencement of a storm clearing the snow as it falls. A similar course of procedure at each successive storm will keep the track open during the entire winter....A crust forms during the snow which prevents its drifting badly....The only point where we shall encounter a level surface of snow is in Summit Valley for about two miles. By elevating the tracks at this point no trouble will be anticipated...."

Mr. Judah died soon after and was succeeded by S. S. Montague, and in 1865 with construction still west of Gold Run he reported that—"The heavy snowfall in the immediate vicinity of the Summit, amounting in the aggregate to 10 or even 12 feet in depth and a much heavier accumulation at some points by drifting will render it necessary to provide a substantial protection, either of timber or masonry to insure the successful and uninterrupted operation of the road during the winter months. The principal points requiring such protection occur upon the eastern slope and within 2 miles of the Summit."

Over the Sierra

Then commenced in earnest the historical race between the Central and Union Pacific for Utah. The U. P. set for its goal the California State Line, while the Central pushed its surveys on through Nevada, into Utah, and finally, a considerable distance east of Ogden. The grade reached Summit in 1867 and construction started on the 1650 foot tunnel through the ridge. Deep winter snows stopped all but the tunnel, and Chas. Crocker hauled men and materials, and even locomotives, 40 miles through the winter snow on sleds and wagons over the Summit to carry on the work in the Truckee River Canyon, where the snow was light enough to shovel aside. At the first breath of spring 2500 Chinamen were put to work shovelling out the deep drifts west of Summit, and no expense was spared to hasten the onward moving track.

Active snow shed construction was started in 1867 and Mr. Montague's report for that year says in part as follows:

"From the valley of the South Yuba across to the Truckee River the deep snow line of 35 miles is met. For the greater portion of this distance the road follows what is called a side hill line and for the most part is sheltered so as to be available in winter. Therefore the route across the

Summit will be but little affected by deep snows....At troublesome spots along nine miles of the highest portion a substantial overhanging shed will be constructed so as to shoot the snow clear of the track...."

The race between the U. P. and the Central was approaching its climax in the winter of '68-'69. With track laying gangs far out in Nevada, any delay in the delivery of supplies would have meant defeat. By this time 23 miles of shed had been constructed. 13 continuous, and with this protection, reinforced by 2500 snow shovelers and six construction trains, the line was kept open through the winter except for two weeks.

Shed construction continued until 1873, when Mr. Montague reported as follows:

"Snow Galleries"

"A novel and important feature of the work on the Sierra Nevada has been the construction of galleries for the protection of the track from the heavy snowstorms incident to that region. The experiment was first made of covering the track in the cuts only, depending upon snow plows for clearing the embankments, but experience soon proved that where snow was liable to accumulate to a great depth, its removal even from the high embankments would incur great expense, and often delay the movement of trains. It was therefore deemed best to make the covering through the deep snow-belt continuous. More than thirty miles of these galleries were built, consuming 44,639,552 feet, board measure, of sawed timber, and 1,316,312 lineal feet of round timber, equivalent in the aggregate to 52,537,424 feet, board measure, of sawed timber; and 721 tons of iron and spikes. Two general styles of construction were adopted. One, intended for localities where the weight of the snow only had to be supported, and the other for such places as were exposed to 'slides' and the slower, but almost irresistible 'glacial movement' of the snow on the steep and rocky slopes, along which a great portion of the road near the Summit was built."

A detailed description of the manner of constructing these galleries would occupy too much space for this article, but I need only say that they ha, proven a complete success. Though frequently covered with drifted snow to a depth of ten or twenty, and, in some places on the slope of Donner Peak, of more than fifty feet, they have afforded a safe passage for trains through this inclement region at all times and seasons, without any noticeable detentions. From that date there was but little change in snow shed mileage until the recent second track construction program was undertaken.

The early snow sheds were constructed very much after the manner of a house, with peaked roof on a very steep pitch. The logs comprising the posts and braces were felled in the adjacent forests of pines, firs and cedars. The neighboring saw mills furnished the timbers for roof and side. Great care, involving much labor, was taken in removing rock and loose material in order to bury the base of posts and braces to obtain a rigidity of alignment but, rot soon set in and with uneven masses of snow on either side of any portion of the line of sheds a subsequent forcing out of line followed its melting. The braces were neatly axed to fit around the posts and here were required the services of many expert woodsmen. The boards covering the roof were largely of sugar pine laid longitudinally, without battens. With melting snows water came down on the roadbed in torrents, adding greatly to the cost of track maintenance. The snow would pack behind the slopes on the roof and force the sheds out of line. When this happened, men shoveled snow from one side of the roof to the other, the weight thus transferred forcing the sheds back again. Following the peaked roof sheds came those of flat construction, detailed as are the present existing sheds but with the following difference. There were no rafters, 3"x12"s laid flatwise and longitudinally forming the roof. Sag due to heavy snow loads and the factor of direct flow of melting snows onto the roadbed lead to the development of the last and present existing types of snowsheds.

The clearance of the first type was 17 feet above top of rail; with the second type an increase to 18 feet was made With the present existing types we find through the process of renewals and repairs a change to the present prescribed standard clearance of 22 feet. This increase of clearance has obviously caused a steady increase of lumber necessary for any fixed type of shed, and has as its underlying cause the increase of dimensions of locomotives and cars since the initial construction of the sheds.

The flat type of shed early proved its efficiency and necessity of adoption. Its ease of restoration to original alignment and the enormous snow loads it has proven capable of bearing in times of abnormal snowfall has marked it as the most efficient type for use in the snow belt of the High Sierra.

The early sheds were constructed with wood footing blocks which quickly rotted out and were in turn replaced with stone footings, the blocks being cut in the granite quarries which were then operating near the base of the Sierra. With the extensive development of concrete however, it has been found more satisfactory to substitute concrete footings, these being cast on the ground in portable forms.

Cost of Snow Sheds

The cost of the original peaked roof showshed construction is not clearly recorded. However, records indicate extremely high material prices during the construction period in marked contrast to prices 20 years later, and it is probable that the snowshed cost was at least $2,000,000. In the reconstruction of peaked roof to flat roof sheds, costs were much lower, so that with the reconstruction completed, and amplified by the additional construction of seven miles of double and three-track shed with expensive truss construction, necessary to cover the sidings and yard tracks built as business increased, the total cost of the reconstructed shed was $2,159,313.00.

Incidentally, it might be of interest to know that since 1902 we have spent

Two scenes of early day methods of removing snow from the Company's lines over the Sierra Nevada mountains are shown in the upper and lower photographs. Inset shows modern rotary snow plow which makes quick work of throwing the snow clear of the tracks, keeping the mountain line open at all times.

in renewals and repairs $3,834,585.00, or nearly double the original cost.

From records available for the years 1896 to 1899 the average labor cost of erecting sheds was $4.00 per 1000 F.B.M. Lumber cost $9.00 per 1000 F.B.M. Contrast with these present day costs labor $20.00 per 1000 and lumber $22.00 per 1000 feet board measure.

EMPLOYE CASUALTIES SHOW DECREASE IN 3 MONTHS

Casualties to employes reported to the Interstate Commerce Commission were decreased 20.2 per cent during the first three months of this year as compared with the same period in 1924. This good showing indicates the keen interest employes are taking in accident prevention work and benefits being derived from the monthly safety committee meetings.

East Bay Electric Division has a clear record for both 1924 and 1925 in regard to train and train service casualties. Stockton Division is second with only one injury as against three in 1924, while Coast Division is third with four injuries as compared with eleven last year during the three month period. There have been four deaths and 100 injuries on the 13 divisions in train and train service casualties as compared with 8 deaths and 145 injuries last year.

Telegraph Department has a clear record for the first quarter this year in non-train service casualties. Sacramento Division has the next best record with 11 injuries as against 17 last year. There has been one death and 364 injuries in this class of service as compared with one death and 434 injuries for the three months last year.

P. F. E. PEOPLE HOLD DANCE IN SAN FRANCISCO

As a part of a series of "get-together family" events being carried on by Pacific Fruit Express employes, a dance was given at Whitcomb Hotel in San Francisco April 24. About 350 men, women and children from several points in California enjoyed the event. In addition to dancing the program included a "kiddies revue," staged by Miss Sylvia Hansen, and vocal numbers by Miss Ellen Corlett, Mme. Petrotta and Thomas Kearney, the Scotch comedian. Talented acts were also put on by some of the employes. Especially notable among these were the square dance enacted by Los Angeles shop employes dressed as '49ers, and the humorous singing and acting by the Colton boys.

There was plenty of delicious punch and varied colored balloons were distributed to the kiddies. Although the dance was scheduled to stop at midnight, so enthusiastic and merry were the P. F. E. folk, that it was necessary to continue the festivities another hour.

"Mamma," said a child recently, "am I descended from a monkey?"

"I don't know," replied mamma, "I did not know your father's people very well."—Exchange.

Fire Brigade Sets New Speed Record

Members of Sacramento General Shops fire hose cart No. 5 which set a record of 31 2-5 seconds in a nine man test, starting 50 feet from plug, running out 300 feet of hose, breaking joint, applying nozzle and getting water. Back row, left to right—Fire Chief H. A. Adams, N. Pendleton, A. Sena, G. Toffee, A. Logan (ass't captain). Front row J. Caldroni, J. Battaglia, L. Jorgenson, F. Parisi, B. DeRiso, J. Hernandez and C. Schmitt.

NOW comes the Sacramento General Shops fire brigade with a record of 31 2/5 seconds in getting water through 300 feet of hose, starting 50 feet from the plug and breaking one joint.

Fire Chief H. A. Adams read in the Bulletin about the 33 2/5 seconds made by Dunsmuir fire brigade in a similar test, and was inclined to doubt the possibility of making such fast time. In fact, after the first attempt of his team, he was convinced that the Dunsmuir boys made their record while running down the Siskiyou mountains. After a short period of training, however, the team was ready to make another attempt, which resulted in the new record being established. Following were the conditions of the run.

The course was over a level, bumpy road, with cart carrying 500 feet of standard fire hose with a crew of nine men. The cart was placed 50 feet from the fire plug and on a signal given by the chief, the crew started and ran past the plug, laying out 300 feet of hose, breaking hose joint, applying standard nozzle and getting water in 31 2/5 seconds. Water pressure was slow and the crew waited 1½ seconds for water to respond.

YARDMAN HELPED THROUGH GROUP INSURANCE

Notwithstanding our continuous efforts in the field of safety education, it is regretable that serious casualties still too often occur. A sad case recently occurred on the Western Division, when one of our yardmen, a vigorous and well-liked young man, met with an accident which resulted in the loss of both legs.

To this employe and his wife the immediate future indeed looked dark, as it always will when a wage-earner is unexpectedly stricken.

Fortunately, he was insured under our Group Plan, but he didn't realize how much protection this insurance afforded him, overlooking the fact that it carried total and permanent disability benefits.

His case was reported to the Metropolitan Life Insurance Co. and within three days their representative delivered to his wife a check for $86.25, an amount sufficient to provide for the family's most pressing needs. This is the first of a series of thirty monthly payments that will be made by the Insurance Company, and during this period of time they will have an opportunity to arrange their affairs for the future.

Have you read your group insurance certificate? Read it carefully today so that you and your family may know the full value of the protection it gives you. It will add to your piece of mind to feel that you have guarded your loved ones against the uncertainties of life.

GENERAL OFFICE PICNIC TO BE HELD JUNE 21

Employes of the General Office at San Francisco, together with their families and friends, will hold their annual picnic June 21 at Fernbrook Park in Niles Canyon. Three special trains, two from Oakland Pier and one from 3rd Street, San Francisco will carry the crowds to the picnic grounds. Hundreds will also make the trip by automobile.

In addition to foot races for young and old, fat and slim, other interesting events will include a tug-of-war, greased pole climb, and pie eating contest. One hundred gate prizes will be given away. Dancing in the large open air pavilion will be a feature of the program.

Problems of S. P. Line Over the High Sierra

By T. AHERN

Assistant General Manager, Northern Division

Part Two

FIRE is the greatest agent of destruction of the snow sheds. The origin of these fires is largely due to outside causes. Some of them are due to adjacent brush fires, originating in the immediate vicinity, or to an extension of neighboring forest fires.

The fire risk with the original type of locomotives which were wood-burning, was great. This has been practically eliminated with the present type of oil-burning locomotives. The exhaust due to a use of full steam pressure loosened the boards of the sheds and a continuous rattling of loose boards accompanied the noise of exhaust gases during the passage of a locomotive through an upgrade portion of the sheds. This is now eliminated by placing a "V" shaped removable casting over the top of the stack, known as a "splitter," thus deflecting the exhaust gases to the right and left where escape is allowed from the sheds by omission of the top boards on the sides of sheds. This deflection of the exhaust gases prevents any possibility of sparks being driven into the roof of the shed with a probable fire resulting.

The necessity of fire protection was early emphasized and in 1870 fire trains were installed. The present train consists of a locomotive, with pump, and two water cars, each of 12,500 gals. capacity. Trains are stationed at Emigrant Gap, Cisco, Summit and at Andover, the eastern terminal of the snow shed line.

On Red Mountain, elevation 7860 feet, is maintained a lookout station for shed protection, with observers on constant duty during the summer fire period of each year. Located in this lookout is an ordinary engineer's transit to which, rigidly attached, is an aluminum indicator which travels over a chart, etched in copper and attached to a stand directly in front of the tran-

sit. On this copper plate are shown the stations, fire alarm boxes, and district telegraph boxes, etc, with the vertical and horizontal angles of their location. Thus by the reading of the vertical angle, it is at once determined if shed itself is on fire. This is a necessary procedure due to smoke obscuring the sheds during a fire in their immediate vicinity. The indicator over the copper plate gives at once the nearest station or district box. Etched on the glass of the windows comprising the front bay of the lookout house is a line following the exact line of the sheds as seen through the transit. Intersection of this line and

Two views of Southern Pacific's fire fighting equipment in the Sierra Nevada mountains. During the summer months four fire trains are usually kept in this district. The two-tank cars each hold 125,000 gallons. The pump, shown in inset, sets on top of boiler just back of steam dome and draws steam from locomotive. The pump has a capacity of about 250 gallons per minute. The trains are each equipped with about 1000 feet of 2½-inch hose. Pipe system is arranged in such a way that water can be had from front or back of locomotive.

a line on the fire gives location where the sheds are burning. At only two portions of the sheds between Summit and Blue Canyon (visible from Red Mountain), is the view obstructed by spurs of the mountain range. These obstructions were utilized in placing the casing composing the bay window of the lookout.

This station reports, by phone, to Central District Station at Summit, every half hour. If a fire is noted the exact location is immediately determined and if such proportions as to require fire train service, as a large brush fire or the sheds themselves, or merely the services of a nearby section crew, advice is phoned to Summit and from here further action is directed.

Extending over the snow shed belt are a series of district telegraph boxes, less than one mile apart, at which watchmen patroling equally given portions of the sheds report every hour

or give an alarm if fire is discovered. This is registered automatically on a tape at Summit. The tape is under continual observation and on the appearance of an alarm, the chief dispatcher is notified and he in turn notifies the nearest fire train and the main line is cleared of traffic. With the sidings existing at the various stations, no great trouble or loss of time is encountered in its clearance. Of course, the main line through the sheds is blocked while the fire train is in service and the consequent loss of time sustained depends entirely upon the magnitude of the fire. The fire train is equipped with an individual high pitched whistle which can be heard for miles and which is blown almost continually en route to a fire. This enables small equipment as gravity and section hand cars to get in the clear. In the cab of the engine is a chart giving the number and location of each alarm and dispatch box, the location being given with reference to the mile posts.

Alarm boxes are located in several of the more important parts of the sheds. When these are used, bells are automatically rung at Summit and other points, serving the same purpose as a general alarm and calling into action all available forces. This alarm is very rarely used, the district telegraph boxes serving the purpose desired.

For watching the sheds not visible from Lookout Mountain, stations are maintained during the fire period on the tops of Tunnel No. 6 and Tunnel No. 3.

It was early recognized that a fire break at intervals in the shed line was imperative. Galvanized iron sheathing of the sheds was first tried, but this proved a failure. Then followed the telescope shed, consisting of a 96-foot piece of shed mounted on wheels so arranged that in summer it can be pulled back within the adjacent shed which is of an enlarged section. Thus an efficient fire break is provided. They are spaced approximately one mile.

In addition to the telescope, we are now taking down 100 feet of shed approximately every quarter mile, particularly wherever the track is on fill,

This is the second part of an address delivered by Mr. Ahern before the Pacific Railway Club at San Francisco on Southern Pacific's operation of traffic over the Sierra Nevada mountains. First installment appeared in the June Bulletin.

and by so doing fire breaks are provided which effectually confine any conflagration to a relatively short distance.

Removing Snow

Prior to 1889 snow fighting outside the sheds was done with Bucker Plows, flangers, and the strong arms of many shovelers. With the Bucker Plow and six to twelve engines, we would back away some distance from a drift and take a run at it, with every engine "down in the corner." When we hit the drift, snow literally spurted aside until stopped by the pressure, when back we went for another run. So it went until we bucked through—that is unless plows were derailed.

"Bucking snow" was a term that carried much meaning in those days and to say that it was thrilling work puts it mildly. As the banks were built up with snow on each side, they became too high for the plow to get rid of the accumulation and after each trip of the plow, it was necessary to excavate chambers into the side of the snow banks into which the snow from the next trip of the plow could be deposited. This necessitated an army of shovelers to move snow back by hand with long-handled, square-pointed shovels. Attending to the physical requirements of the men was no small task as provisions to feed them had to be moved through the storms and we had also to build quarters to house them. At Shady Run (now Midas), we used ten monster ranges for cooking meals in our camp kitchen.

The flanger was, and still is, used to clean out and throw aside the ice and snow packed between the rails which would otherwise fill the flangeways and result in derailment. Riding a flanger was in those days anything but pleasant. The operator rode on an open flat car raising and lowering the plows, which were arranged beneath, by hand. It was run at fairly high speed to throw the snow aside, and the operator had to be constantly alert.

Subsequently, the type of flanger was materially improved, the operator's quarters being housed and an air lift for the plows installed, valve being operated by locomotive engineer.

First Rotary Plow

The first rotary was received in 1889 and its coming was hailed with delight by those charged with the responsibility of keeping the line open.

In 1908 we first developed the use of the spreader, designed by our then Superintendent of B & B, C. F. Green, which was followed next year with the steel spreader commonly used for leveling embankments and ballast. This machine will spread the accumulated snow banks to a width of 20 feet from the track, leaving the snow level with the top of the rail. It does its most effective work when the depth of snow does not exceed four or five feet; however it has been used successfully for an initial spread with a snow depth of nine or ten feet.

We have also adopted the practice of installing pilot plows on a number of locomotives, this being a small "V" shaped plow placed on the pilot. We have from time to time added to our equipment so that we now have six rotary plows, four of which are of the largest and most modern design; also seven flangers and three spreaders which are used continuously during the snow season.

During light storms the line is kept open with pilot plows and flangers. Pilot plows are more efficient under these conditions than the rotary as they can make better speed. Both the pilot plow and the flanger deposit the snow close to the track and at frequent intervals, particularly at the close of each storm, spreaders are run to level off the accumulated snow banks.

During heavy storms the rotaries are run. These, with one or two Mallet engines behind, eat steadily through the drifts, casting the snow far to one side. The rotary is followed a short distance behind by the flanger which cleans out the accumulation of ice and snow between the rails and behind this comes the train.

Snow Sheds

The present day sheds are a sturdier development of the early flat roof sheds with all the little refinements of details developed through years of experience.

Single track sheds have overhead clearance of 22 feet, and side clearance 8 feet 6 inches. There are 326 board measure feet of lumber in each track foot of this shed and its cost, in place with concrete footings and all, is $25.00 per foot.

There are a series of openings in the boards of the shed between the posts at about the height of car windows. These serve the double purpose of ventilating the sheds in summer and providing view for the passengers. In winter a solid board frame is placed over the open space.

Double track snow sheds are of two types—one, where the tracks are on 17 foot centers, in which case a row of center posts between the two tracks is provided, and the other with tracks on 13 foot centers where the span over the two tracks is trussed. The first type is usually known as double track shed and the second type as double track truss shed.

The double track shed requires 442 board measure feet of lumber per foot of shed and costs approximately $33.00 per foot in place; the double track truss shed requires about the same amount of lumber, but is more expensive of erection on account of the truss construction.

The staff stations are built in the side of the shed at stations, the shed being somewhat widened for a short distance on each side to provide for a platform in front of the office. The office has a bay window for the operator and the name of the station is shown in front of the office by an illuminated sign.

Operation in the sheds has its own peculiarities. The powerful Mallet locomotives are built with the cab foremost, the tender being attached to the smoke box end on account of the restricted vision within the sheds, particularly due to smoke. The flagman hangs two Dietz lanterns under the rear platform of the caboose, where he sits watching the track. If he sees freshly splintered ties he knows something is wrong and he stops the train with the emergency valve.

The rear helpers are cut in 13 cars ahead of the caboose, and as whistle signals cannot be heard, the helpers keep working steam until fully stopped by the head engine. When stopped, the head engineer releases brakes, allowing the slack to run back on the helper who sets his independent straight air as soon as stopped. When ready to go, a heavy application of air is made from the head end, which when released, is a signal to the helper to start.

"Railroading in Barn"

This brings to mind the remark of the boomer brakeman who made his first trip in the sheds. "I've railroaded all over the United States," said he, "but this is the first time I ever railroaded in a barn."

A word should be devoted to the problem of shed maintenance, which, on account of the extremely heavy traffic during the summer months when the work must be done, is difficult in the extreme. Lumber yards are established at convenient points throughout the sheds, and as soon as possible in the spring work trains are used to distribute lumber up and down the line. Sheds are repaired by splicing posts, placing new roof boards, etc., as required, except that where the general condition of the shed justifies it, it is more or less completely renewed. The lumber for the beams, rafters and roof is hoisted by means of an "A" frame and a horse, and our snow shed horses, the product of years of training, have grown remarkably wise in their ways and can almost foresee the desires of the foreman. Each gang erecting shed is, of course, fully protected by flag; however, as trains can usually be heard approaching, the delay to operation through shed construction is remarkably small.

The snow sheds have frequently carried a load of from 20 to 25 feet of snow. As a matter of general practice, however, we do not allow the snow to remain on the sheds to a depth of over ten feet, snow to greater depth being shoveled off by hand.

Fewer Snowsheds Now

The development of improved equipment has made possible a new policy in the maintenance of snow sheds, as it is now possible to plow open and keep clear stretches of track where sheds were formerly essential. The condition is being still further improved through the construction of second track over the mountain.

During the past two years we have started to take down sheds wherever the track is on fill or in slight cut where the lower side can be daylighted. Sheds will be retained in all through cuts, side hill cuts where slides are possible and over sidings and switches. Wherever possible sheds will be eliminated in connection with second track construction, and wherever this is done, all cuts will be daylighted. Nearly ten miles of the present snow sheds will be permanently abandoned when second track is completed.

"HOUSE OF A THOUSAND CURVES" SAFEST SPOT ON EARTH.

By W. L. MASON
(Staff Operator at Gunter)

Did you ever hear of the "House of a Thousand Curves?" It is forty miles long, built on a railroad track and the safest place for a traveler in the world.

Whenever I hear people raving over the beauties of the Southern Pacific's route over the Sierras, past the blue lakes, the snowy mountain tops, the miniature Switerland of crags and glacial formations, I always say to them: "Have you ever given a thought to the wonderful system which makes it possible to send trains over a single track and around short curves without the slightest possibility of a collision? Think back sixteen years and you cannot recall having read or heard of any life being lost in the snowsheds through collision of trains."

That gets them interested, whereupon I sketch briefly for them a description of the staff system which furnishes Southern Pacific passengers absolute insurance against accident while traversing the beautiful Sierras through the "House of a Thousand Curves."

Those of you, who are not familiar with the staff system and who have the opportunity of traversing this region should look out of the window on the righthand side as the eastbound train approaches Blue Canon station. You will note a sign reading "Beginning of Staff System."

With the approach of your train, the operator at Blue Canon has stepped up to a machine about six feet high and a foot in width. The machine has a zigzag slot containing twenty or more steel staffs, each about a half inch in diameter and ten inches long. These staffs correspond somewhat to a Yale key and are grooved so that they cannot be removed without the help of the operator at the next station. The Blue Canon operator presses a button at the lower end of the machine, which informs the operator at the next station that he proposes to take a staff out of the machine. The second operator has two staff machines, one for the station on either side. He holds the electric button connected with Blue Canon just long enough to release the staff. The two machines then become locked automatically, so that no more staffs can be removed and no more trains pass. There is only one way by which either of these machines can be unlocked. The staff that was removed from the Blue Canon machine must be replaced at the end of the block.

To return to the operator at Blue Canon. When he takes a staff out of its place he puts it in a leather pouch attached to a steel ring about two feet in diameter and hangs it on a crane, somewhat similiar to a mail crane, where the engineer may catch

The Staff System and the "House of a Thousand Curves" which it protects. Upper photos show operator and machine (at left) and method of delivering staff to engineer in motion. Below, a winter scene in the snowshed area.

it as his train goes by. There is a "catcher" on the engine for this purpose.

The staff is the engineer's authority to move from one block to the next. Under no circumstances will an engineer proceed to the next block without the staff. The staff, you see, takes the place of train orders. There are staff stations all through the sheds, approximately two miles apart, where there are sidings so that opposing trains can pass one another.

As a train moves out of a station, the operator there informs the operator at the next station over the staff telephone that a train has entered his block. Then that operator repeats the process, signalling the next station in advance that he has a train coming and wishes to remove a staff from the pair of machines governing that block. This operation is repeated over and over again all the way through the sheds from staff station to staff station. The operator takes the staff that is left at his station by the train and replaces it in the machine connecting with that from which it was removed,

thus synchronizing the pair, leaving that block clear for another train in either direction.

When your train is to meet another train at any one of these stations the signals in the staff office are set by the operator for one train to take siding and one to hold main line track. If the superior train arrives first, it stops at the office and waits for the other train to get clear into the siding, and then for the operator to put staff back into the machine and replace it before the train can move on the next block.

The signals for both the eastbound and the westbound trains are located at forty feet and again at fifteen hundred feet from the switch on either side of the station. The distant one is called the "distant" signal, the other the "home" signal. When a block is clear for a train, the lights in these signals are green, showing that a train may proceed with safety. When a train is to take siding the distant signal shows yellow and the home signal shows red. All signals are made to read "clear" when a staff is hung on

THE BULLETIN

the crane. In order to get these signals to clear, the staff must be run through a controller which governs them. As was mentioned before the staff is grooved and as it is put through the controller it locks the signals.

In addition to the staff system, further protection is afforded by fire trains, track walkers, shed watchmen and a small army of other employees whose duty it is to get you through these forty miles of snowsheds.

In short, nothing that human ingenuity can suggest nor money pay for is neglected to make your ride both enjoyable and safe.

Insurance statistics show that you are in less danger in a railway coach than out of it, so I can truthfully say that when you leave Blue Canon, eastbound, you enter "The House of a Thousand Curves," the safest place on earth.

STEAMER DIVISION NOTES.
By C. C. SANDELIN

Fred R. Smith, assistant chief engineer ferry steamer Solano, has returned to Benicia after several months' absence on a trip to Australia.

Mr. Smith stated that he enjoyed his trip immensely and found much to interest him in that country. However, he is more convinced than ever that the renowned Garden of Eden was located in California.

William Hamilton, fireman on the ferry steamer Alameda, has also been absent on a trip to Australia, returning to San Francisco on the same steamer which brought Mr. Smith.

The ferry steamer Edward T. Jeffery, which was operated by the Southern Pacific on its Alameda ferry during government control, was returned to the Western Pacific, March 1st, together with the tug Virgil C. Bogue and the car floats.

The Western Pacific steamer employees who were employed on the Southern Pacific ferry steamers during the time the unified ferry service between San Francisco and Oakland was in effect, are no longer under our jurisdiction.

Captain H. H. Holmes, of the ferry steamer Melrose, is an earnest advocate of Safety First, and never loses an opportunity to make suggestions that will correct unsafe conditions He has already won two medals for his interest in safety work, and is among those selected for medal to be bestowed for distinction in this work during period October 1, 1918, to February 29, 1920, having submitted 44 excellent suggestions between these dates.

A WASTE OF ACTION.

Pat—O'im not such a fool as t' fight wid ye."
Rastus—What you mean?
Pat—Shure, if Oi give ye a black eye it wouldn't show.

SALT LAKE DIVISION NEWS ITEMS.

By NORBERT J. THOMAS
(Superintendent's Office, Ogden)

Fifty-two employees attended the Division Fuel Oil meeting at Carlin, March 12th. Many important aspects of conservation of fuel oil were brought before the meeting and discussed at length. It was shown that all concerned on this division are co-operating creditably in the saving of oil.

J. W. Clark, chief of Fuel Oil Bureau, San Francisco, explained a graphic chart reflecting oil consumption on the system. The Salt Lake division made a saving of fuel oil in both freight and passenger service during the month of February, 1920, over February, 1919, of approximately $1,852.

Our last regular safety meeting was held at Ogden, February 17th, when many excellent suggestions were offered and discussed.

The committee has made it a practice to talk safety at every opportunity and it developed that a total of 326 talks were made since the last meeting.

Mr. H. G. Valleau has been appointed chief dispatcher, Sparks, Nev., vice Mr. J. T. Bell, who is on an indefinite furlough.

Mr. H. F. McDonald, former trick dispatcher, has been appointed assistant chief dispatcher, vice H. G. Valleau.

Mr. O. M. Barlow has been assigned the duties of trainmaster as well as roadmaster on the Fernley and Fallon branches.

Mr. Otis Weeks, division engineer, went to Chicago to attend the American Railway Engineers' meeting, March 16th.

SUPERINTENDENT WILSON PRAISED BY EDITOR.

Superintendent William Wilson of the Tucson Division has a warm friend and admirer in Editor Willard E. Holt of the Lordsburg "Liberal," who made public note recently of the fact that the Southern Pacific official had given up his office car to an injured Mexican laborer in order that the latter might be hurried to the hospital in comfort. Mr. Wilson followed later in a motor car.

"The Creator evidently picked up pretty good material when Wilson was built," concludes the journalist.

EXPERIENCED.

Reader—This paper says that 85,000 women are now employed by the railway systems of the United States.
Speeder—Hardly proper work for women, I should say.
Reader—Why? Who's had more experience in looking after trains and switches than women. I'd like to know·
—Yonker's "Statesman."

FORTY INDUSTRY TRACKS ON WESTERN DIVISION.

By R. J. HOCKETT
(Superintendent's Office, Oakland Pier)

During 1918 and 1919, forty industrial tracks were constructed on the Western Division, their total length being 37,156 feet, or seven miles. This new work demonstrates the industrial development that has been going on in the east bay cities and on the Western Division. There are numerous projects under way and new ones coming up all the time.

From indications the fruit canning industry in the east bay region will enjoy an extremely heavy season. During the past few months several extensive additions have been made in the way of canning plants, among them being the large plants erected by Libby, McNeill & Libby and the Jones Canning Company, as well as a branch plant of Biseglia Bros. of San Jose at Fruitvale. The Pacific Coast Canning Company is likewise providing for a heavy run, a big extension to their present plant at the foot of Fourteenth Street, Oakland, being under way.

Ray Le Mieux, relief ticket agent at Oakland Pier and Berkeley, has been promoted to city ticket office, Thirteenth and Broadway, Oakland.

L. T. Bayley, clerk, Sixteenth Street Station, Oakland, and T. A. McGeough, clerk at Fruitvale Station, have been promoted to ticket clerks, city ticket office, Thirteenth and Broadway, Oakland. They entered on their new duties a few days ago.

G. A. Erickson, clerk, Fruitvale Station, has been promoted to a responsible position in the office of L. Richardson, D. F. & P. A., Oakland.

J. D. Rippy, relief agent, Western Division, has been appointed chief clerk to C. H. Jasper, D. F. & P. A., Fresno.

A. J. Lloyd, ticket clerk, Oakland Pier ticket office, has been appointed assistant chief clerk in the office of Mr. Jasper, D. F. & P. A., Fresno.

J. R. Spurgeon has been appointed agent, Crocket Station.

C. A. Woodson has been appointed agent, Walnut Creek.

P. R. Witt has been assigned position as second trick operator at Santa Rosa.

H. A. Hargis has been assigned position as second trick operator at Hayward.

J. S. Silva recently assumed his new position as assistant agent (Mare Island), South Vallejo Station.

REDUCING TELEGRAPH TOLLS.

Superintendent of Telegraph E. L. King advises that in order to eliminate unnecessary telegraph tolls the name of the railroad and the official title of the sender should henceforth be omitted in messages sent to other railroads via Western Union. The revised official roster contains this information.

KING SNOW: *Railroad's Snow Fighting Battalions*
Match Wits and Courage with Sierra and Cascade Storms

When winter comes and "King Snow" begins his reign in the high Sierra of California and the Cascades of Oregon, there are battles aplenty in the offing for the "mountain boys" on the Sacramento and Portland divisions of Southern Pacific.

The beautiful white blanket that spreads itself over these regions, beckoning to the fun-loving thousands who revel in the thrills of snow sports, simply brings another seasonal job to the men who keep the tracks cleared for the safe and uninterrupted movement of passenger and freight trains over these main transcontinental routes.

At the first flurry of snow the fight is on. The battery of eight rotary snow plows (3 on Cascade), four spreaders (2 on Cascade), and ten flangers (3 on Cascade), is in position at Truckee, Norden, Emigrant Gap, Crescent Lake, Klamath Falls and other strategic points in the two battle areas. Powerful locomotives are primed, ready to couple onto the snow fighting equipment. Trained crews await the trainmasters' calls any hour of the day or night. Bridge and Building and track men are already on the job. Linemen look after the telegraph and telephone wires, and signalmen are alert to catch anything amiss with their circuits.

"Front" of 97 miles

The railroad's troops and heavy artillery stand ready for action over a "front" that extends for 70 miles in the Sierra and 27 miles in the Cascades. That's the answer to the challenge hurled by "King Snow" every winter.

The courageous and determined accomplishments of these snow fighters, year in and year out, form one of the most dramatic episodes in the operation of our railroad.

The Sierra summit of the Company's line reaches its maximum elevation of 7017 feet at Norden. Highest point in the Cascades is 4844 feet at Odell Lake.

Beautifully blanketed in its ermine mantle, the Sierra region presents a glorious spectacle—one that even the railroad snow fighters pause to enjoy after the tracks have been cleared and the storm gives way to sunshine. East portal of the long Sierra Tunnel is seen at the right, and a portion of the snow sheds on the left. The location is near Donner siding overlooking Donner Lake where 36 emigrants to California were trapped in the snow and perished during the winter of 1846-47.

A total snow fall of 45 feet during a winter is not uncommon at these two summits, with 10 to 20 feet on the level at a time. In many places the snow drifts to much greater depths. High winds, blizzards, and sub-zero temperatures, multiply the difficulties of the task in keeping the lines open. Storms often follow at quick intervals, the heavy fall of one being no sooner cleared away than another sweeps the mountains.

Such are the conditions in a fairly normal winter.

Yet, so well has the railroad fortified itself with powerful equipment, sturdy snow sheds, and trained forces working under the direction of leaders skilled through years of experience, that trains are seldom delayed even a few minutes, which is remarkable considering the gigantic task the operating and maintenance crews are confronted with at times.

The present winter has been comparatively mild. Total snowfall on the Sierra summit amounted to 276 inches

up to February 1, with 9 feet 9 inches the maximum on the ground at any one time. The Cascade line had a heavy storm during the first two weeks of January when the snow level was increased from 48 to 104 inches at Abernethy near Cascade Summit.

But, in light of the experiences our snow fighting champions have battled through in former years, there has been hardly enough snow this season to give them a really good work out. The winter, however, is by no means over, and before this issue of *The Bulletin* reaches its readers the boys on the mountains may have fought through one of those storms the old-timers tell about when snowfall has averaged as much as a-foot-an-hour.

How Fight Is Handled

Here's what happens, for instance, at Truckee, headquarters of the Sierra snow fighters, when a snow storm begins.

The first defense used against "King Snow" is the V-shaped snow plow at-

The Cascade region in Oregon matches the Sierra in the depth and beauty of its snow, as well as in the severity of its storms through which the railroad snow fighters battle. Picture on the right was taken at Cascade Summit station, and the other one at Abernethy where the Cascade snow usually reaches its greatest depth. The Bulletin gratefully acknowledges the assistance of several Portland Division employes in loaning pictures and negatives of snow scenes.

Some of the Operating and Maintenance Men Who Direct Sierra Snow Fighters

Eric Gram
General Foreman
Norden

Claude Gangler
Ass't Signal Sup'v
Norden

August Olson
B&B Foreman
Norden

H. Montgomery
B&B Foreman
Emigrant Gap

Bat Riordan
R'ndhouse F'man
Truckee

Henry Williams
Night Y'rdm's'tr
Truckee

Harry Oliver
Day Yardmaster
Truckee

MERLE L. JENNINGS
Trainmaster, Truckee . . . supervises the snowfighting job.

tached to the pilot in front of a locomotive. More than 50 locomotives operating "over the hill" are equipped in this manner. The pilot plows are capable of pushing the new, light snow off to the sides of the track sufficient to allow standard clearance for train operations.

As the snow fall increases, the flangers are brought into action. This type of snow fighting equipment gets its name from the fact that it is used to scrape the snow and ice from between the rails so that the flanges on the car wheels will meet with no obstructions.

Each flanger has two pairs of cutting blades which are raised and lowered by air, the valve being operated by the engineer of the locomotive pulling the flanger, or by one of the crew on the flanger. Each pair of blades is at the base of a flare-shaped metal moldboard which serves to throw the snow and ice to the side of the track; and each is operated independently of the other, depending if used in double track area or on downhill side of canyon.

Blades of flangers used in yard service extend the full width of the track inside the rails, but with automatic train control in use over this portion of the railroad in the Sierra, blades of the road service flangers are divided into pairs so as to clear the track magnet boxes located at intervals in the center of the track. This arrangement avoids the necessity of the engineer having to

raise the flanger blades except when passing over switches, road crossings, etc.

Flangers are used more extensively than any other type of snow fighting equipment and are operated continually during the storm until the snow banks they have built up alongside the tracks are so high the moldboard flares will not throw the snow clear of the tracks.

Then the rotary plows, the most powerful and effective of all snow fighting artillery, are called into action. With huge fan wheels whirling at great velocity, the rotary is pushed forward at a moderate speed by one of the big 4100 class locomotives cutting through the snow and throwing it far to the side of the tracks or down the mountain side. The fans are propelled by a power engine in the rotary operated by an engineer and fireman, while an operator and assistant manipulate the controls of the rotary itself. The rotary fans in various types of plows have diameters ranging from 9 feet 11 inches to 11 feet 2 inches, and are housed in a hood that can be raised or lowered, but which is ordinarily operated about three inches above the rails. The fan blades whirl

REPAIRS: A. "Bimbo" Esola (center), machinist at Truckee roundhouse, assisted by E. J. Falltrick (right) and J. Lipschenberger in adjusting a new blade at base of moldboard on one of the flangers used in cutting snow and ice from between the rails.

SCOTT E. GORDON
Roadmaster, Truckee . . . wide-wing rotary is his pet.

and cut the snow back of the blades into the cones, which in turn throw the snow through an opening in the top of the hood.

During a heavy storm, the flangers and rotaries are operated continually, always going ahead of passenger and freight trains to insure a clear path.

When the storm subsides the "cleanup" job begins immediately. This means the cleaning out of the cores built up by the flangers or left by the rotary plows between the double tracks, widening the path through the snow to a maximum clearance of 8½ feet from center of the track, clearing sidings, and shoving down the banks on the down-hill side or where ever else it is possible. This operation calls into use the rotary, plus a wide-wing rotary plow and the spreader.

The spreader, pushed by a locomotive, is equipped with wings that can be extended to 20 feet on each side of the track, thus clearing a 40-foot path. Each wing is divided in three sections, and under the skillful manipulation of the spreader operator who regulates them up-and-down and in-and-out by using an air valve, the wing sections are made to function like the powerful steel arm of a robot, jointed at the shoulder, elbow and wrist. Moving along at a fair speed, the spreader wings level the snow off even with the tops of the rails.

One of the requirements of a com-

Continued on Page 12)

"King Snow" Meets His Match in the High Sierra . . .

Business-end of wide-wing rotary snow plow . . .

kicks up a rumpus like this when in action . . .

(Continued from Page 4)

petent spreader operator is to be thoroughly familiar with the location of every narrow cut, high bank, rock abutment, signal box, switch, or other obstruction that comes within the 20-foot radius of the spreader, so that even though the object may be completely covered from sight by deep snow, he knows when to lift or pull in the wings to avoid damage to the spreader or to operating facilities along the tracks.

Certainly the spreader operators, as well as the rotary operators, have to know the A, B, C's of their rights-of-way, even in darkness, or else up-root expensive signal and track appurtenances, to say nothing of damaging the spreader wings by digging into earth and rock through narrow cuts.

The Clean-up Job

In handling the clean-up on double tracks, the spreader and rotary work together. The spreader, with both wings extended, is moved forward on the outside track. The inside wing levels off the snow bank between the tracks, shoving the snow over onto the inside, or hill-side tracks. The outside wing shoves down the snow bank on the canyon side. The rotary then comes along on the inside track clearing off the core shoved over by the spreader, throwing the snow clear of the outside track and entirely off the right of way.

The finale in the clean-up act is the performance of the wide-wing rotary. This especially equipped plow has hinged wings on each side, which cut into the snow bank on the hill-side, providing an additional clearance of four feet.

After the wide-wing has done its duty, the railroad line is clean. At least it's clean until another snow storm. Sometimes that storm comes most disgustingly quick. Then the job's to be done all over again.

The day seems to have passed when snow can fall fast enough and heavy enough to blockade the railroad. Modern equipment is too powerful and effective. Yet, such a thing happened occasionally in days gone by.

Veteran railroaders tell the fiction-like story of the winter of 1889-90 when a Sierra storm piled snow to depths of 30 and 40 feet on the level, blocking rail traffic for ten days. The rotary plow had not yet come into use; it was introduced later that winter. There

PICTURES: Above—Right: Crew of spreader. Left to right—J. E. Henderson, locomotive engineer; E. E. Smith, locomotive fireman; C. H. Ray, conductor; A. Dominguez, laborer; R. T. Henderson, laborer; T. B. Wagner and C. E. Carl, brakemen; H. N. McColl, spreader operator.

Below—Right: Crew of wide-wing rotary snow plow. Left to right—Harry Bell, hostler; Nick Zucke, locomotive engineer; Scott E. Gordon, roadmaster; W. Nixon, brakeman; Walter Aske, rotary engineer; Jess Fisher, machinist; W. W. Zell, rotary fireman; C. H. Preston, conductor; F. Spanger, locomotive fireman; W. A. McElhiney, brakeman.

was no equipment then that could throw snow clear off the roadbed.

Snow fighting outside the snow sheds was done with "push" plows, a pioneer type of flanger, and the strong arms of many human shovelers. The push plow was exactly what its name implies. It was used to push the snow off the tracks, being pushed at the head-end of six to twelve of the small, wood-burning locomotives of that day. The battery of locomotives would back away some distance from the snow pack and then take a run at it with the throttles of all locomotives "wide open." The snow was pushed and tossed aside until resistance became too great and the whole battering ram came to a halt. Then the locomotives backed up and took another run at it. So it went until the plow was pushed through, or was derailed.

As the snow banks were built up alongside the tracks they became too high for the plow to get rid of the accumulation. To take care

... *Rotary and Spreader Clear Path for Fast Trains*

sweeps the railroad clean, like this . . .

for the transcontinental trains.

of this situation, chambers, or trenches, were cut in the sides of the banks at frequent intervals to provide a dumping place ahead of the push plow. This operation required an army of men wielding long handled, square-pointed shovels.

During the storm and blockade of 1889-90, the railroad had more than 5000 shovelers in the Sierra. Attending to the physical requirements of these men was no small task under the extreme storm conditions. Housing quarters were built at Shady Run (now Midas), where ten monster ranges were kept busy almost constantly cooking food that had been brought in by dozens of dog sled teams.

"Cyclone" and "Headlight" Plows

The "cyclone" plow, a contraption resembling a huge screw driver, was also tried out about this time, but was of little use. The rotaries, however, stood the test, and their efficiency has been greatly increased from time to time through new developments and certain changes in design.

Another type of plow used quite extensively in the Sierra up to about 1916, was the "headlight" plow. This plow was attached to the pilot of the locomotive, as are the present-day pilot plows. Its mold-boards extended all the way up to the locomotive headlight, and was capable of sweeping a path 10 to 11 feet wide through the snow that was too deep for the ordinary pilot plow.

The wide-wing rotary plow, reconstructed this last year at the Company's shops in Sacramento, is the finest piece of snow fighting equipment yet developed by any railroad. Needless to say, it is the pride and joy of the boys on the Sierra "hill."

The snow sheds were the railroad's principal protection in the early days.

The Company's pioneer engineers were well aware of the vagaries of a Sierra winter. Their construction of the first transcontinental rail line was halted completely during the winters of 1866-7-8, except for the underground work in the tunnels, and they were quick to adopt the sheds as the most practical means of keeping snow off the tracks in the most exposed sections of the mountains.

Construction of the sheds was begun in 1867. By 1873, four years after the first transcontinental line was opened to traffic, 40 miles of these snow protections had been erected, extending in almost a solid, unbroken stretch from

---

CREDIT: *Bulletin Pictures Only Few of Snow Fighters*

The employe pictures in connection with this article on snow fighting, show only a few of the men who take care of the job in the Sierra. There are dozens of others involved in the work, but it happened to be their day off when *The Bulletin's* camera man showed up for pictures. Also there is another group of employes on the Cascade Line who perform an equally fine job each winter. At another time they will come in for their share of credit in *The Bulletin* with pictures and all the trimmings.

---

just west of Truckee to a short distance west of Blue Canyon. This introduced a strange phase in customary railroad operations, which once prompted a boomer brakeman to remark: "I've railroaded all over the world, but this is the first time I've ever railroaded in a barn."

The early sheds were constructed very similar to a house, with peaked roof on a very steep pitch. Logs for the posts and braces were felled in the forests adjacent to the rail line, and were of pine, fir and cedar. Neighboring saw mills supplied the timbers for roof and sides. Posts and braces were buried deep to give rigid alignment to the sheds. But this type of shed proved impractical. Posts rotted in the ground and offered little resistance when snow packed against the peaked roof facing the mountain slopes, began to force portions of the sheds out of alignment. When this happened, men shoveled snow from one side of the roof to the other, transferring the weight so the sheds were forced into line again.

These difficulties have been overcome. The sheds are now flat roofed and have proven capable of supporting heavy loads of snow. The posts and braces are anchored in rock or concrete bases to prevent rot, and clearances have been increased from 17 to 23 feet. In many places concrete walls have replaced the wooden sides facing the mountain.

As the efficiency of snow fighting equipment increased, there became no need for sheds over the tracks in many locations. The original 40 miles had been reduced a few miles even before the memorable storm of 1889-90. When double-tracking was extended from Colfax to Blue Canyon during 1913, the sheds in that district were abandoned, and by the time double-tracking was resumed in 1924, there was slightly less than 30 miles of sheds remaining. A successive program of abandonment has been carried out since the latter date and at the present time there are 10.208 miles of sheds on Sacramento Division and 0.522 miles on Portland Division.

HONORED: *Chief Engineer Is Awarded Prize for Best Paper*

Chief Engineer W. H. Kirkbride was recently honored by the American Society of Civil Engineers, in being awarded the Thomas Fitch Rowland Prize for outstanding paper submitted on a construction subject. His paper was on the Southern Pacific Martinez-Benicia bridge across Suisun Bay near San Francisco. The award was endowed by Mr. Rowland, honorary member of the Society, in 1884, to be given for papers describing in detail accomplished works of construction.

Mr. Kirkbride, who is a member of the society and former president of its San Francisco Section, attended the recent annual convention in New York. He is also a member and former director of the American Railway Engineering Association, and a member of the rail committee of that organization; a member and vice-president of the Seismological Society of America; a member of the American Wood Preservers' Association; and president of the Pacific Railway Club.

CONSCIENCE: "In the summer of 1931 I was hitch-hiking through Texas trying to get home," wrote a gentleman to Superintendent Fairbank's office at El Paso. "I was unable to obtain a ride, so I rode the freight train from El Paso to Phoenix. Enclosed is my check for $1.50, which I believe is about the cost of 250 pounds of freight for that distance."

"We can never say too much in praise of S. P. service. . . . When any of our friends make a trip West we hope it will be over your lines . . ." wrote the Misses Orlena and Mary Futrelle, of Louisville, Ky., in thanking F. A. Taylor, investigator, Roseville, for efforts taken to locate baggage that had strayed.

"The railroads have made traveling a relaxation and a real rest. . . . The S. P. people were just as kind to me in El Paso and elsewhere as our own Bob Welton (agent) is here at Mt. Angel," wrote Rev. Fr. Alcuin Heibel to the Mt. Angel, Ore., News, on his return from an extended trip.

"I congratulate you on your Tray Service in the coaches and tourist cars. The food is good and the price within reach of everyone," wrote Dorothy Meyers, of San Jose.

F. A. Allen, chair car porter, Pacific Limited, was commended by a passenger who wrote to Superintendent G. E. Gaylord, commenting on the courtesy and efficiency with which he conducted his work.

"Southern Pacific is my favorite road in crossing the continent. . . . I have just completed the 18th round trip in less than eleven years. . . the best food for the least price of any road . . ." writes Mrs. Geo. B. Snow, of Long Beach.

Mrs. F. V. Stevens, Vermontville, Mich., writes commending the service received from all Southern Pacific employes, and in particular the "cheery,

CONFAB: *"Get More Business" L.A. Traffic Meeting Theme*

Plans for increasing traffic, coincident with the improvements in train service through the San Joaquin Valley and to Portland, were discussed by representatives of the Southern Passenger Traffic District at a conference held in Los Angeles, January 22 by General Passenger Agent Henry P. Monahan.

District Passenger Agents met with Mr. Monahan and Assistant General Passenger Agent Nels Kinell during the afternoon. This was followed in the evening by a general meeting attended by more than 125 men from many offices throughout Southern California, at which Chas. A. Pestor, district passenger agent at Los Angeles, presided.

President D. W. Pontius of the Pacific Electric was a guest at the afternoon meeting and discussed pending matters of importance. A feature of the evening session was the showing of the Associated Oil Company's motion picture "Friendship Lives Here." The picture strikingly emphasized the good that can be accomplished by friendliness in dealing with the public and with one's fellow employes.

THANKS: *Courteous and Attentive Services Commended by Customers and the Company*

courteous manner of Lewis Peeples, news agent."

John Hicks, waiter on the Daylight, is commended by F. F. Gualano, of Monterey Park, Cal., for "courtesy, excellent service, and detailed attention to the children's needs."

"On a trip East I rode on six different roads and I found your dining car service the best," Mrs. H. V. Matthews, of San Francisco, writes. "I talked with a lady who was returning East. She said S. P. service was grand, and that when she changed roads during the night she could always tell the difference."

". . the best meal I have eaten in many a day," wrote Anna M. P. Barnes, R. N., Tulalip, Wash., who also commended the "gracious and courteous services" of Steward A. E. Robinson and his co-workers.

"As a patron of your road over a period of fourteen years, I take this opportunity of expressing appreciation of the fine, efficient and courteous service received from H. J. Taggart, F. H. Works, and John Esterak, in the ticket office at Fresno," wrote Miss Edith A. Demorest of that city.

REWARD: *$50 Check to Girl Who Found $680 in Wallet*

New Year's Day was a really happy one for Miss Margaret Slot, attendant at the Union Station news stand in Ogden.

During the few minutes stopover of the Overland Limited at Ogden December 23, W. M. Brewster, oil company executive of Paris stepped up to the news stand to make a few purchases. Returning to the train he discovered his wallet was missing. A hurried search was futile, so he left the matter in the hands of the railroad people while he continued on to California.

Margaret Slot

In the meantime Miss Slot noticed a wallet left on her counter and laid it aside while she continued to take care of the rush of customers from the train. When she had a moment to examine it she was startled to find it contained $680 in bills. She turned it over to Special Agent H. H. Cordon, and in a few days the wallet and its contents was again snug in Mr. Brewster's pocket.

And on New Year's Day Miss Slot received a $50 check from the oil company official.

Trains and crews of the twenty-first century crossing the Sierra in winter are equipped with technology that provides communication with other trains and yard personnel, wind shield wipers, motors that detect the slippage of wheels, dynamic breaking which helps reduce the wear on the shoe breaks of each car, the ability of the engineer in the front of the train to control the locomotives at the rear of the train, a system that provides near instantaneous application of the brakes, and a cab that can be easily warmed. Trains of the current era run over heated switches that depend on a hot steel cable that runs along the rail, heats the rail near the switch and melts the snow in contact with it. In contrast, an entire four-car train crossing the same summit in 1875 likely weighed less than a modern diesel-electric locomotive and contained none of the features that are standard on locomotives one-hundred-twenty years later. Passenger cars were heated with a wood stove. Brakemen on the cars applied the brakes by turning a wheel about two feet in diameter at the end of the car. They needed to stand on top of the cars, in some cases, which was a demanding task given the cold, wind, ice and snow. The signal to brake came from the engineer's whistle signals if they could be heard over the sounds of a storm.

164

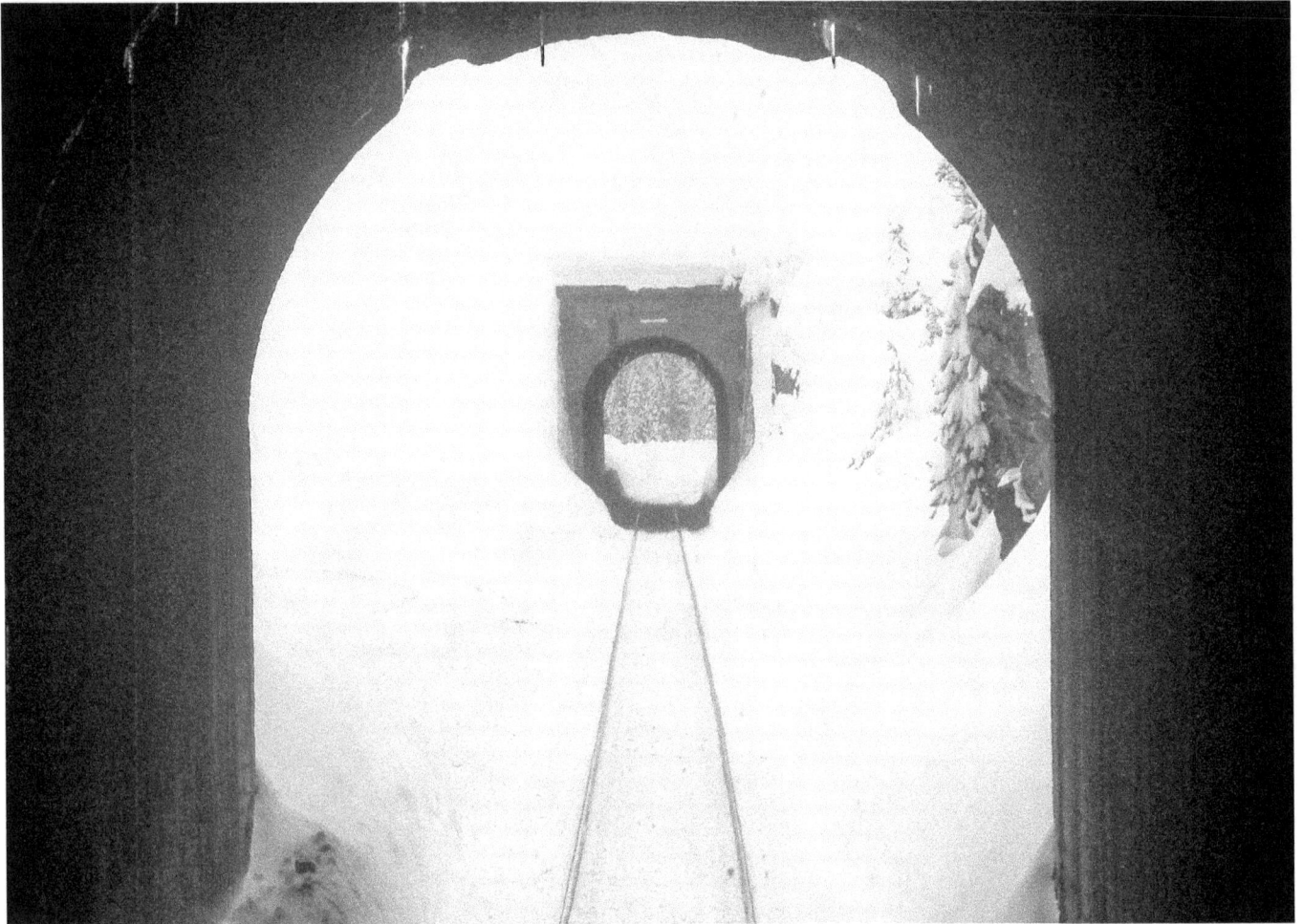

Many of the tunnels over the Sierra were made with the vertical walls sixteen feet high. Above that height, an arch with a diameter of sixteen feet curved over the track. Initially, tunnels constructed in soft material were lined with timber. Later, some tunnels were lined with concrete. The railroad learned quickly that snow sheds were needed between the tunnels. This led to tunnels of wood connecting tunnels through mountains. The location shown above, near Yuba Gap, is a likely location of this construction.

The concrete lining the tunnel is not impervious. Water dripping through cracks can lead to icicles. Small icicles, as shown above, can grow to be over ten feet long. Some locomotives in the diesel era are equipped with metal grids, called antlers, on the roof to break off the ice.

In *High Road to Promontory*, by George Kraus, John R. Gillis, who worked on the tunnels, is quoted from a speech made to the American Society of Civil Engineers. He said, "A single foreman, with a gang of thirty to forty men, generally constituted the force at work at each end of a tunnel; of these, twelve to fifteen worked on the heading, and the rest on bottom, removing material." He added, "As soon as each heading became sufficiently advanced, the center line was secured, generally by small holes drilled in the roof, with wooden plugs and tacks. In most cases the entrances were afterwards so blocked up with snow that it was impossible to accurately refer to the line outside."

The advent of double-stacked container cars created a major problem for trains: They didn't fit. The image above shows two chiseled notches that extend the entire length of the tunnel to accommodate the higher cars.

Chapter Twelve
Fuel Savings – A Concern from the Beginning

Imagine a situation on January 1, 1900 where someone is sitting on a horse looking at a steam locomotive en route from Sacramento to the Sierra foothills. As viewed from the side of the road, the fireman throws coal into the firebox, the fire gets hotter, more steam is formed and the train goes faster. That is the elementary series of events that, in no way, paints the true picture of the physics, timing, and mechanics that are in play to make a fuel-efficient trip for the engine.

The first piece of information that the engineer and fireman must have before departure is the type of coal that is being used. Coal contains carbon, hydrogen, ash, water, sulfur, and nitrogen. Carbon makes up the highest percentage of coal and is the part of coal that burns as a solid on the grates of the locomotive firebox. The type of coal is determined by the proportion of each constituent.

Anthracite coal has about 85 percent "fixed," or solid carbon, about 5 percent gaseous matter and moisture, and about 10 percent ash. This composition allows it to burn with a small flame. Bituminous coal has 50 percent fixed carbon, 40 percent gaseous matter, and about 10 percent ash. When this coal burns, there is more flame and smoke. Anthracite coal is further categorized into semi-anthracite, which is less dense, has a greater amount of volatile combustible matter, and is more easily ignited. There is also semi-bituminous coal, which is not as hard as the regular bituminous. It also ignites and burns quickly.

In a solid fuel burning locomotive, the fire lies on a grate. The grate acts like a screen. It holds the burning wood or coal while allowing air from outside the engine to circulate up through the openings in the grate. An ashpan below the grate catches and collects hot embers and ashes as they fall through. The grate may be shaken to clean the ash from the bottom of the fire. They are shaken either manually or (in larger locomotives) by a powered grate shaker. Why shake the grate? By doing so, more openings in the grate become exposed, air is allowed to move up in to the fire area and combustion of the coal on the grates is more efficient. So, a watchful fireman monitors the air moving through the grates and shakes the grates as needed.

Because too much steam pressure in the boiler can cause a disastrous explosion, the design of the locomotive has a relief valve at the top of the boiler. When the boiler pressure reaches the danger level, the valve opens and releases steam. Letting steam exit without going through the cylinders is a waste of the fuel it took to make the steam. Calculations indicate that one-quarter to one-half pound of coal is wasted for each second the pop valve is open. If, during a trip, the valve is open ten minutes, 225 pounds of coal is wasted.

One of the potentially large fuel losses that a fireman needs to minimize concerns coal dust. When a shovel of coal is put in the firebox, the draft picks up the fine coal dust and sucks it through the tubes, into the smoke box and out the stack. If a large amount of coal needs to be burned under a high demand situation, the draft is greater and so is the loss of unburned coal dust which leaves the engine through the stack having never been burned. For this reason, the fuel consumption record of an engineer's locomotive in light service on flat terrain cannot be compared, per mile, to one whose engine is laboring uphill with a four thousand ton load. Some crews soak the slack coal with water before departure so that it sticks together when ignited. This must be done at such a time to allow excess water to drain.

The seemingly small amount of savings that a fireman and engineer can make needs to be multiplied by hundreds of locomotives operating 365 days a year to understand the huge savings that wise management of water and coal can realize in a solid fuel burning locomotive. The efficient consumption of oil and coal in Sacramento and the West was especially important. Those items needed to be delivered long distances compared to eastern yards where coal might have been a locally-produced commodity. Sacramento crews also had the seven-thousand foot climb to Donner Summit that created a heavy demand for fuel.

The seven articles on the next pages illustrate the importance placed on fuel economy in the early part of the twentieth century, and illuminate the challenges faced and cooperation required from all areas of railroading. These include:

"Fuel Economy Issues of 1916 – Saving Oil Over 100 Years Ago," July 1916

"Fuel Use – One of the Railroads Major Costs," June 1916

"Monetary Savings From Fuel Consumption," October 1916

"Oil and Coal – the Railroad's Second Largest Expense," March 1921

"Thrift in the Locomotive Cab – What the Fireman Can Do," May 1921

"Oil Burning Locomotives - Care and Operations," December 1921

"Handling Superheated Locomotives," February 1919

The Bulletin

ISSUED BY THE SOUTHERN PACIFIC COMPANY BUREAU OF NEWS

VOL. IV.   SAN FRANCISCO, CAL., JULY 1, 1916   No. 13

Notes Along Company's New Coos Bay Branch.

Through its building of the Willamette Pacific Railroad, the Southern Pacific has opened up in Southern Oregon one of the richest territories in the West. As comparatively few Southern Pacific employes are familiar with the conditions and possibilities of the line, The Bulletin prints herewith a few notes on the Coos Bay branch trip from Eugene to Marshfield and through to Powers, for which it is indebted to General Passenger Agent John M. Scott.

Eugene, home of the University of Oregon and county seat of Lane County, is substantially built with good hotels, neat business streets and fine residences. Population, 9000.

The new branch line to the Coos Bay country leaves the main line at this point. From Eugene to Veneta is flat farming country. Veneta to Noti is some flat farming country, but gradually changes to foothills of Coast Range Mountains.

Veneta is agricultural center and railroad station for Elmira, two miles north.

Noti, center of agriculture and logging district, is near junction of Elk Creek, Noti and Long Tom rivers, excellent fishing streams.

West of Noti you pass Horse Shoe Curve and Noti Tunnel at summit of the Coast Range. This tunnel is 2480 feet long. We now follow and cross at intervals Chickahominy Creek, Wild Cat Creek and, near the station Austa, come to the Siuslaw, which stream is followed to tidewater.

At Swisshome on the south bank of the Siuslaw is a curious rock formation called "The Old Man Siuslaw." Scenery is excellent along the Siuslaw. There is good fishing in this stream and its tributaries.

Mapleton at tidewater on the Siuslaw River is center for dairying, fishing and lumbering.

Cushman is the railroad station for Acme, Florence and Glenada.

Acme, population 200, is located on the north bank of the Siuslaw River or Bay, a mile or mile and a half from Cushman station. Communication with Cushman is by boat. Has a small sawmill and creamery. Depends on dairying, fishing and some lumbering.

Florence, population 200, is located on north shore of Siuslaw Bay, about four miles from Cushman. Communication is by boat. Launches make run in about thirty minutes. Has a sawmill, hotel, two newspapers, bank and

Interesting Fuel Economy Talks

In view of the importance of economizing in the use of fuel oil, and the increasing interest taken in the subject by Company employes, The Bulletin prints herewith extracts taken from minutes of recent fuel meetings.

Extract from paper read by Master Mechanic Merry of Los Angeles Division at fuel oil meeting in Los Angeles on May 5th:

"A saving of fuel oil can be accomplished through the co-operation of the different departments. Conductors by recognizing the cost of unnecessary moves and switching, yardmasters by demanding that brakes of all cars be released while switching, station and train crews by working quickly around stations to prevent delays which have to be made up on the road, the roadmaster and his foremen by not having too many slow orders out causing trains to slow down, etc., dispatchers by putting light tonnage trains on sidings for heavy tonnage trains regardless of ruling direction and by giving steam orders far enough in advance so that engines can be fired up without forcing and then not be held around under steam unnecessarily long awaiting orders. I would recommend that all these conditions be brought to the attention of those concerned. Co-operation between engineer and fireman is essential to economical fuel consumption, and so is proper inspection and comprehensive work reports from the engineer. Great care should be exercised by the engineer in making an intelligent work report for the information of roundhouse foreman, as it is only through the closest co-operation between the engineman and those supervising the roundhouse repairs that many of the defects which vitally affect the steaming and other qualities of the locomotive are remedied and which are essential to give us the highest degree of fuel efficiency together with a reduction in unnecessary inspection and labor on the part of roundhouse employes."

AT OAKLAND PIER

The following remarks were addressed to train and enginemen by Road Foreman of Engines House of the Western Division:

"I do not believe that the entire responsibility rests with the man who is running the engine or the man who is firing it, but there is great responsibility resting upon him. There are others concerned, as well as the mechanical men, in the economical operation of a locomotive. I refer to train

dispatchers, train crews and station employes. Every minute of time that a train stands at a station unnecessarily means a draft on the fuel supply to overcome this delayed time, and every time a train is stopped unnecessarily for train orders that might be put out in advance or handed up on a hoop, another tax is put on the fuel supply and so on down the line.

"I believe it would be a good proposition if every engineer and fireman would keep a memorandum of the fuel that he burns on each engine that he operates, for the reason that if mistakes are made, it is a very easy matter to go back to the engineer and fireman and get the correct figures, and if a man gets in the habit of watching the fuel performance as he should, and if he notices that on a certain trip he has burned more fuel with about the same conditions as before, he will try to figure out what was the trouble and call the attention of the master mechanic to it, so he may be able to do something with the engine to better the condition."

THE USE OF STEAM

From Fireman Cirby's paper on fuel economy read at the Coast Division fuel meeting on Saturday, May 27th:

"Economy in the use of steam is important in regard to saving fuel oil and should receive very careful consideration from engineman, because an engineman's reputation as an economical fuel performer depends to a great extent on his knowledge of this subject. In order that an engineer may get the best results from an engine, he must understand how to handle the reverse lever and throttle to the best advantage. There are several methods of increasing or diminishing the power that is being exerted by a locomotive and the engineer should understand these methods and endeavor to learn which method will be the most economical to use under a given set of conditions. With this knowledge at his command, the engineer's or fireman's chance of a good position on the performance sheet is greatly increased. Expansion of steam produces power and if it is not given a chance to expand there is a loss of power and there will be a corresponding loss in fuel.

"It is important that the oil in the tank be at the proper temperature. If it is too cold, the oil will not flow as freely to the burner as it should, the fire will lag and the exhaust will have a greater effect on the fire than

Interested in Fuel Economy? Read These Recent Remarks on the Subject

(Continued from Page 1)
it should, due to insufficient oil supply at the burner and to poorer atomization of the oil. Overheating the oil makes it hard to regulate the fire at the burner, the oil 'gases' in the burner, and makes the flow of oil to the burner uneven. This causes the fire to fluctuate in intensity and wastes considerable fuel, because a large percentage of the gas escapes from the burner. Another bad effect of overheating the oil is that the asphaltum separates from the lighter oils and precipitates to the bottom of the tank, causing an increased amount of the lighter oils to be driven off as vapor."

SAVING BY BEING ON TIME

Engineer Jefferis gives views on the economical use of fuel at the Western Division fuel meeting on April 29th:

"I have found that by maintaining an 'on time' schedule considerable oil can be saved, as when a train is behind schedule the consumption of oil is greater on account of the efforts to make up lost time. Considerable oil is saved if, instead of shutting off too much when approaching curves, the train speed is regulated with the braking power. I figure it is economy to use steam on curves, rather than shut engine off and pick up again after curve has been passed. Enginemen should study track conditions, location of slow orders, study the condition of the engine, and, in fact, keep constant observation of all conditions along the road to aid in the handling of a train in such a way that oil can be saved. Engines are different; some will use more oil than others under the same conditions. It is necessary to be acquainted with the engine you are running. Also, there is a great difference in the grade of oil; sometimes it is heavy and you have to use the heater, and again we get oil when it is not necessary to use the heater so much. A constant observance or study of physical conditions while running over the road affords the greatest opportunity for the economical use of fuel oil."

GREATER EFFICIENCY

By Fireman H. J. Ogilvie at fuel oil meeting held at Los Angeles on May 5th:

"Greater efficiency in the fireman is the thing that will accomplish what we are trying so hard to do. There are three things very essential on an engine. They are co-operation, investigation and attention being paid to those operating the engine. To reduce fuel consumption co-operation is essential and for that reason the engineer and the fireman should work together to get the better results."

Remarks of Engineer Phillips at Western Division fuel meeting on April 29, 1916:

"To use fuel oil economically the engineman must be watchful. Often times we can avoid a water stop if train is handled properly; this saves fuel. Another feature is 'Do not be in too big a hurry leaving town,' but be sure you have your air all picked up so your brakes are not dragging. To get an engine working good and hard keep a good fire in her so she will not leak, but let your fire down gradually, not cut it all down at once. A study should be made of the conditions and the enginemen should work accordingly."

In a series of tests made on Beaumont Hill by Road Foreman of Engines W. E. Stoermer, Los Angeles Division, to indicate the relation of increased fuel consumption to increased speed of train it was shown that several trains weighing one thousand tons each consumed between Colton and Beaumont fuel oil costing $16.11 at speed of 10.91 miles per hour which gradually increased until at a speed of 25.53 miles per hour the cost was $35.65. This would reflect the cost of unnecesary delays to both freight and passenger trains which are compelled to recover lost time with increased speed.

ON THE COAST DIVISION

The following is from a paper by Road Foreman of Engines Holdredge and read at a fuel meeting held on the Coast Division recently:

"We all understand that improvement in the design of boilers whereby more heating surface is provided and also the application of the superheater, will produce a saving, but this is not the question. The question is to get the best results possible with what we have to do with, and regardless of what improvements are applied, the careless man will still be behind.

"Co-operation between the engineer and fireman is necessary. In starting out of a terminal, the engineer should work the engine easily and gradually increase the speed, recognizing the fact that it is just as detrimental to expand the boiler too fast by forcing the firing from the start, as it is to cool the boiler too suddenly after having worked engine hard. A great many hard trips are due to this one cause as the sudden expansion causes leaking which is disastrous to fuel performance.

"Firemen should maintain the steam pressure in starting as well as while running and not allow pressure to drop back ten or twenty pounds and figure on picking it up after starting. Considerable fuel, in excess of that necessary to maintain pressure, is consumed in forcing fire to regain pressure that is allowed to drop. It can readily be seen that if pressure is allowed to fluctuate, the boiler is being forced beyond requirements which results in damage to the boiler and entails considerable loss of fuel.

"A record should be kept of each trip in order that excessive consumption over some previous trip can be noted and the cause remedied."

AT BAKERSFIELD

The following is from Oil Burning Inspector Morgan's remarks at San Joaquin Division meeting at Bakersfield:

"Co-operation between the engineers and firemen is a very essential feature and between enginemen and roundhouse men. If the engineer makes an intelligent report so the roundhouse men will know where to look for the trouble it helps greatly. I see a great many reports of 'Engine not steaming,' 'Engine smokes,' no cause given. This entails considerable expense, as the roundhouse forces have to look all over the locomotive, where if the men would state in the report the symptoms of the engine and give their views as to what they think is the matter, it would help in shop expense as well as in fuel economy."

AT LOS ANGELES

From remarks by T. O'Leary, roundhouse foreman, Santa Barbara, at fuel oil meeting at Los Angeles:

"Black smoke at all times should be avoided, as when prevailing it usually indicates faulty drafting or firing of the locomotive. Care should be taken by enginemen when first starting out to see that flues are properly sanded on account of liability of flues becoming more or less coated with soot when being fired up in roundhouse. Soot is known as a great nonconductor of heat. By sanding the engine when first starting out it will avoid smoking the locomotive and make her a better steamer. When engineers are preparing to shut off they should notify the fireman. The fireman should always be ready to respond quickly. He should keep close observation of how the engineer is working the throttle so that fire may be properly regulated with draft produced by the exhaust."

"Particular care should be given to the heating of fuel oil. Fuel oil should not be heated to more than 100 degrees Fahrenheit. Overheating fuel oil causes it to separate, allowing the asphalt or creosote earth to settle to the bottom of the tank, the incombustibility of which results in fire being extinguished in the fire-box. Overheated oil causes fire to kick back, oftentimes working the brick work loose in the fire-box."

THE BULLETIN

Fuel Issue one of Livest Problems Railroads Have to Face

"The fuel issue is one of the livest one the railroads have to face today, and in order to handle it in an economical way they must have the co-operation of the men who use the fuel and who are responsible for its consumption," said Superintendent T. Ahern of the Coast Division at a recent fuel oil meeting held at the Third Street station, San Francisco.

"The problem requires the co-operation of everyone," he continued, "but you have to have the machine in good shape to get good results. This applies to the Motive Department in the use of fuel; to the dispatchers in seeing that trains are not unnecessarily delayed and, where there is a meet between heavy and light trains, in seeing that something is done to expedite the movement of heavy trains over the road rather than having them tied up at sidings, where there may be a waste of oil. In getting trains in and out of stations the conductor must be prompt in the handling of his business and return to the train as soon as possible. It is up to the trainmen, the trainmasters and to proper station supervision to see that trains are moved with as little delay as possible, for train delays cost fuel."

"The cost of fuel oil on the Coast Division is roughly $700,000 per year," said Master Mechanic C. R. Burroughs, in a paper read by Mr. Ahern. "This money, after passing through the hands of officials, whose duty it is to see that every dollar entrusted to them for disbursement is economically and judiciously expended and full value received therefor, assumes the form of fuel oil and passes out of their hands and into ours, through the firing valve, directly under control of the hands and the brains of the engine crews and those responsible for its consumption. It is up to us to drive as good a bargain with it and get as full value as the men who handle it before it reaches us.

"We are not going to attempt to lay down a set of cast iron rules and instructions to follow, or try to show you how, for I do not think we have an engineer or fireman on the division but who knows the economical and the wasteful methods of fuel consumption, but, of course, we can all learn and improve, and it is the purpose of this meeting to exchange ideas and learn and improve, and not only get back to our former performance, but better it. I want first to call your attention to the matter of making reports of defects on engines in order that shopmen may keep the machine in the perfect condition that it should be, otherwise the best efforts on part of engine crew cannot overcome the loss due to defective machine. Engine crews are always in

the best position to detect defects, such as blows, etc., and should use every effort to locate and report them, in order that the House Doctor may know what medicine to prescribe and not use the wrong pill when he should be doing something else, for it costs something to pull a pair of pistons, looking for a blow in the valves. I have noticed many times where a crew, as long as they had no particular trouble themselves, although they were aware of the fact that the engine was not doing just as well as it had been, made no report of same. For example, I will cite an instance that happened a few days ago. One of our engines cut out of a fast train on account flues leaking. On application of hydrostatic test, found pipes leaking, which was the primary cause of the engine failure. On inquiring among engineers and firemen, found that this engine had been lagging for a number of trips and no report made of same. Had this been reported, it would probably have saved the delay as well as the excessive amount of fuel consumed to overcome leaky pipes. Crews must not feel, although their reports and suggestions are not always acted on, there is any intention to ignore them and even though reports are not absolutely correct, they ofttimes give a clue to the real trouble.

"The matter of drafting engines properly to get perfect combustion is one of the most difficult problems the roundhouse foremen have to contend with. Not seeing the engines in operation he practically works in the dark, with only the assistance of such reports as he can get from engine crews, and many times makes unnecessary changes before reaching the cause of the trouble. This not only adds to the cost of repairs, but results in loss of fuel until engine is steaming properly. Of course, when the road foreman of engines is available and can spare the time from his numerous other duties, etc., he can ride engines and locate any irregularity, but with 170 engines and 547 miles of division and half of his time devoted to other than mechanical matters, it is impossible for him to cover the entire ground and it behooves enginemen to make a study of their engine and give all assistance possible to the roundhouse, in order that oil-burning apparatus may be kept up to the highest state of efficiency.

"Roundhouse foremen should follow the matter of the condition of the apparatus closely and know that when the engine leaves the house that everything is in first-class shape for the trip, and when reports are made that engines are not steaming properly, use every effort to find out why not. You are the Doctor and it is up to you to take the numerous symptoms de-

More Increases in Prices of Railroad Materials.

Referring to Vice-President W. R. Scott's circular issued April 25th, calling attention of employes to the increases in prices of material, a portion of which information was shown in a subsequent issue of The Bulletin, employes are advised of further increases since the issue of the last circular. They are published below, with the hope that employes will use the greatest possible economy in their use and, where possible and consistent with safety, use substitutes.

Marking crayons	13%
Rags	7%
Track spikes	45%
Paper car roofing	47%
Insulators	24%
Wool waste	125%
Automatic block signals	13%
Station signals	15%
Headlights	15%
Railway and marine fire extinguishers	20%
Drop forged steel wrenches	50%
Carpets	16%
Postal bag racks and fixtures	12%
Mechanical interlocking appurtenances	7%
Boiler tubes	78%
Glue	17%

One S. P. Envelope That Has Been Doing Its Duty.

Roadmaster's Clerk E. W. Stone at Carlin, Nev., recently mailed one of the reuse envelopes, Form 5558, which bore the following destinations, the point of origin being unknown:

1, Brooklyn, Ore.; 2, Springfield, Ore.; 3, Brooklyn, Ore.; 4, West Oakland, Cal.; 5, Ogden, Utah; 6, Montello, Nev.; 7, Carlin, Nev.; 8, Golconda, Nev.; 9, Carlin Nev.; 10, Palisade, Nev.; 11 Carlin, Nev.; 12, San Francisco, Cal.

scribed and diagnose the case and prescribe the medicine."

A lively interest in the fuel problem was shown at the meeting, most of those in attendance participating in the discussion. Those who attended were:

T. Ahern, Superintendent; E. R. Anthony, Assistant Superintendent; C. R. Burroughs, Master Mechanic; H. H. Carrick, Asst. Master Mechanic; C. H. Holdredge, Road Foreman of Engines; E. C. Morrison, Division Engineer; W. H. Phelps, Asst. Division Engineer; I. J. Onyon, Chief Dispatcher, San Francisco; P. Slater, Trainmaster; W. H. Curran, Trainmaster; J. Hall, Master Car Repairer; P. N. Nelson, Supervisor B. & B.; S. J. DeGraeff, Division Storekeeper; H. P. Buchenery, General Foreman Store Department; E. A. Millsap, Station Supervisor; H. R. Hicks, Chief Clerk, Superintendent's Office; F. W. Veaco, Fuel Clerk; J. B. Kelly, Roundhouse Foreman, San Francisco; A. T. Brennan, Roundhouse Foreman, San Jose; J. C. Dolan, Roundhouse Foreman, Watsonville Junction; W. C. Buttle, Roundhouse Foreman, Pacific Grove; W. H. Utsinger, Roundhouse Foreman, San Luis Obispo; W. G. Fifield, Engineer; B. D. Silsby, Engineer; W. C. Airey, Engineer; J. T. Moore, Engineer; G. A. Van Sickle, Engineer; C. C. Benbow, Engineer; R. A. Bush, Engineer; J. A. Burke, Engineer; W. H. Thompson, Engineer; F. V. Cox, Conductor; R. P. Summers, Fireman; L. A. Hamlin, Fireman; V. M. Cave, Fireman; J. T. Lynn, Fireman; R. McHenry, Fireman; H. E. Russell, Fireman.

VISITORS PRESENT

F. W. Huller, Chairman Efficiency Committee; J. N. Clark, Chief of Fuel Bureau; R. E. Doyle, Fuel Inspector; F. Shalda, Chief Clerk, Fuel Bureau.

For Those Interested in the Greater Economy in Use of Fuel

Gratifying evidence of the interest taken by Company employes in the subject of greater fuel economy is found in Chairman Kruttschnitt's letter transmitting the annual report for the fiscal year ending June 30, 1916. Mr. Kruttschnitt's reference to the subject is in these words:

"The greater efficiency in the use of locomotive fuel, shown in last year's operations, not only has been maintained but has been increased, as 5.16 gross ton miles were moved per pound of fuel in passenger service, an increase of 2.18 per cent, and 5.98 gross ton miles in freight service, an increase of 2.42 per cent. The money value of this gain is $217,396.08, compared with 1915; $740,395.92, compared with 1914, and $1,515,645.12, compared with 1913."

Road Foreman of Engines E. E. House of the Western Division, in a recent address at a division fuel meeting at West Oakland, made the following remarks:

"There are thousands of gallons of water and consequently considerable fuel wasted daily, by the thoughtless haphazard method employed by the majority of engine crews in blowing off boilers. It seems to be a religion with some of them to blow off the boiler before beginning a trip, regardless of whether or not the boiler has been washed since making the last trip.

"The correct manner of blowing out a boiler when such action is necessary, is a matter that enginemen should give close study. It has been found by practical experiments made by competent persons that in order to obtain the best results when blowing out a boiler, the blow-off cock should be opened for about ten seconds at a time, and in no case should it be held open longer than twenty seconds. Then close it, and after standing for a few seconds open it again an equal length of time. This operation should be repeated three or four times, when it will be found that the mud will be blown out of the boiler, and at the expense of comparatively little water and fuel. The method generally in vogue in blowing out boilers is to fill up the boiler to the whistle while at the water column, then after pulling down the get away track, open up the blow-off cock and leave it open until the water is barely in sight in the water glass, thus blowing away several hundred gallons of clean water, heated by the burning of fuel, when a far better effect could have been obtained with less than one-half of the water and fuel expenditure by opening and closing the blow-off cock at short intervals three or four times. By this means mud is allowed to settle around the outlet to the blow-off cock after the cock is closed, and when it is opened again it is forced out to the atmosphere, and the boiler is thus rid of it.

"A thorough inspection of the engine should be made at the end of each trip with the view of locating defects and having repairs made. Some employes are falling into the habit of inspecting your engines while the station work is being done at the last station before arriving at the terminal point. An inspection carried on at a station while you are looking for a proceed signal from the conductor is perfunctory at best, and generally results in something being overlooked that should have been discovered by the engineer; therefore, the engine should be inspected at the roundhouse where time and opportunity is furnished for such work.

"The fireman should give particular attention to the manner in which water is supplied to the boiler, care being taken to furnish no more water per minute than is used by the engine, for if the boiler is allowed to fill up, the steam pressure will fall and the fire will be crowded in the endeavor to maintain a full head of steam, with the result that fuel is wasted in smoke, the temperature of the firebox is changed considerably by the fall of steam pressure and leaks are liable to result from contraction and expansion."

Railway Age Gazette Nails Lies About Railway Salaries.

So many wild ideas exist exist about the salaries paid railroad officials that it is interesting to see the Railway Age Gazette nail them in a recent issue. Under the heading, "More Lies About Railway Salaries," the Gazette denounces as untrue the estimates made by the American Railway Employes' Journal that the pay of officials amounts to $113,577,000 a year, and says:

"The Interstate Commerce Commission has published sufficient data to show exactly what are the average salaries paid to railroad officers, and to demonstrate conclusively that even taking into consideration the high salaries paid to a very few, these averages are exceedingly modest. According to the Commission's latest detailed statistics on this subject there were 5740 general officers on all the roads of the country in 1914, and their total compensation was $21,338,995, an average of only $3717 a year. This includes all the classes of officers embraced in the Employes' Journal's list and, in addition, some of the more highly paid classes not mentioned by it, such as vice-presidents, general counsel and the heads of the engineering and and mechanical departments, as well as the majority who, of course, receive less than the average."

A Real Lesson in This Tale, Says the Visalia Delta.

"Railroad men are inclined to believe that they are an unappreciated body of men and that their best efforts to please the public meet with little encouragement," says the Visalia Delta of October 5th. "They are wrong sometimes, as the following little story will show, and will prove also that they are not operating always in ungrateful communities. At least not in Visalia. A gentleman who was visiting our city recently was met by some friends at the new station and among those who met him was a little girl. Before she said, 'How do,' she remarked to him:

"'Don't you think we have the prettiest depot in the whole world?'

"Now this little girl showed great civic pride, and she showed also that she had a proper appreciation of art and of Southern Pacific architecture. At any rate if this little girl was pleased and was proud that ought to go a long way in satisfying Mr. William Sproule that he is making friends in territory which is not quite all his own.

"The trouble is that we are all inclined to believe that we are not appreciated. Napoleon often said to his followers at lonely St. Helena, 'I shall have no place in history.' When a man or woman thinks he is not appreciated then is the beginning of decline in efficiency. 'Believe in yourself and care not at all,' is the one motto we should follow. If we are doing the right thing, even in building depots, and though we are not thanked, think, each one of you who is discouraged and downcast, that in this wide world there is some little girl saying, 'Don't you think we have the prettiest depot in all the world?'

"That is a heartening thing. The human being is very much like that noble animal, the dog. Both must have companionship. There is no more lonely thing in the world than a homeless dog, and the next most pitiable object is a lonely man. But if that man believes he has not a friend, let him think of that little girl in Visalia who said, 'Don't you think we have the prettiest depot in all the world?' There is always some one in this wide world who thinks now and again of us, especially if one owes money, but not that. None is so disconsolate as not to have at least one well-wisher in the world. And, thank God, let us all try and think that each has a friend and of that little girl who said, 'Don't you think we have the prettiest depot in the whole world?' Someone appreciates us—though we know it not."

16   THE BULLETIN

Our Organization For Fuel Economy

Rising cost of Fuel Oil such that Strict Economy in its use must be observed at all times.

By R. J. Clancy, Assistant to General Manager

Aside from wages, fuel is the largest item of expense in railway operation. The importance of this item becomes apparent when we consider that the annual fuel bill of the Southern Pacific is about $30,000,000 and of all the railroads in the United States about $600,000,000. With a full realization of the ratio of that amount to total operating expenses it should automatically follow that intensive economy should dominate an expenditure of such proportion.

Ever increasing consumption and cost of fuel prompted many of the more progressive roads to study more intensively the subject of fuel conservation, resulting in the establishment of an organization for checking and analyzing fuel performances and in turn placing such statistics in the hands of operating officials for guidance in locating and eliminating sources of waste and excessive fuel consumption.

The Southern Pacific (Pacific System) is one of the larger roads that early instituted a system of fuel checking and control, resulting in the effectation of substantial economies. As inaugurated in 1911, the plan is based on the theory that fuel conservation is a subject of prime instead of secondary importance, that in many respects it is a problem of mental attitude on the part of officials and employes having to do with the handling and consumption of fuel, that fuel consumption and tonnage hauled, while dispatcher, trainmen and others play an important part, for the purpose of guidance should be allocated to the engineer, firemen and the individual locomotive, and that in this as in every other successful undertaking there must be the fullest co-operation on the part of those whose duties are directly related to fuel consumption. With view of having all departments assume their share of responsibility, this organization is under the supervision of a general operating officer. Briefly stated the working plan of the organization is as follows:

A daily report is sent to superintendent of each division showing fuel consumption and work done by trips for each individual engineer, fireman and locomotive. From this report superintendent is required to bring to attention of enginemen any instances where fuel consumption appears excessive, and if investigation develops locomotive is responsible the mechanical department is called upon to remedy the cause.

Consumption is Watched

The daily performances are accumulated for the month for each engineer, fireman and locomotive and segregated by class of service, type of locomotive, district, direction, etc., and reduced to average amount of fuel consumed per 1,000 gross ton-miles per trip, average gross tons and average time terminal to terminal. Comparison is then made with average of all men or locomotives in the pool. These sheets are issued shortly after the close of each month and copies issued to all transportation and mechanical department officials on each division in addition to copies posted on bulletin boards at all terminals and roundhouses for enginemen's inspection. These reports serve to encourage good natured rivalry among enginemen, each striving to obtain position at top of list.

From the monthly report these performances are entered by superintendent in loose-leaf books, a separate sheet being used for each engineman and locomotive on which their work is accumulated for a period of one year or more, giving quick reference to past fuel records of men and locomotives. These books are made in duplicate, one copy being furnished the Road Foreman of Engines and the other is retained by Superintendent.

The monthly sheets together with the loose-leaf book are checked each month by superintendent and instances of excessive consumption, after allowance is made for locomotive load and time on road, are called to attention of those responsible; Road Foreman of Engines using his book record to locate locomotives and enginemen consuming excessive amounts of fuel.

Letters to enginemen about excessive performances are not written in a spirit of criticism but rather the men are encouraged to tell their troubles with view of helping them to overcome their difficulties. Good performances are likewise noted and commented on with letters to enginemen.

Fuel Committees Active

Each superintendent has a fuel committee on his staff consisting of four to eight engineers and firemen in addition to one conductor and one brakeman. Committees' duties are to study methods for reducing fuel consumption, increasing efficiency in train operation and create interest in subject of fuel economy among their fellow-workers.

Fuel meetings are called bi-monthly on each division for purpose of discussing fuel economy. These meetings are attended by superintendent's staff, fuel committee, roundhouse foremen and all engine and trainmen available. Superintendent acts as chairman. The meetings are addressed by experts on locomotive handling, air-brakes, superheaters, etc., and ways and means of effecting fuel economy are freely discussed by the men and officials. Employes are encouraged to submit papers for discussion. Suggestions and recommendations made at such meetings are either immediately disposed of or carried over as unfinished business to next meeting through appointed committees.

Stationary plants are also given close attention through a special committee which has saved large sums of money through improvements and correction of mechanical defects in boilers and furnaces and proper instruction to employes operating the plants.

In the organization and operation of a bureau or department for checking fuel consumption full co-operation between transportation and mechanical departments is of first consideration, for no matter how much effort be made no practical achievements can result without co-operation. To mechanical department officials is entrusted the improvements in design and efficiency of the power, the installation of super-heaters and other capacity-increasing factors so essential to any scheme for fuel economy.

The plan under which the Southern Pacific's Fuel Bureau is operated covers the whole field of fuel handling and consumption and in the movement for economy embraces everyone whose duties relate to transportation from the general officer in charge to the call boy. Superintendents, trainmasters, dispatchers, master mechanic, road foreman, trainmen, enginemen, station agents, roundhouse foremen, yardmen, carmen, operators, section foremen, water foremen, signal maintainers, engine dispatchers and callboys all play their respective parts in the elimination of waste and the effectation of economy.

A GOOD SOLDIER

The captain, walking between ranks, saw that one of the men had a shine on the toe of his shoes, but that heels were caked in mud.

"How is this?" he asked. "Didn't you notice that your heels were dirty?"

"A good soldier never looks behind," was the reply. A week of K. P. was the result.—California Electric Ry. Journal.

tral figures in the history of Imperial Valley's development were Edward H. Harriman, President of the Southern Pacific Company, Epes Randolph, (Mr. Harriman's personal representative), George Chaffey, C. R. Rockwood, the man who held his faith in the project despite all adverse circumstances, A. H. Heber, the man who succeeded Mr. Chaffey and H. T. Corey, General Manager of the California Development Company, appointed by Mr. Randolph, president of this company and representing the Southern Pacific, who finally accomplished the great undertaking of shutting out the Colorado River from flooding the valley. In the promotion and colonization work, Dr. J. W. Oakley and W. F. Holt were conspicuous figures in 1902.

A. H. Heber, W. F. Holt and their associates organized the Imperial Gulf Railroad and in May, 1902, Mr. Julius Kruttschnitt, then General Manager of the Southern Pacific Company (Pacific System) had a conference with Mr. Heber in San Francisco, as a result of which the Southern Pacific Company assumed construction of the railroad, and on October 16, 1902 the work began. On February 21, 1903, the railroad was completed to Calexico. Thus transportation facilities were assured and the real boom in the Imperial Valley began.

The Colorado River runs along the edge of the valley at an elevation of from 25 to 200 feet above the valley's floor. In 1904 the original intake for irrigation became choked with silt, checking the flow of water into the valley and causing a water shortage. To remedy this situation, Mr. Rockwood, as General Manager of the California Development Company in October, 1904, cut a new intake, 50 feet wide and 6 to 8 feet deep. This had to be enlarged again in November and December, and in January, 1905. Preparations were being made to close this emergency gap in February, 1905, an unusual flood of the Colorado River broke through the intake and poured into the valley, making two great channels across it, the Alamo on the east and the New River on the west, both emptying into the Salton Sink now the Salton Sea. Thus the Salton Sea was born, a body of water, covering an area of 400 square miles. The Southern Pacific Lines were flooded and had to be moved to higher ground on the mesa.

After an unsuccessful struggle, through lack of sufficient finances, or large enough organization to combat the Colorado and to close the break which permitted the river to flood the valley. the citizens of Imperial applied in June, 1905 to the Southern Pacific Company for aid, and the California Development Company was given the backing of the Southern Pacific resources. Mr. Epes Randolph was made President of the California Development Comapny and Mr. Corey was appointed General Manager of the company by Mr. Randolph April 19, 1906. Conditions

Thrift In The Locomotive Cab

Fireman Tell How Economy May Be Applied to Fuel Oil Consumption

By F. W. WINBERG
Fireman, Western Division

Petroleum with its manifold uses is one of this country's great resources and is increasing in demand by the industrial world. Thus, as it has increased in value, and taking this fact into consideration, with reports that the supply is failing to meet the demand, it is apparent that there is need for endeavor to effect its conservation.

The resources of this country were ample to meet the demand of the nation a score of years or so ago, hence as there was little need for economy little thought was given in that regard. During the recent past though, there has been occasion for conservation and fuel economy is now a live question with us all.

The advent of petroleum as fuel proved a great convenience to the railroads, in as much as it increased the efficiency of the locomotive is handled and stored easily, and did away with considerable labor mainly on the part of the firemen.

The railroads have inaugurated different plans to conserve fuel oil. On our own particular road, the Southern Pacific, fuel oil meetings are held on each division at regular intervals. Suggestions are submitted and acted upon whereby much has been accomplished.

Manners of conservatism are various concerning those who are responsible, for handling of the oil among whom the fireman is one of the leading principals, hence his part will be considered in one particular.

First experience in firing oil gives one the impression that there is little possibility of waste, as it seems that all that is required to maintain proper steam pressure is to supply a sufficient amount of oil. But there are other factors that the fireman must consider, on o fwhich is sanding out the flues to rid them of carbon.

Carbon collecting gradually on the flues retards steaming, being a nonconductor of heat. Engines are provided with a box of sand and the sand is poured through a hole in the firedoor with a scoop as means of cleaning the flues of carbon. The exhaust when sufficiently heavy draws the sand through the flues.

The flues should be sanded out as often as necessary and at least a half dozen times on a 100 mile trip using sufficient scoopsfull each time till there is little evidence of carbon coming out of the stack. Five or six scoops are generally necessary under normal conditions.

Though the engine steams well and can be fired with little or no smoke, the sanding out process should not be neglected, as many engines drafted well are not so apt to smoke or use excessive oil if they be sanded out.

It is safe to state if this practice were not resorted to over the entire trip of 100 miles an excessive amount of about 50 gallons of oil would be

Excessive smoke under some circumstances has been necessary, but an effort should be made to overcome it through proper firing and use of sand.

When the engine fails to steam free a good practice is to sand the flues just before stopping, especially with a heavy train, because very often in starting it takes some time before the exhaust becomes sufficiently accelerated to permit using sand. This, of course, would help to maintain full steam pressure which otherwise would drop back.

The flues should be sanded out just before reaching a terminal; this will facilitate firing the engne up at the roundhouse.

were then deperate. A considerable portion of the Imperial Valley had been flooded and the town of Calexico appeared to be doomed.

By working every quarry between Los Angeles and Nogales and giving the rock cars right of way over all other classes of trains, the gap in the river was closed February 14, 1917. From time to time in the last 14 years the Colorado has broken its banks and has threatened to inundate the valley again, but each time the Southern Pacific Company has come to the aid of the Imperial Irrigation District and has held the waters in check.

The cities and towns of Imperial

Valley are bustling, rapidly growing places and should be considered here. El Centro is the county seat and is the distributing point of the valley. It has a population of 5,464 and is the social and commercial center for the district. It is also the valley's railroad center. It has good hotels, parks and fine municipal buildings. Four banks there have assets of over $4,000,000.

Calexico is the center of the cotton industry. It has a population of 6,223. Heber has a population of but a few hundred, but is surrounded by rich agricultural lands and ships to market hundreds of cars of the vari-

Oil Burning Locomotives' Care and Operation

Close Teamwork Between Engineer and Fireman is Essential. Practical Suggestions Given in Handling and Maintenance of Engines.

By J. N. CLARK, Chief of Fuel Bureau

At the annual convention of the Traveling Engineers Association, held in Chicago, Sept. 7 and 8, 1921, J. N. Clark, 4th Vice President of that organization and Chief of the Fuel Bureau, Southern Pacific Company, presented the following treatise on "Operation and Maintenance of Oil Burning Locomotives." Mr. Clark's discussion of the subject has brought forth favorable comment and his paper has been reprinted in a number of technical publications, the accompanying article being a reprint from the Railway Review of September 8, 1921.

THE first attempt in the United States to burn crude oil in the firebox of a locomotive was made at Santa Paula, Cal., in October, 1894. The subsequent study of combustion soon led to a better understanding of the firebox and boiler design and realizing the many advantages to be gained by the use of oil as fuel, every effort was made to develop the most efficient burner and drafting arrangement. Most obstacles were overcome and in a comparatively short time a high degree of efficiency was attained.

Crude oil possesses many advantages over other fuels and in the West its use soon became widespread, being employed in steam making in power plants and floating equipments until the degree of popular interest grew to such proportions that the economic need of the fuel oil situation became quite apparent. At no time during the past few years has the production of crude oil in the United States been sufficient to meet requirements and records for June, 1921, show the consumption of 81,000 barrels per day in excess of production, which deficit was and is being supplied from the Mexican fields.

Daily production, United States,
June, 1921 ....1,346,833 bbls.
Imports from Mexico .. 340,175 bbls.
Total ....1,687,008 bbls.
Daily consumption ....1,427,367 bbls.
Amount placed in storage
daily .... 259,641 bbls.

At the present rate of consumption we are told by geologists that the crude oil supply in the United States will be exhausted within a few years as there now remains but about 60 per cent of the original supply under ground. With this prospect is it small wonder that strenuous efforts are being exerted to conserve the remaining supply by maintaining and operating oil burning locomotives as economically as possible. At the present time there are about 41 railroads operating in 21 states which burn oil for fuel.

Enginemen and Upkeep

The maintenance of the locomotive rests largely with the enginemen. The medium through which the maintenance is accomplished is the work report and its correct rendition. This means that if the locomotive is to perform economically at its maximum capacity specific information must be given to the roundhouse force as to just what defects were noted under actual operation. Maintaining an oil burning locomotive would be a very simple task if a thorough and accurate report of defects noted by engine crews were reported at the end of each trip. "Don't Steam" covers a multitude of sins, but the worst sinner of all is the man who writes it on his work report, with no further detail, and then has writer's paralysis.

Generally speaking there are but few points of difference in maintaining an oil burning locomotive as compared with a coal burner and these points of difference affect the design of tender, oil piping and firebox arrangement. The fuel oil tank is constructed of 3/8-inch material and the capacity is approximately 3,000 gallons — a coal equivalent of about seventeen tons on the basis of 42 gallons per barrel and four barrels of oil equal to one ton of coal. On filling the square tank two-

inch space is left in top on semi-cylindrical six inches to allow for expansion when oil is heated. The ratio of expansion is one per cent for each 25 degrees increase in temperature. Fuel tanks are provided with a measuring rod designed to show inches on one side and gallons on the other so that accurate measurements may be obtained at all times. Oil flows to burner by gravity on all types of locomotives except the Mallet, in the tender of which is used six pounds air pressure. The oil piping arrangement conveys the oil from the tank to the burner, from which it is sprayed into firebox by steam. This conduit is two inches in diameter and passes through a superheater four inches in diameter which is used, in addition to the tank heater, to heat the oil before it reaches the firebox.

The tank heater should be so constructed and maintained as to keep oil at proper temperature, the oil feed line should be free from leaks, with as few elbow joints as possible, metal with flexible joints is preferable to rubber; cut-out or safety valve on tender, blow-back valve, superheater and firing valve should all be in perfect working condition. Burner should be of suitable size, varying from one inch to three inches according to size of firebox, clean and lined up about 60 inches from flash wall to insure perfect combustion. Six to nine and one-half inches—depending on draft arrangement—is the correct distance from burner to fire pan floor and particular attention should be given to see that burner is so placed that oil strikes flash wall in center. Care must also be exercised to see that fire pan is free from air leaks with no obstruction on floor and all damper controls are in perfect working order.

Drafting Oil Burning Locomotives

There are two ways of drafting an oil burning engine; one, the horizontal draft and the other the vertical. Both are in general use on oil burning roads. The arrangement of the brick work on both drafts is practically the same. The horizontal draft, by which air is admitted through fire door, is the most economical, but in poor water district, from the standpoint of boiler maintenance, the vertical draft arrangement with the flat door, by

---

High Lights of Report on the Use of Oil

CRUDE OIL has been used as a locomotive fuel less than 25 years, but there remains only 60 percent of our original supply under ground.

* * *

If the locomotive is to perform economically, the roundhouse force must be advised by the enginemen of defects noted under actual operation.

* * *

The closest teamwork between engineer and fireman is necessary, for the fire must be changed to meet the operation of the throttle and the reverse lever.

* * *

A slight color of smoke at the stack is better than no smoke at all, for a clear stack indicates too great an amount of air. Losses from too much air are frequently three times as great as from insufficient supply. Firemen should avoid black smoke at all times if possible; it is a loss of fuel and should not be tolerated.

---

which air passes to rear of firebox, prevents cold air from reaching the staybolts, thus reducing boiler maintenance. The horizontal draft in good water districts has proven very successful both from the standpoint of boiler maintenance and fuel consumption.

Air openings from firepan should vary, depending on size of firebox. From sixteen square inches air admission around burner and 85 square inches at rear of firebox to 225 square inches around burner to 224 at rear. It has been found that the best results obtained on an oil burning engine, with respect to front end arrangement, is the use of an extension stack extending down to center line of boiler, being twelve to sixteen inches in diameter. In order to insure economical fuel performance fire pan must be maintained in good condition. To secure the best results the fire pan should be welded to the mud ring and rigidly secured so as to obviate all possibility of air leaks at sides and front—also behind brick work. The maintenance of fire pans on oil burning locomotives is a very important feature and too much attention cannot be given them.

Front end air leaks and outside steam leaks should not be tolerated. In the roundhouse organization one man should be assigned to make torch tests and repair all air leaks into front end and around outside steam pipes. There are many recommended practices for the elimination of air leaks around outside steam pipes, the best one in use is the application of a casting made of ¼-inch steel plate with a welded seam. This casting is riveted permanently to the smoke box and then caulked or welded around the edge to make it perfectly airtight. It is large enough in diameter to allow removal of the steam pipes without disturbing it. The lower end of the casting is flanged outward and to it is riveted a wrought iron ring. A cast iron flange made in halves is fitted together and bolted snugly around steam pipe, which is machined true at this point. The cast iron flange is secured to the wrought iron ring on the bottom of the casing with eight studs and a copper wire gasket used between them to obtain an airtight joint. A copper gasket is then caulked into a dovetailed groove in the cast iron flange around the steam pipe.

Closest Teamwork Necessary

Operation of oil burning locomotives call for the same attention from the engineer as on a coal burner but the fireman has no manual labor to perform in delivering fuel to firebox. He has to be alert at all times for there is no bank of burning oil to aid him in keeping an even firebox temperature when engineer is working a light throttle or drifting after having forced the locomotive to its capacity. The closest teamwork between engineer and fireman is necessary to meet the fire must be changed to meet the operation of throttle and reserve lever.

A supply of clean, gritty sand

WRITE FRANKLY AND SIGN YOUR NAME

A number of letters, signed "Employe," "Engineer" or some similar designation, have recently been received by Southern Pacific officers. These letters were courteous and dealt with matters of such interest both to employes and the Company that the officers addressed have expressed regret because they were unable to make full and friendly replies, inasmuch as the writers' names and addresses were not given.

The officers have requested this space in the Bulletin to say that no employe need hesitate to write them frankly. Letters marked "personal" will be opened only by the person addressed.

should be in sand box of locomotive with sand scoop so fireman can clean soot from flues. Flues should be sanded while running by placing a small quantity of sand in scoop and by inserting through opening in fire door while engine is working hard, allowing the exhaust to draw the sand through the flues, thus cutting the soot in its passage and discharging it from the stack. It is good practice to sand frequently in order to remove soot which is a non-conductor of heat and causes steam pressure to drop rapidly. Engineers should take care to give the valve sufficient travel, opening throttle far enough so that exhaust will carry sand through the flues and do the work for which it is intended. The sand scoop should be held as far in fire door as possible in order to prevent sand from falling on brickwork or flash wall.

Firemen should watch the temperature of oil and endeavor to keep it just warm enough to insure an even flow to the burner—this temperature will vary from 100 degrees for light California or Texas oil to 180 degrees for heavy Mexican oil. The fireman has direct control of amount of steam used to heat oil in tender and unless he is careful he may not only damage the oil by overheating it so that it loses its lighter gases, but also increase the hazard of explosions from the escaping gas. Where excessive steam is turned into tender there is excessive condensation and unless this water is drained off it will go to the burner, causing trouble and often putting the fire out.

Heating oil in tender is not for the purpose of aiding in burning it but to cause it to flow in a steady, even stream from tender to oil feed line, where it passes through a superheater box or joint before it reaches burner. The purpose of the superheater is to raise the oil to a temperature where it will be easily broken up by atomizer as it goes from burner tip into firebox.

Since it takes less than one-fifth of a second for a particle of burning oil to travel from burner tip to flue sheet it is of utmost importance to have oil so atomized that no time is lost in burning it.

Firing Valve and Blower

The use of the blower is to create sufficient draft to keep the firebox clear of smoke and gases and to produce artificial draft. The misuse of the blower causes waste of fuel and damage to the boiler. Very light applications should be indulged in under all conditions. The strong use of blower draws cold air into firebox damaging flues and flue sheets, causing them to leak. A slight color of smoke at the stack is better than no smoke at all, for a clear stack indicates too great an amount of air. Losses from too much air are frequently three times as great as from insufficient supply. Firemen should avoid black smoke at all times if possible; it is a loss of fuel and should not be tolerated. Black smoke is either the result of faulty firing or the condition of the engine.

Firing valve by which the flow of oil to the burner is regulated requires constant attention. Lost motion frequently occurs in this apparatus and much fuel is wasted due to the fact that fireman loses the correct adjustment of this valve account too much play either in the rods or in the valve itself. When an engine is worked to full capacity on long grades and has been fired heavily it requires very careful attention to prevent flues from leaking as the small tongue of flame does not fill firebox sufficiently to keep out the cold air. Engineers cannot be too careful at such times, they should start gradually to ease off on throttle, allowing fireman plenty of time to adjust his firing valve, injectors should be shut off at once, if only for a short time, and then worked at short intervals. In most cases engines start leaking while descending grades after having been worked hard. Careful teamwork between engineer and fireman will prevent black smoke when shutting off or pulling out of station.

FOREMEN OF BROOKLYN SHOP GIVE BAND A DINNER

The Brooklyn shop foremen gave the Portland S. P. band a reception and dinner, Friday evening, November 4th, at Waverly Hall, Portland, in recognition of their faithful volunteer service in boosting the interests of the employes to the best of their ability, and to strengthen plans for making the S. P. band one of the best bands in the city.

The dinner was preceded by selections by the band, and several solos by Mrs. Duva. After dinner, the several foremen made a number of interesting talks. The solos of Mr. Kemnitzer gave special pleasure, as did also the unaccompanied comical songs of Herb Croft, the band's basso.

Handling Superheater Locomotive
on Road Explained by W. L. Hack

How enginemen can do their part in getting all the success that the superheat engine will deliver under economical and efficient operation was explained recently by Mr. W. L. Hack, assistant superintendent on the Sacramento Division, at a meeting of the Pacific Railway Club.

In part, Mr. Hack said:

"Skillful manipulation of the superheat engine is an act that must be learned; the supervision force from a mechanical standpoint on a railroad should be vitally interested in the performance of the superheat engine, as when properly handled by an engine crew they will deliver great efficiency, and when improperly handled are a very wasteful machine.

"The only difference between a superheat and saturated engine is temperature. Some seem to think that the added efficiency comes from pressure. It does, but from an absence of back pressure. We will take Mr. Engineer on his journey with this engine from the time he takes charge at the turnout track until he arrives at the completion of his trip.

"The first thing the engineer does after making out his report as to the condition of the locomotive and inspecting her is to oil around. We, on the Southern Pacific, have done away with swabs on valve stems and piston rods, for this reason:

"The piston rod is entering and coming out of the cylinder, being exposed to a temperature of 690 to 700 degrees in the cylinder, and if engine oil, which is of a low flash test, is put on to this piston rod the result will be a carbon deposit; this carbon deposit rapidly cuts away the piston rod packing, giving us a blowing piston. This is one of the reasons why the engineers should not put engine oil on piston rods. The other reason is:

"Valve oil lubricates a cylinder; although entering with steam it sticks to the cylinder wall and piston rod, because they are the coldest surface that the oil comes in contact with, and the piston rod, if no oil is applied by the engineer, will be an absolute indication of the condition of the cylinder wall, because the piston and cylinder wall are practically the same temperature. Thus the engineer can always tell the condition of his cylinder walls by observing the piston rod.

"After the engine has been oiled around, some oil fed through lubricator, he backs out his train. The lubricator should be started ten minutes before leaving time, set to feed to the cylinders and valves an even amount, five or six drops a minute being the usual amount fed. However, that could be regulated by the owners of the engine. We average about fifty miles to a pint of Perfection valve oil on the Southern Pacific. Our lubricators on the Southern Pa-

cific are equipped with a Booster attachment, this attachment admitting the steam at boiler pressure to the tallow pipes.

"Experiment on the Southern Pacific has shown that engines without Booster valve, after accumulating a pressure in the steam chest of 170 lbs., with 200 lbs. boiler pressure, lubricator ceases to deliver oil by gravity, but glass installed in tallow pipes shows current from the steam chest toward the lubricator. This pressure in the steam chest after 170 lbs. has accumulated keeps the superheat engine from getting lubricant in the valve chest. The cylinder will always be lubricated, because the tallow pipe is tapped into the top of the cylinder in the center, and after the engine exhausts on either end of the cylinder, the cylinder lubricator is fed from the choke plug from the tallow pipe against no pressure, and always insures that the cylinder will be lubricated.

"The Booster should be worked wide open at all times, putting boiler pressure in the tallow pipe between the choke plugs, insuring the delivery of oil to the steam boiler under any and all conditions.

"The engine is now lubricated, the water in the boiler down to the level of where it should be and we are ready to start. The cylinder cock should be open with the superheat engine just the same as the saturated engine, to take out the condensation that has accumulated. An engine equipped with a pyrometer shows how quick the superheat equipment gets in action. We would take No. 20, with twelve cars, engine 2460 of Pacific type, 77-inch wheels, 22x28 cylinder engine, leave Oakland Pier—the first stop being Sixteenth Street, Oakland, we say two miles away. The temperature of the steam on leaving in a saturated engine with 210 lbs. pressure is 392 degrees. Knowing the temperature on leaving, the first question Mr. Engineer asks is when does my superheat efficiency commence? The answer:

"When the temperature reaches about 500 degrees or 100 degrees of superheat. Now we come back to the saturated engine.

"It doesn't make any difference how large the nozzle tip is in the saturated engine, you will always get back pressure, due to the fact that re-evaporation takes place along the cylinder walls, and when exhaust opening comes ahead to piston, it is necessary for piston rod to shove out the re-evaporation that has taken place along the cylinder walls through the exhaust port to the exhaust nozzle to the atmosphere.

"The first point to determine: When the superheat engine really gets her efficiency is due to the fact that after the steam gets heated to a temperature

of 500 degrees, re-evaporation ceases to exist. Thus—when the exhaust opening comes ahead of the piston there is no back pressure adding to the efficiency of the steam that is shoving on the piston doing the work. That is the reason why low superheat gets you no efficiency and your efficiency does not start until the steam is heated to a temperature of 500 degrees or over.

"Now, as we have started out, how hot will the engine get or how high will the temperature get in the steam pipe, the valve temperature and the cylinder in going this two miles from Oakland Pier to Sixteenth Street? It is surprising to note a pyrometer will go to 690 degrees in this short distance.

"Now as to the quality of valve oil furnished: We have Perfection valve oil of about 535-degree test. The bit of instruction issued to engineer is to consume this lubricant without leaving deposit of carbon on valve rings or cylinder pack rings, which causes a grinding deposit that soon wears the rings out. In the live steam pipe will be found a vacuum valve for the admission of air to destroy the vacuum that always exists ahead of the retreating piston when throttle is closed. Our instructions to engineers on the Southern Pacific are:

"As you are approaching Sixteenth Street, temperature in engine, as you will observe, is 690 degrees; if you shut the throttle tight you will get carbon for this reason: The walls of the cylinder and the walls and rings of valve chest are well lubricated. Ahead of the retreating piston and with the throttle closed would be a partial vacuum; when the throttle was closed air would be admitted through the vacuum valve by the live steam ring into the cylinder to destroy this partial vacuum. With this cylinder heated to 690 degrees, with air admitted into partial vacuum—all of the conditions are there that will burn lubricant. We do not want this lubricant burned, we want this viscosity utilized for the purpose for which it was intended, and the only way the engineer can fully get all the lubricating properties is to leave steam in the engine until practically the time to stop, excluding the air and applying enough steam to break the vacuum and heat of the retreating piston.

"Early in my talk I mentioned the necessity of the piston rod not being supplied with outside lubrication, and here is another point where the piston rod tips off the engineer as to the excellence of his handling. For instance, if the piston rod appears black and smoky, there is a reason for it. We will say approaching Sixteenth Street, Oakland, that he was working the engine of an 8-inch cut-off, temperature of 690 degrees, with full throttle opening. He did not change the position of the lever, but eased the throttle down until steam just showed at the cylinder cocks. For some reason or other he smells the oil burning in the cylinder, and he says: 'She shows steam at the cylinder cocks—why does she burn the oil?'

"When you get over to Sixteenth Street, you take the engineer down on the ground and show him this piston rod and we explain in this manner what has happened:

"When the engine was at an 8-inch cut-off and he left the lever in that position he eased the throttle down until the volume of steam supplied at the cut-off was not sufficient to destroy vacuum ahead of retreating piston. After the piston travels eight inches on its journey the valve cuts off the amount of live steam supplied and that amount of steam must expand from the 8-inch point the full length of the cylinder to destroy the vacuum ahead of retreating piston. The throttle has been eased down until the volume admitted to 8-inch cut-off is small; exhaust port is open ahead of the retreating piston and enough steam has not been supplied to break the vacuum. Thus the vacuum will be broken by admission of smoke arch gases reaching the cylinders through the exhaust port.

"To overcome this feature we advise equipping the lever with Stephenson valve gear to the point where the engine rides free or generally about one-half stroke. Where steam is admitted from a drifting throttle with a half-stroke cut-off, although the volume is small coming through the throttle valve, it only has to expand one-half the length of the cylinder to break the vacuum; thus we get better success and our instructions to our engineers are when the throttle is eased down to drifting throttle to change cut-off to Stephenson about one-half stroke with Stephenson gear and to follow stroke with Walschart gear.

"This explains to the engineer the why for every move you expect him to make. There are conditions that arise on our mountain districts where we do not advocate the use of drifting throttle for a great distance. For instance: Leaving the summit of the Sierra-Nevada mountains where district is covered with snowsheds the speed is low—twenty-eight miles an hour being passenger restriction—we leave summit working drifting throttle for approximately two or three miles, until engine, the cylinders and valve chest cool down; then throttle is closed and we drift at this low speed perfectly. The reason it is altogether reasonable to believe that a superheat engine will drift when she is cooled down to approximately the temperature of the saturated engine better than the saturated engine is because the superheat engine is supplied with cylinder lubrication.

"The superheat people who furnish this equipment, our Government who controls it, our shopmen who maintain it, are all necessary factors in getting the efficiency, but to the locomotive engineer is due the severest task of all in getting the full efficiency, as co-operation is absolutely essential and necessary, and on the Southern Pacific we have held meetings at all the vari-

California System for Testing Railroad Weights Is Approved

California is believed to have the most satisfactory system for testing railroad weights of any State in the Union. An interesting account of how the railroads, co-operating with the State, were able to bring such a satisfactory condition about is given in the biennial report of Charles G. Johnson, superintendent of the State Department of Weights and Measures.

George Easton, general scale inspector for the Southern Pacific, has been in close touch with Superintendent Johnson in the work of perfecting the system now in vogue, and has reported to the latter the result of all tests of platform and track scales. The Southern Pacific's master scale and scale test car is practically under the jurisdiction of the California State Department of Weights and Measures in that the car is sealed with the State seal. In case it becomes necessary to break the seal the car is not used until it has been again sealed by the State. This test car with the California seal attached is used not only in California but is accepted as the official car in the States of Arizona, Nevada and New Mexico, and is also used at Brooklyn for weighing in and checking the scale car of the States of Oregon and Washington.

Sixteen 150-ton scales for the weighing of loaded freight cars were built by the railroads of California under the supervision of the State Department of Weights and Measures. In addition thirty such scales were rebuilt and about fifty large commercial platform scales were installed. By strict adherence to a policy of testing all scales frequently complaints have now been reduced to a minimum. Formerly ten or twelve such complaints reached Superintendent Johnson's office daily. Now if there is one a week it is an exception.

All railroad tariffs or charges, Superintendent Johnson will explain in his report, are based on the weight or measurement of the commodity hauled, and consequently the commodity hauled must be weighed or measured, and such records, forming the basis of settlement, must be true. There is no weighing apparatus employed by the carriers today that has not been inspected and sealed as a guarantee of its accuracy.

During the past two years the number of scales repaired and tested at stations on various divisions are as follows: Western Division, 586; Sac-

ous terminals and round-houses continually to tell the engineer the why for every move we wish him to make, and I am pleased to say results are accruing from that method of handling. Numerous tests have been made; the results of such tests are given to the supervision force and in turn given to the engineer to stir up interest."

ramento Division, 488; Portland Division, 436; Shasta Division, 106; Salt Lake Division, 154. This was accomplished through the use of scale repair cars. About three hundred scales were tested on those divisions in the same period and 350 portable and dormant scales were rebuilt at the scales shop. Superintendent Johnson recommends that all freight cars be weighed empty at least once every six months, as question on the point of such weight often occasions controversy. The equipment for testing these gigantic scales consists of several cars constructed for that purpose. The report describes them thus: These cars, together with other standard weights employed, are tested on master scales and sealed by the department. The master scales, together with as many more as time will permit, are tested annually by the equipment furnished for this purpose by the Federal Bureau of Standards. This equipment consists of an over-size railroad car equipped with electric cranes and 90,000 pounds of legal test weights. Each lever and part of the platform is carefully checked and all nominal errors noted are corrected. When the inspection is completed the apparatus must be correct within the allowed tolerance of one pound in one thousand. All scales which are found inaccurate and which require new parts or extensive repair are condemned for a period of time in order that these repairs may be made, when they are again tested for accuracy. With the employment of this apparatus it is just as easy to place a 10,000 standard weight on the corner of the platform as it is to place a one-pound weight on a counterbalance.

The service is without cost to the State. The railroad companies build the scales and provide for their inspection, having developed a special organization with experienced crews for this important work.

Important Pamphlet.

Bureau of Explosives Pamphlet No. 8 covers the I. C. C. regulations for the transportation of explosives and other dangerous articles by freight. Freight conductors should study the instructions contained therein and agents see that all employees at their respective stations whose duties require knowledge of transportation of explosives are given opportunity to thoroughly familiarize themselves with these regulations.

Director Carter Names S. A. Bramlette as New Representative.

W. S. Carter, director Division of Labor, has announced the appointment of S. A. Bramlette as representative of the Division of Labor, effective December 16, 1918.

Expert Explains Working of Superheater Locomotives

The Los Angeles Division recently held a fuel oil meeting where economical locomotive handling was discussed from the viewpoint of fuel conservation. Mr. C. N. Cross, efficiency engineer, U. S. Fuel Administration, gave a very interesting talk on the relation of co-operation to fuel economy. He was followed by Mr. W. L. Hack, who exhibited a chart explaining cylinder temperature, the booster valve and the proper method of lubrication. Mr. W. G. Eawse, traveling engineer of the Locomotive Superheater Co., followed Mr. Hack, and his remarks are quoted below in full for the information and guidance of all enginemen handling the superheater locomotive:

"I have been invited to attend this meeting by Mr. J. N. Clark, Chief of the Fuel Bureau, to make a talk on the Superheater locomotive, touching on maintenance and operation, and its relation to the subject that has brought us here today. Fuel economy is inviting more attention than any other subject connected with transportation.

"The superheater applied to locomotives consists of a header and a given number of elements called units. The top header fire tube type is universally adopted in locomotive practice, while the construction af the superheater is no doubt familiar to all railroad men. The header being located in the upper part of the smoke arch and the units extending into large flue, 5½ inches in diameter, to within 24 inches of the fire box end of large flues. An auxiliary mechanism in the form of a damper and cylinder is used, which operates automatically with the flow of steam from the steam pipe. The function of the damper is to protect the units from the hot gases when the engine is at rest or drifting. When the damper is open the units are in a direct course of the gases from the fire box, and are exposed to very high degrees of temperature.

PURPOSE OF SUPERHEATER.

"The purpose of installing a superheater in a locomotive boiler is to impart heat to the steam in addition to that which it already has as saturated steam. The saturated steam when entering the dry pipe from the steam displacement of the boiler contains a certain percentage of moisture. This moisture is evaporated into steam and then superheated.

"The superheated steam is a poor conductor of heat. This is one of its valuable qualities in the operation of a superheater locomotive. The average losses from condensation in a saturated engine run from 22 to 27 per cent of all the steam that enters the valves and cylinders. The greatest proportion of losses takes place during the admission opening, as at the time when there is the greatest difference between the temperature of the steam and that of the cylinder walls, while condensation does not cease during the initial admission,

but continues during the expansion period to near that of the exhaust port opening, or when the pressure in the cylinder becomes less than corresponding temperature of the cylinder walls. When a second heat transfer takes place the heat from the cylinder wall condensation and is termed re-evaporation, which continues until the exhaust port closes.

LESS LOSS OF HEAT.

"When highly superheated steam is admitted to the cylinders, some of the heat is given up, but not in proportion to that of saturated steam, on account of the poor heat conducting qualities of superheated steam. There are two reasons explaining this fact: The first is that since it is a poorer conductor of heat it loses less of the energy by the transfer of its heat to the cylinder walls. Second, because of its higher initial temperatures no energy is lost by cylinder condensation.

"Saturated steam has the same temperature and pressure as the water from which it is evaporated and with which it is in contact within the boiler.

"When heat is added to saturated it become superheated and increases in volume in proportion to the increase in temperature; when 200 degrees of superheat is obtained the increase in volume is 32 per cent or in other words there is practically one-third more steam with 200 degrees of superheat per pound of water evaporated.

HEAT INCREASES VOLUME PER POUND.

"The addition of heat to steam increases its volume per pound at any given pressure. This characteristic is exactly the reverse in action of saturated steam, as the temperatures and pressure are increased, the volume per unit of weight is decreased; for example, 100 cubic feet of saturated steam at 150 pounds pressure weighs 36.2 pounds, 100 cubic feet of steam 200 pounds pressure weighs 46.8 pounds. Superheated steam having a pressure of 200 pounds with 260 degrees of superheat, 100 cubic feet weighs 33.8 pounds. These data may not be understood by all who are operating superheated locomotives, but an understanding to some degree can be applied to more ideal locomotive operation which in turn will be responsible for more economical fuel consumption.

"The operation of a superheater locomotive is quite similar to that of a saturated with the exception of using steam when drifting to prevent carbonization of the lubricating oil, the lubricating oil having a certain flash point. It can always be relied on that the cylinder wall and valve chest temperatures are higher than the flash point of the oil, and where no steam is admitted to cylinders under those conditions carbonization and excessive wear of valves and cylinder rings take place as valve and cylinder rings make a tight joint between negative and

positive sides of the piston. Where improper operation is practiced, the results are increased fuel consumption per output of work performed.

"One other point to be considered in locomotive operation is that of carrying water too high in boiler. It should be carried to a height identical to that of a saturated locomotive when the best results are to be obtained; when water is carried too high it is carried over into the superheater and must first be evaporated before being superheated; for an example—assume a boiler carrying 200 pounds pressure and developing a temperature of 650 degrees F. and the water carried over into the superheater in quantities to reduce the total temperature to 450 degrees F. The pounds of water that would be evaporated for 100 cubic feet of steam would be 42.2 pounds, as against 46.8 pounds for a saturated locomotive or a saving of only 10 per cent instead of 30 per cent, with the high temperature; in other words, it would require 20 per cent more fuel to perform a given amount of work, and such improper operation as referred to can be avoided and the fuel consumption reduced per output of work.

HANDLING OF THROTTLE.

"While as a general rule a superheater locomotive is handled in practically the same manner as a saturated locomotive and to correct some impressions as to the handling of the throttle to obtain the maximum efficiency therefrom. There is no set rule other than using a throttle opening most conducive to obtain the best results; a wide-open throttle is only recommended when necessary under certain operating conditions, as pyrometer tests have proven that the greater the velocity of the steam through the unit the higher the steam temperature will be obtained and the higher the steam temperature the increase volume per unit of water evaporated.

"The proper maintenance is of equal importance in obtaining the maximum efficiency from the fuel consumer. Care should be taken to keep all the units tight at the header joints, and where a reasonable degree of workmanship has been applied with the recommended material for the manufacture of header bolts there should be no trouble in keeping the units tight at the header.

An interesting and helpful fuel oil meeting was also recently held at Tracy, on the Stockton Division. C. N. Cross, of the United States Fuel Administration; R. E. Doyle, company fuel inspector; A. M. Weston, air brake inspector; W. L. Hock, general road foreman of engines, and others made talks on fuel conservation.

Chapter Thirteen
Railroad Watches are a Crucial Tool

Imagine a train, in 1869, leaving Sacramento late in the evening with an expected arrival time at Colfax about three and a half hours later. Not having exclusive trackage rights due to other trains heading west, the engineer is expected to reach sidings to allow passage of an oncoming train. Did the phrase, "Timing is everything" originate in the railroad industry? The emphasis on scheduling and timekeeping increased with the number of trains. So, as the years passed, the engineer leading a consist to Colfax became increasingly reliant on his watch and the watch of the other engineer. The headlight in the distance at the wrong time was an adrenaline rush for sure.

The importance of an accurate watch to railroading cannot be overstated. The congestion in any railroad yard, especially large ones like the Southern Pacific's in Sacramento, required synchronized movements of equipment for the efficiency of the business and to assure a safe work environment for those on and off the train. Over the years, the understanding of metals and physical principles were incorporated into the railroad watch.

Even though the history of western railroading in regards to Sacramento and the transcontinental railroad begins in the 1860s, dedication to accurate and coordinated timekeeping in the railroad business existed prior to 1850. The Pennsylvania Railroad, in 1849, published a rule that stated: "Each engineer will be furnished with a watch which shall be regulated by the Station Agent at the commencement of each trip and must be deposited with him when the engine returns. If not returned in as good order as it was received, the engineer must pay the expense of repairs." Four years later, the Boston and Providence Railroad published their own rules to control the quality of watches. They included the statement, "Conductors will submit their watches to Bond and Sons, 17 Congress Street, Boston, for examination, and procure from them a certificate of reliability which will be handed to the Superintendent" (National Association of Watch and Clock Collectors Bulletin, November 2000).

In the late 1880s, the General Time Convention, which later became the American Railway Association and was renamed again to the Association of American Railroads, made attempts to certify watches and create a standardization of performance across the industry. During the 1890s, numerous railroads started efforts to incorporate specific design features into railroad-accepted watches.

The questions that needed to be answered at the beginning of the twentieth century focused on the mechanical features of a watch that enabled the mechanism to tolerate the demands of a non-office work environment.

By 1908, most railroad watches needed to meet several significant requirements. One feature was that the marking "Adjusted 5 Positions" be engraved on the watch plate. This phrase indicated that the watch had its accuracy calibrated while the watch was in each of five positions. This was done with the knowledge that certain parts of the watch, such as springs and wheels, can operate differently depending on the orientation of the watch. Calibration was eventually measured in six positions: dial up, dial down, pendant (winding stem) up, pendant right, pendant left, and pendant down.

Another required feature was that new watches be fitted with a double roller escapement. The addition of the second "safety roller" reduced the chance of the watch stopping after being jarred. The transition to include this feature occurred over a relatively short time and across most of the industry.

Railroad crews working between Sacramento and Reno were faced with the challenge of working in summer temperatures that could approach 110 degrees during the summer and 10 degrees in winter. Because a watch needed to perform well in a variety of temperatures, most railroads required that a crewmember's watch be temperature compensated. There was a known physical principle that metals change shape slightly as the temperature changes. Higher temperatures cause an expansion of the metal. (Railroads of the twenty-first century still deal with expansion of rail and the resulting "heat kinks" that can cause a slight ripple to the horizontal alignment of the track.) So, watch springs and wheels that depend on a near perfect balance had a tendency to have varied performance as the temperature changed. The hairspring, for example, weakened as the temperature increased. The result was not a good one; the watch lost time. This spring behavior was the main source of error in watch accuracy and, in general, watches were good to about five minutes per day without temperature compensation. The temperature correction process included the addition of bimetallic rims to the perimeter of the balance ring. There was an outer ring of brass and an inner ring of steel. As the temperature increased, the brass expanded more than the steel and caused the balance ring to bend inward. Like an ice skater spinning faster as the arms are drawn in, the wheel spun slightly faster and compensated for the weakened hairspring. The problem of metals changing shape with temperature was minimized with the revelation that Elinvar, made of iron, nickel, and chromium could be used. This alloy changes shape very little with temperature and was introduced into balance wheels and hairsprings. So, the need for the bimetal features of the balance ring were eliminated.

Even with these new precautions in place, the railroad employee was expected to have the watch checked for accuracy. The maximum error per week was +/- 30 seconds as measured by an authorized jeweler with a timepiece considered the "correct" time. Some jewelers, at least in the 1940s, used a ship's chronometer that was calibrated against Western Union time received from Washington D.C. After being checked by a railroad inspector, the employee received a card showing that the watch met the qualifications for design and accuracy.

The general public's usual method of setting a watch was to pull out and turn the crown of the pendant. The pendant was the nearly spherical piece just above the twelve on the dial. When the watch was wound, there was the possibility that the pendant could be pulled out accidentally and the watch time inadvertently changed, the problem going unnoticed for some time. For a conductor, this risk was unacceptable. To solve this problem, an official railroad watch needed to have a lever-set mechanism. With the lever set, one needed to unscrew the glass (bezel) covering the dial and pull out a lever before the pendant could be used to set the watch. When the watch was set, the lever was returned to its usual position and the bezel re-attached.

A watch has no use unless it can be read properly. In the poorly lit, nighttime operation of a locomotive, the engineer needed to be sure that the watch was oriented correctly so he knew where the hands were pointing. To help the crewmember avoid the potentially disastrous, and at least embarrassing, mistake of not reading the time correctly, a railroad watch was required to have the winding stem at twelve. With the stem at this position on the watch, the employee knew where twelve was located on the dial.

The following section contains four articles. The editions of Southern Pacific's publication *The Bulletin* in which the articles appeared are:

"A Story of Time – Regulation of the Railroad Watch,"

June 1928

"A Careful Man Has an Accurate Watch," September 1920

"Ask the Conductor," August 1923

"On the Dot – What is the Correct Time," April 1939

A Story of Time

....how the universal system of regulating our watches came into being and its importance in railroad operations....

By S. A. POPE
Supervisor of Time Service

TIME may be likened to a staircase upon which mankind has ascended out of barbarism through remote ages to our present stage of civilization and progress. In this staircase minutes are the steps. Upon them must stand every thought, action and even every spoken word.

As life has become more involved and society more highly organized, it is safe to say that time has become the greatest single factor in any line of endeavor. By accuracy of the modern chronometer navigators cross oceans through storm and calm, maintaining great ships on schedules that seldom vary. Present day manufacturers govern the production of every article by the clock. Even the housewife would find it difficult to boil eggs without the aid of watch or clock. In fact the three familiar black hands which measure moments and minutes have come to govern every standardized activity of life.

Nowhere is the value of time more emphasized than in the transportation industry, and no other industry has done so much to promote its standardization. Just as American railroads are prominent in standardization of facilities and equipment, so they are chiefly responsible for our present orderly nation-wide time system, commonly known as "standard" or "civil" time.

Were you ever curious as to why our day is divided into 24 hours and the hours into 60 minutes, etc? Why could it not just as well have been 30 hours, of 25 minutes to the hour, or any other combination?

It is probable that our time divisions have come down to us from the Babylonians who first divided the circle into 360 degrees. They had a very definite fondness for the figure 12 with all its multiples and subdivisions, and so we have 12 months to the year; 24 (12 x 2) hours to the day; 60 (12 x 5) minutes to the hour, and 60 (12 x 5) seconds to the minute. Can you think of any other number that would serve us so well, enabling us as it does, to divide hours into intervals of 5, 10, 15, 20 or 30 minutes?

In pre-railroad days when travel by horse-drawn vehicle was the popular mode of transportation and when a stage journey required preparation of weeks and perhaps months, the need for accurate time measurement

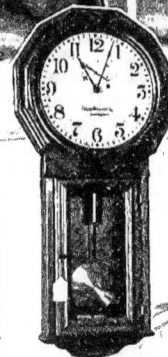

The passenger above knows that Stationmaster J. Gernant has the correct time and is safe in setting his watch before boarding the Gold Coast at Oakland Pier. (Right) Standard clock of today and the ancient timekeeper used by the railroad in the early '70's. (Below) Watchmaker M. A. McLelland at his bench in the Company's repair department.

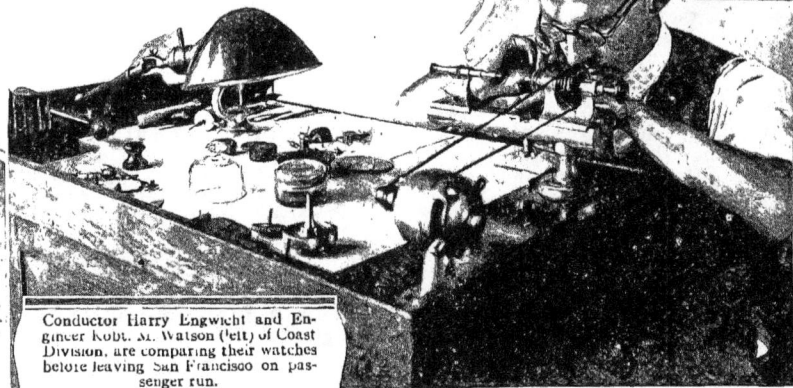

Conductor Harry Engwicht and Engineer Robt. M. Watson (left) of Coast Division, are comparing their watches before leaving San Francisco on passenger run.

did not exist. Long journeys were only undertaken account of some extraordinary emergency. In 1797 clocks were considered so much of a luxury that the English Parliament levied a tax of five shillings a year on them. So it is clear to see that any sort of time was considered sufficient for the rank and file and presumably only those in affluent circumstances were fortunate enough to own time-pieces. It should be added, however, that the act was repealed the following year.

The coming of railroads introduced an entirely different situation. Trains had to be run on a schedule more exact than those of the mail coach. Travelers moving over long distances found that their watches constantly disagreed with the local time in cities and towns through which they passed. It is recorded that as many as five different times were in evidence at one railroad center, indicating the arrival and departure of trains north, south, east, and west, and local time. The confusion may well be imagined, and it became imperative that something be done to find a solution to the problem. This was overcome to some extent by the railroads adopting the time of the central or larger cities in the districts through which they operated. But even this was only a temporary measure of relief as a difference still existed between railroad time and local time.

Suggests World Standard

In 1870, Charles Dowd, of Saratoga Springs, made the suggestion that an international time standard be established for the timing of all civil pursuits. The suggestion was not acted upon until 1879 when it was taken up anew by Sandford Fleming, at that time chief engineer of the Canadian Pacific Railroad, who, through the agency of the Canadian Institute, introduced the subject officially to the leading governments of the world, with the view of bringing about a uniformity in designating the hours of the day for civil use.

In 1882, the United States Congress authorized the president to call an international conference for the purpose of adopting a uniform, common prime meridian to be used in reckoning longitude and in universal regulation of time throughout the world. Present at the conference were representatives of 26 countries, but an agreement on the question of a prime meridian could not be reached. The majority, however, favored establishing the longitude of Greenwich, England, as the origin of longitudes.

Before the conference met, the United States and Canada had selected a series of standard meridians to be reckoned in multiples of 15 degrees east and west of Greenwich, 15 degrees being equal to one hour of time, and on October 18, 1883, a convention was called by W. F. Allen, secretary of the General Railway Time Convention, at which it was decided

to introduce such a plan of standard time to take effect at noon November 18, 1883.

Local time, with all its accompanying complications and difficulties, so far as railroad operation was concerned, then became a thing of the past. Thus, Standard Time varies from that of Greenwich in whole hours. When it is 12:00 noon at Greenwich, it is 11:00 a. m. 15 degrees west and 1:00 p. m. 15 degrees east of Greenwich. To carry the illustration still farther, the 120th meridian west (which is Pacific Standard time, and governs the greater portion of Southern Pacific operations) is 8 hours behind that of Greenwich time, and 120 degrees east of Greenwich is 8 hours ahead of Greenwich time.

Four Zones in U. S.

The division of time zones is arranged as nearly as practicable midway between time meridians so that the greatest difference between local time and standard time is not more than one-half hour at any given point. Reference to a map describing the time zones of the United States will give a comprehensive understanding of this arrangement. From this it will be seen that we have four time standards in the United States—that of the 75th meridian, known as Eastern Standard Time; that of the 90th meridian, Central Standard Time; that of the 105th meridian, Mountain Standard Time; and the 120th meridian, Pacific Standard Time. The irregularity of the lines dividing these zones is because certain cities or districts have good and sufficient reasons for using the time of either one zone or the other.

Changes in time standards affect-

Story of 1869 Record Recalls Early Days to President of Santa Fe

The following letter, in part, has been received by F. Q. Tredway from W. B. Storey, president of the Santa Fe:

"I have read with very great interest the article in the May number of the Southern Pacific Bulletin entitled, 'A Railroad Record That Defies Defeat.'

"Many years ago, and in fact during my younger railroad life, I was employed by the Southern Pacific in the Engineering Department and recognize the correctness of the description given in the article as to the method of track laying.

"There is one incident connected with the work of April 28, 1869, which I think is of interest. Mr. Strobridge personally told me that he had provided relays of men to handle the work and that when the relay for the iron gang came to relieve the original force the latter refused to be replaced and therefore the same men handled the entire ten miles of iron rails in that one day. You will note from the facsimile of the time book page shown in the May issue that the men were given four days' time for that one day."

ing the movement of Southern Pacific trains, occur only at Yuma, Ariz., at which point passengers set their watches backward or forward one hour. East and west-bound passengers on our Overland, Golden State and Sunset routes set their watches backward or forward one hour when reaching or leaving our rails at Ogden, Tucumcari and El Paso.

Every department in a railroad's activities is benefited by accurate time. For this reason Southern Pacific has a department which specializes in time service. It is the business of the Time Service Department to protect train movements from hazardous situations arising as a result of time variations. Not only is time accuracy necessary in closely adhering to schedules but in negotiating meets and passings, which are often made on short time, watches of train crews must be accurate to prevent unsafe conditions arising.

The watches of 12,000 employes on Pacific Lines are subject to our Time Service requirements and comply with very exacting rules and specifications. At regular intervals these watches are carefully examined by inspectors appointed by the Company. All train and enginemen visit an inspector twice a month and watches are set if a variation is shown of twenty seconds or more from Standard Time.

Haste Makes Waste

The need for a greater interest in accurate time is well illustrated to those of our readers who are daily commuters to suburban points. Note the frantic rush of the people who patronize our ferry boats during the peak hour between 5:00 and 6:00 p. m. You will see commuters running for the gate when, as a matter of fact, there may be several minutes to spare. All the attendant worry and excitement in making connections could be eliminated with correct Standard Time in the pocket of each individual. If the public would take the same interest in having precise time that is the habit of railroad men, uncounted hours, now wasted, would be saved, and probably a good many heart ailments could be avoided too.

A Scotch story recently appearing in a popular weekly publication, tells us that: "Sandy MacPherson, after being shown to his room in a hotel, looked from the window and noticed a large illuminated clock in a tower across the street. He stopped his watch."

In all walks of life many people are not unlike the Scotchman. They do not give their watches as much care as they devote to the lawn-mower or the family washing machine. For lack of competent attention their watches are seldom in the condition in which they left the manufacturer and are constantly depreciating as time-keepers. This is false economy. The few dollars of expense involved in having one's watch or clock cleaned

These clocks and watches represent some of the many different types used on Pacific Lines. 1. Time is an important factor in punching baggage checks. 2. Types of watches that have been used by railroad men, and men long before them, during the last hundred and seventy-eight years. 3. Before starting his run the conductor compares his time with the Southern Pacific Standard Watch —the symbol of accurate time. 4. Clocking "in" and "out" is a well known procedure. 5. Patrolmen and watchmen punch a clock to indicate their location at all hours during the night. 6. Clock dial that is familiar to thousands of travelers through the busy Oakland Pier terminal. 7. Section of the repair shop at San Francisco where clocks are sent from all over Pacific Lines

and oiled annually, is money wisely invested and the owner is repaid many times over by the personal satisfaction and joy to be derived from a reliable time-piece. If given this attention, there is hardly any limit to the life of a good watch. Railroad men are required to have their watches cleaned and oiled at stated intervals and the date each watch is due for cleaning and oiling is carefully recorded by the Company. A great many employes give their watches this attention once a year, but eighteen months is the longest period watches are permitted in service without this attention.

The following letter recently received by the Company from Headquarters Ranch, Quihi, Texas, pointedly illustrates how Southern Pacific time is generally depended upon by the public:

"Gentlemen:—When our last reliable time-piece failed us recently we did not know for a while what to do since the nearest watch was several miles away. One of the boys remembered that your limited train would be along past the place at 11:36 in the morning so we waited around and when she blew by we set all clocks and watches accordingly. On my last trip to the city my watch tallied exactly with that of the best jeweler there."

The Time Service Department does not confine its activities to the supervision of watches only. It encompasses all types of time-keeping mechanism. It may surprise many to know that we have approximately 3,000 clocks which are used for every purpose in the railroad's operation. Leading in importance are Standard Time Clocks which, for the sake of brevity, are designated "Standard Clocks." These are of the highest type of workmanship and are found at all terminals and turn-around points as a reference for train and enginemen, enabling them to compare their watches with a positive standard of accuracy. Standard clocks receive closer attention than any other timing equipment and are as nearly accurate as human ingenuity can make them.

The "Boss" of Time

The seat of authority from which our time is obtained, is that of the Government Naval Observatory at Washington, D. C. By astronomical observations, degrees of accuracy are obtained that are almost unbelievable. The performance of the precision clocks in the observatory may be checked to within 1/1000 of a second. The majority of radio fans are accustomed to the method of sending out time signals. Commencing at 11:55 a. m. Eastern Standard Time of the 75th meridian, the seconds beats of the distributing clock in the observatory are sent out by wire and wireless simultaneously. The 29th second is omitted and the 55th to 59th inclusive. Prior to the final signal at exact noon, the 50th to 59th inclusive are omitted. Telegraph operators on duty check their standard time clocks with these signals and correct them when the variation accumulates five seconds. For an almost infallible source of information, be sure to check your time-piece with Naval Observatory signals.

Among other Southern Pacific timing equipment which should receive mention are our marine clocks which guide our navigators on ferry and river steamers in adhering to their schedules. They are placed in pilot houses and engine rooms.

Employes' Time Recorders have an important function. These are commonly known as "punch clocks" and are mostly used in our shops and larger offices. They eliminate disputes on the part of employes with regard to the number of hours worked, by indicating on a card, which employe inserts in the clock, the exact time he "punches in" and "punches out." On this record he is compensated.

Clock equipment also insures proper protection of property during the hours of darkness, against fire and theft losses. Watchman clocks, as they are known, are carried by watchmen and patrolmen who, in making their rounds, insert a key placed in station boxes at desired points, imprinting on a paper dial the time at which these stations are visited, thus giving a complete record as to whether or not the watchman attended to the duties assigned to him.

There are many other types of time equipment too numerous to mention in detail.

This department now has a fully equipped clock repair shop through which all time-keeping mechanism, irrespective of design or purpose, is serviced and repaired. Some clocks come to the shop in the last stages of dilapidation and would probably be scrapped if sent outside for attention. By the combined skill of the cabinet maker and the watchmaker, they are reconditioned at a very moderate cost and sent back to their allotted tasks as timekeepers.

All our clock equipment will now be serviced periodically and the more important types, such as Standard Time clocks, watchman clocks, pilot house clocks, and time recorders will be given annual attention. This will prevent depreciation and the unreliability which is bound to result when time-keeping mechanism is neglected.

What Is "Time"?

Ask the average man if he understands the meaning of the word "time" and he will assure you that he does. But ask him to define the word, and ten chances to one you will have him non-plussed.

Although without question the most remarkable word in the English vocabulary, it's every day use in our business and social contacts has made it common-place. We often say that we "make time," "save time," "lose time," etc., but these terms are only figures of speech. One thing, however, we do know, not one of us can be separated from its influence for we are all most vitally concerned with one period of it—what we refer to as "life time."

Here is the dictionary definition: "The general idea, relation, or fact of continuous or successive existence, infinite duration or its measure."

Rather abstruse you will say, and rightly so, for a comprehensive understanding of it is impossible. It is one of the mental twisters which has baffled scientists and philosophers. One writer puts it this way: "Time is naturally divided into three most unequal parts—whereof the Past includes all that has happened until now from that far-distant period when 'Heaven and Earth rose out of Chaos';

Death of R. P. Kyle at Phoenix Comes as Great Shock

Russell Parsons Kyle, district freight and passenger agent at Phoenix, died April 28 at the Arizona Deaconess Hospital, following a brief illness caused from spinal meningitis. He was stricken suddenly April 25 and was removed from his home to the hospital the next afternoon. He was unconscious practically the entire time of his illness.

Russell P. Kyle

Mr. Kyle was one of the most prominent figures in business and civic circles of Phoenix and was very popular socially. He was a past president of the Kiwanis Club and at the time of his death was a director of that organization. He was a director of the Phoenix Country Club, vice president and director of the Phoenix Chamber of Commerce, chairman of the Phoenix District of the Roosevelt Council of the Boy Scouts of America, director of the Arizona Club, an exclusive business men's organization, director of the Arizona Health league, director of the Central chapter of the American Red Cross and during March he was elected president of the Phoenix Community Chest Association and was a member of the Board of Governors of the Merchants and Manufacturers Association. He was a captain in the Arizona National Guard, prominently connected with the Hiram International Club, a Masonic organization, actively associated with the Phoenix civic center charitable institution, a member of the Ma-

the Present is but a moment, expended in a breath momentarily renewed; the Future is, as the Past,—'a wide unbounded prospect' an 'undiscovered country,' into which Prophecy itself penetrates but partially."

We can only associate time with some physical phenomena recurring with regularity. To us, therefore, time is really a measurement of action, motion or human effort. With so many things to learn and so little time allotted us in which to put to good purpose, should we not apply to ourselves the lesson which the railroads are constantly giving us of the advantage of "on time" performance?

Paraphrasing an old saying—"Save the minutes and the hours will take care of themselves."

The minutes saved in the course of a year by correct measurement of our action and efforts will soon repay the original cost and maintenance of good time pieces, whether they be clocks or watches.

sonic Order, also of the Loyal Order of Moose, senior warden in the Episcopal Church and took active part in Sunday school work in that denomination. He was chairman of the Greater Phoenix and Salt River Valley Committee at the time of his death.

Mr. Kyle was born in December, 1885, at Boston, Mass. He was married at Tombstone in September, 1912, to Miss Lucy Wolcott. He is survived by his widow and three sons, Warren Wolcott, Russell Parsons and David Mason Kyle, also his mother, Mrs. W. O. Kyle of Boston.

Private funeral services were held on the 30th, from the Chapel of A. L. Moore & Sons and at the same hour memorial services were held for the general public at Trinity Cathedral. Rev. B. R. Cox read the Episcopal funeral rites and cremation services were held at the Greenwood Crematory immediately following the services at the Chapel. The speaker at the Memorial service was the Honorable Judge E. S. Clark.

Mr. Kyle came to Phoenix in 1922 from Morenci, Ariz., where he was serving the Morenci Southern Ry. Co., as general manager. He previously had long service with the El Paso & Southwestern as clerk, brakeman, material accountant, train inspector, safety supervisor, trainmaster and was general agent at Phoenix at the time he was appointed district freight and passenger agent for Southern Pacific in November, 1924.

All Phoenix and the Salt River Valley mourned his passing.

Veteran Surgeon Is Honored at Dinner In Oakland

Men prominent in the medical profession from all parts of California and civic leaders attended a dinner at Hotel Oakland, April 27, given in honor of Dr. O. D. Hamlin, division surgeon of Southern Pacific with headquarters at Oakland. The occasion was in recognition of the many years Dr. Hamlin has served in his profession and the speeches during the evening, in addition to dealing with health subjects of general interest, were laudatory to the career of the veteran doctor. Dr. Hamlin is chief surgeon of the emergency hospital in Oakland.

Auditor Is First Vice President of Accounting Association

T. O. Edwards, general auditor, was elected first vice-president of the Railway Accounting Officers Association at the organization's annual meeting held early in the month at Atlanta. For a number of years Mr. Edwards has been very active in the affairs of the association. H. W. Johnson, comptroller for C. B. & Q., is president of the organization.

"A Careful Man—Accurate Watch"— and Why

Wreck twenty-five years ago led to present inspection system by which time exactitude on the country's railways is assured at all times

By S. A. POPE, Assistant General Time Inspector

Over a quarter of a century ago there was a bad railroad wreck near Cleveland, Ohio. The fast mail, known as No. 4, was going east, an accommodation train was coming west. The two trains came together head-on at a small station where the accommodation train was supposed to have taken the siding.

The engineers of both trains were killed and the dead bodies of nine clerks were taken from the kindling wood and broken iron of the postal cars.

Following the official inquiry, the case went into the Federal courts at Toledo. The conductor of the accommodation train admitted he had not taken his watch out of his pocket from the time he had been ordered to take the siding until the accident occurred. He said he supposed the engineer would look out for that. But experts testified that the engineer's watch must have stopped four minutes and then began to run again, a little matter of life and death that he never found out. The fast mail was on time to the dot, the other train four minutes later than the engineer's watch showed.

The disaster proved two things: that the watch of one engineer was inaccurate and that the conductor of the accommodation train was negligent. Therefore, "a careful man and an accurate watch" became a recognized fundamental principle in safe railroading, and the accepted motto of the Southern Pacific's Operating Department.

To Mr. Webb C. Ball of Cleveland, Ohio, belongs the honor and distinction of having conceived and perfected the system of watch and clock inspection which nearly all the great railroad systems of the country have accepted as their standard. His offices stretch across the continent and his assistants are on the road continually, seeing to it that the watches of thousands of engineers and conductors are so nearly exact month in and month out that accidents due to the element of time are theoretically impossible.

On the Southern Pacific, Pacific System, there are approximately 12,000 employees whose watches come under the Time Service requirements. Local watch inspectors are located at all principal points, these inspectors being appointed by Webb C. Ball, general time inspector, after investigation has proved them competent to be entrusted with the work of inspecting and repairing railroad watches. To these watch inspectors 12,000 employees report on an average of 24 times a year. The rules re-

A CAREFUL MAN AND AN ACCURATE WATCH

quire that they must visit an authorized inspector twice each month, and any variation from correct standard time is noted on a card certificate which the employee carries in his pocket at all times. Watches used in railroad service are permitted a maximum variation of 30 seconds a week. If a watch varies in excess of that it is an indication that it needs adjusting or that there is some mechanical trouble, and the watch inspector makes the necessary correction. As an indication that he has visited the watch inspector the employee signs what is known as a "comparison sheet," and in addition to name and occupation, the seconds fast or seconds slow are shown. This comparison sheet is mailed to division superintendent's office and there checked to see that all employees have complied with the rules.

Before starting a run, engineers and conductors compare watches with a Standard clock—there are 150 of these

clocks on the Pacific System, and the maximum variation allowed is 10 seconds. A daily rate card is kept on each clock, and as the time is received at noon each day over the wire from the Naval Observatory the variation of the clock from the correct time is recorded by party in charge.

The conductor also compares the time shown by his watch with that of the engineer before commencing a run, and by these means every possible precaution is taken to insure against errors in time.

Have you ever noticed what pride the average watch owner displays in the fact that his watch has run five or possibly ten years without any expense for repairs? He has no reason to be proud of the apparent achievement of his watch. When it finally ceases to tick and he is obliged to visit the local jeweler he is astounded to learn that the bill for repairs will be a large one for he is told that his watch has been very unfairly treated. The watchmaker reports that the pivots are cut, cap jewels are pitted, etc., and it will take considerable time and material to put the watch in proper order again. Mr. Watch Owner either absorbs a great deal of useful information on the future care of his watch, or, as is often the case, he classifies the watchmaker as a "porch climber," pirate, etc., and feels that he has been unjustly dealt with. If the repair man properly informs him, he will learn that the balance wheel of his watch vibrates one hundred and fifty-seven million times each year, and that the rim of this wheel travels the equivalent distance from San Francisco to New York and back in that time, and in a year his watch has run 9,000 hours. If he profits by his talk with the watchmaker he will follow the suggestion that in future his watch should be cleaned and oiled every 12 months.

Nothing is left to chance or the whim of the owner of a railroad watch who comes under the time service rules. He is instructed that his watch should be cleaned every year, and under no circumstances must it be carried in service without cleaning a longer period than 18 months. Each of the 12,000 watches subject to watch inspection on the Southern Pacific are recorded and the dates they were last cleaned. When the allotted 18 months have expired the matter of watch cleaning is made the subject of a letter from the superintendent to the employee, thus insuring against the possibility of oversight. A watch permitted to run indefinitely without mechanical attention would be a hazard in the operation of trains and therefore every precaution is taken to make watches in railroad service safety appliances in every particular.

A good many of us at various times ride on the "Overland Limited" or the "Owl Limited." The next time you ride one of our crack trains see that your watch is set to correct time "to the second," before starting, and as your train clicks off the miles you will obtain diversion by comparing the time of arrival and departure at the scheduled stops, and you will be surprised at the precision with which the engineer holds that train to schedule. The principal reason for this is that his watch is a piece of standardized equipment, and he is not taking chances in minutes—his watch is on the second—it must be.

Take personal pride in the watch you carry; by so doing you display your personal interest in maintaining the efficiency of the company you work for; you are promoting safety and contributing toward the proper support of one of the greatest safety appliances in connection with the operation of trains.

SALT LAKE DIVISION NEWS LETTER.

By NORBERT J. THOMAS
Superintendent's Office, Salt Lake Division

At all points on the Salt Lake Division the prospects are good for heavy crops. Wheat, oats, barley and hay will show a splendid average yield per acre. In the vicinity of Winnemucca the hay yield, which will be very good, will not be sufficient to take care of the cattle and sheep raised on the ranges, and will necessitate large shipments of live stock in the fall of the year to feeding points. The stock and ranges are in the very best condition at the present time.

At Winnemucca some new prospects are opening up, which may develop into mines, several of them very promising with good ore. The Silver State Chemical Company is contemplating the establishment of a chemical and small refining plant. There is also a new Court House, a new five-story hotel, and a new Federal building under course of construction. A new Catholic church is to be built soon as well as a number of new residences. Business is generally good in that vicinity.

The Maintenance of Way Department is laying new 90-pound steel rail between Jackson and Groome, Nevada, a distance of 18 miles, and it is the intention to carry this work a distance approximately 57 miles from the east end of the Great Salt Lake Trestle to Lucin. It is worth noting that the old 80-pound steel rail was right on the Trans-Continental line and stood 18 years of hard service.

W. O. Johnson, formerly chief clerk in this office, is now chief clerk in the office of the Regional Engineer, J. Q. Barlow, room 221, Pacific building, San Francisco, and is handling work in connection with railroad claims against the Government. Mr. Johnson is making his home in Berkeley.

A. C. Emerson has been appointed chief clerk, Salt Lake Division, succeeding Mr. Johnson, F. J. Bartonek being appointed assistant chief clerk.

W. F. Turner, assistant division engineer, has been transferred to the Tucson Division as division engineer.

J. H. Harshaw of the Engineers' Force has been transferred to El Paso as assistant superintendent and division engineer.

E. S. Adams has been appointed car distributor of the Salt Lake Division, effective June 1, 1920.

Permanent changes in agencies were made as follows: W. W. Manning, Elko, effective May 8, 1920; J. J. Cullen, Susanville, effective May 13, 1920; W. F. Shelly, Wabuska, effective March 1, 1920.

The clerks of the M. of W. at Ogden accepted an invitation by Otis Weeks, division engineer, for a Sunday outing at Saline, June 13th. It is an ideal beach for lovers of salt water bathing. One can walk out into Great Salt Lake for a mile or more and the water never exceeds in depth five feet.

Leo J. Clark and Miss Edna Ludwig were married on June 9th. Both were employees of this office. Mr. Clark was formerly assistant division accountant and was recently transferred to the staff of Assistant Auditor R. Adams at San Francisco, while Miss Ludwig was one of our transportation timekeepers. Other benedicts of the month were Archie Sheppard, accountant, and Don K. Hastings, timekeeper.

NEWS NOTES FROM LOS ANGELES DIVISION.

By S. C. McCLUNG,
Superintendent's Office, Los Angeles

H. M. Read, agent at San Pedro, reports that a portion of the large concrete warehouse on municipal wharf No. 1 has been bonded by the United States Government, thus meeting a long felt need. The warehouse is served by the Pacific Electric.

Pursuing its usual liberal policy of assisting projects which promote the interests of communities it serves, the Southern Pacific has granted permission to officers and men of the Pacific fleet to use an acre of railroad land overlooking San Pedro harbor as an athletic field. Baseball, football and track sports will amuse the thousands of sailors gathered at San Pedro.

Agent E. T. McNeill reports that 197 carloads of ice were shipped from Colton during the first week of June to Imperial Valley destinations for the protection of cantaloupe shipments.

William Lowry succeeded O. D. Guire as yardmaster at Colton, June 1.

An easterly extension is being built onto the Pacific Fruit Express Co.'s ice house at Colton that will add much to the ice storage capacity there.

Agent J. E. Sloan at Santa Barbara reports that walnuts are setting heavy and the coming crop promises to be a record breaker.

George Donnatin, Ralph Heistand and A. W. Hamilton are the three yardmasters now in charge of yard at Santa Barbara.

"Boyland" will be converted into a first-class hotel, to be known as "Samarkand" with Persian effects carried out as much as practicable. The hostelry will open about the first of next year.

The dam at Gibraltar Lake is to be stocked with rainbow trout to add to Santa Barbara's attractions for the anglers.

Tourist travel over the Los Angeles Division continues to tax railroad equipment and hotel accommodations, but so far everyone is being well cared for. Travel to the Shrine Convention at Portland and the National Convention at San Francisco helped to swell the usual summer traffic and keep all hands busy.

"Ask the Conductor; He Has the Right Time"

S. P. Time Service Supervisor Traces Evolution of Timepiece from Grass Rope of the Cave Man to the Modern Railroad Standard Watch

By S. A. POPE, Supervisor of Time Service

SOUTHERN Pacific train No. 1, the "San Francisco Overland Limited," pulled out of University Avenue Station, Berkeley.

"She's ten minutes late, we're going to miss that boat at the Pier" grumbled a man in the smoking room, looking at his watch, "and I have an appointment in San Francisco."

"I think you're fast," said another occupant of the room, "It's 1:35 by my watch, that makes us only four minutes late."

There followed a comparison of watches among those present. No two showed exactly the same time.

"There's the Skipper" said the first speaker. "Conductor, what's the right time?"

The company's representative pulled out his watch.

"It's 1:31," he said, "we're leaving here on time. You won't miss your boat."

That settled the argument. The train conductor was the last court of appeal, no one questioned his authority on the subject of time.

Those whose business requires railroad travel can all cite similar instances showing that the general public have come to regard time and the railroad as synonymous, and when an employe in train or engine service is asked for the correct time his word is accepted as final. And there is very good reason why the railroad man's watch is the time standard, although, like many other benefits enjoyed in this day and age, we accept it as a matter of course and are seldom curious enough to delve into the reason.

History of Watch

But, having suggested to you the inseparable relation between the railroads and accurate timepieces, and further, that this condition did not "just happen," allow me to tell you some things concerning the history and development of that most wonderful of all machines — the railroad standard watch.

Although an interesting story, it would require more space than I am allotted to tell you of man's earlier methods of recording the passing of time. Briefly we will review them:

There was the prehistoric Cave Man's grass

S. A. POPE

rope. It is recorded that he knotted it, dampened it and set afire, and as it slowly burned it recorded the passing of the day; even he vaguely sensing the value of time. The first water clock, or Clepsydra, attributed to a

Chinese Emperor, 2656 B. C.; the first Sun-Dial, 742 B. C., referred to in the Bible as the Sun-Dial of Ahaz; and the 12 hour Sand Glass of Charlmagne, 807 A. D. In the year 900, King Alfred made a time candle of wax, 12 inches in length with notches one inch apart; each candle burned four hours. Finding that it did not burn with regularity because of the guttering of the candle in the draft, the King devised a lantern of white horn scraped thin, thus making it transparent.

The Oldest Clock

The oldest clock of which definite mention is made in English records, was placed in a turret in New Palace Yard, opposite Westminster Hall in London in 1288, during the reign of Edward I. During succeeding years, vague references are recorded concerning mechanical attempts to construct wheels and pinions that could be assembled for the purpose of recording the passage of time.

In the year 1364, a Dutch inventor by the name of DeVick built a tower clock much superior to anything preceding it for Charles V of France. It was a crude, cumbersome contraption with only one hand and was wrought from iron. Instead of the pendulum control of its power, it had what was known as a verge and balance escapement, taking a horizontal motion.

History records that clocks hitherto had been shut up in monasteries, but now they began to be used for common convenience of cities. In connection with this clock, the following anecdote was handed down. The King ordered DeVick to make this clock for the Palace Tower and when the work was finished DeVick reported to the King, who, wishing to find some fault with it, told its maker that the hour of four o'clock should be indicated on the dial by four vertical strokes (1111) instead of the Roman numerals IV.

"I fear your Majesty is wrong" replied DeVick.

"I am never wrong," said the King. "Change it immediately."

And from that day to this, when Roman numerals have been used on clock or watch dials, the King's own version has been followed.

For the next two centuries, development in

Southern Pacific Establishes Own Bureau of Time Service

ON July 12, 1923, the Southern Pacific Company established its own Bureau of Time Service for the Pacific System under S. A. Pope as Supervisor. This bureau will take over the supervision of time formerly in charge of the Official Bureau of Railroad Time Service.

Mr. Pope has been Pacific Coast representative of the Official Bureau since 1917, as Assistant General Time Inspector. He started his profession as a watchmaker's apprentice in England and continued his trade after arriving in Canada in 1903. He first became identified with railroad time service in San Francisco in 1909. Mr. Pope's railroad experience, however, has not always had to do with watch inspection, for he has served in both the operating and traffic departments of various railroads in Canada and the United States.

4   THE BULLETIN

timekeeping instructions was a slow laborious proposition. The world was emerging from the dark ages and inventors were not in popular favor. It was 1504 before the first portable clock or "watch," as it came to be known, was constructed by Peter Henlein, a young locksmith of Nuremburg. It was a spring-power contrivance, half a foot high, made entirely of iron. DeVick's old clock, although a crude machine was, however, the real ancestor of the numberless timepieces which now govern the lives of our hustling modern civilization.

18th Century "Turnip"

We are now going to pass over a period of 200 years. Accompanying this article you will see a photograph of an old watch in my possession— perhaps it might be better described as an old "Turnip" because it more resembles that humble vegetable than the watch as we know it today. It was about 1750 that this pioneer was constructed and it probably took the artisan who designed and made it several months to complete the job. It measures 3 inches in diameter. If it performed within a variation of ten minutes a day it was regarded as accomplishing its purpose. A matter of ten minutes either way would make no material difference in the quiet easy-going lives of people of that period. Seconds were of no account, in fact, this watch was constructed but a few years after the minute hand had been introduced. There was no commuting, no established eight hour working day; and no transportation systems requiring "time to the second," but this old watch illustrates a very definite advance in mechanical construction from the time of DeVick's work, paving the way for the modern wonder which now makes possible the "on time" performance in which the Southern Pacific takes such justifiable pride.

Some really marvelous inventions were made during the Eighteenth Century. Craftsmen pitted their ability one against the other endeavoring to gain supremacy in mechanical achievement. With this idea in mind their work was focused on ornamentation and complicated mechanism, rather than timekeeping excellence. Watches were expensive toys and ornaments for the rich—in no sense of the word articles of utility for the general public. Wonderful specimens, the construction of which illustrate infinite patience and skill, may be seen

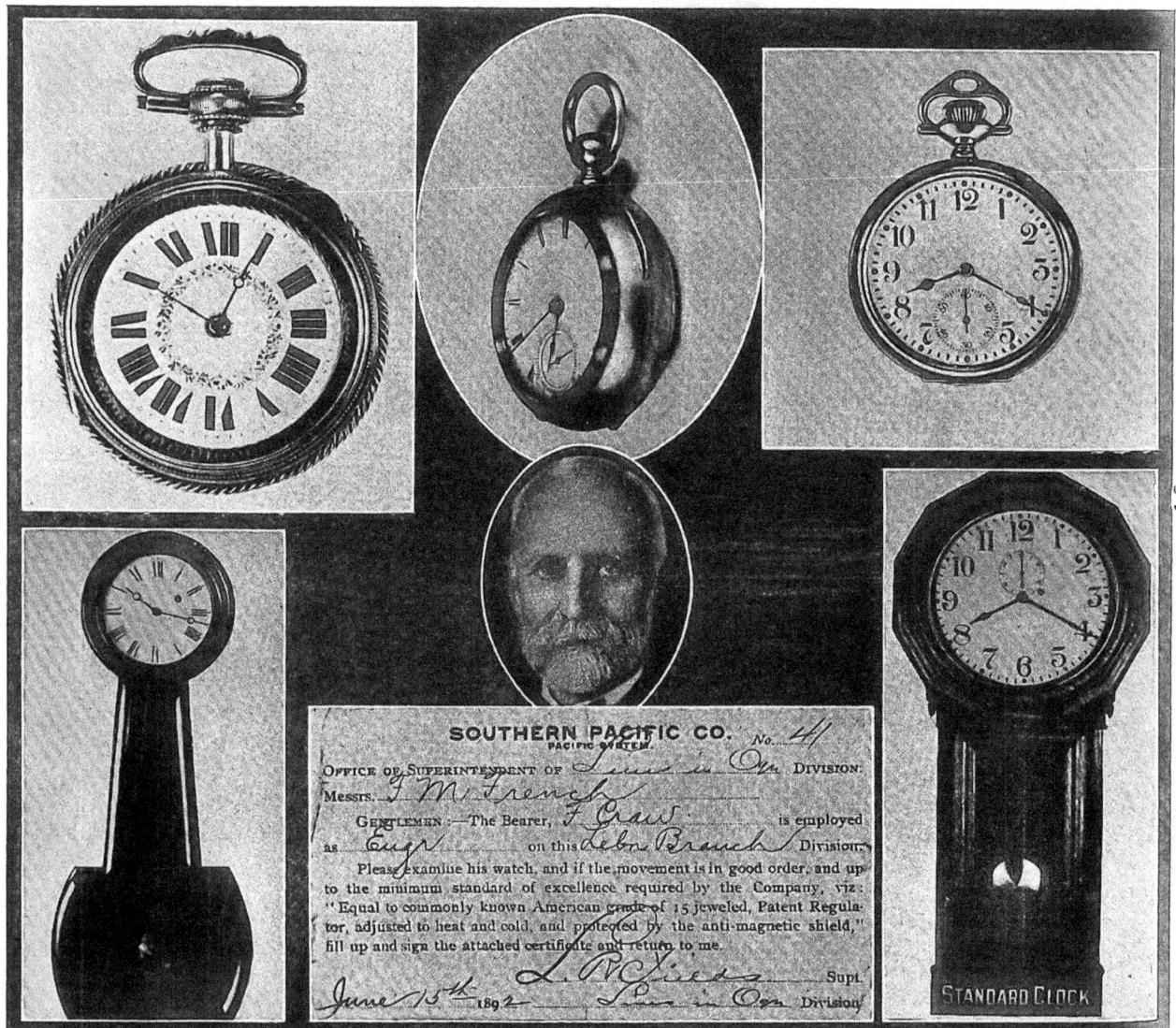

Evolution of the watch is well illustrated by the above watches, showing from left to right, "turnip" of the 18th century; watch in common use on the railroads in the early '60's; and one of the modern standard railroad watches. Below at left is a clock common to our "grandfathers" in the '60's, and at right the standard clock that is familiar to all Southern Pacific employes engaged in train operations. The insert is of Webb C. Ball, "Father" of the time inspection service. The early watch inspection form is interesting, being that of Engineer Craw who is remembered for his excellent record in safety, and also having the signature of L. R. Fields, who for many years was superintendent of the lines in Oregon.

in our museums, and there are some priceless private collections. Those who have access to the Memorial Museum in Golden Gate Park, San Francisco, may see some excellent examples of the work I refer to.

During the Nineteenth Century rapid development was made in all branches of mechanical invention. Watchmakers had been keeping abreast with this progress and were developing a timepiece of real practical value. Let me direct your attention to a period still to be remembered by some of our railroad pioneers—back in the Sixties. The railroads had stretched their ribbon of steel across this country opening up tremendous productive areas. Industries were developing rapidly and calling upon the railroads for larger and better facilities, continually increasing the density of traffic and necessitating more accurate timepieces that would make possible speeding up of train schedules. Accidents due to variation in time were by no means uncommon, often resulting in loss of lives and destruction of railroad equipment. With the knowledge that watches did not, as a general thing keep accurate time, and in the absence of any system of watch inspection and standardization the railroad managements permitted five minutes margin at meeting points to allow for variation in watches of the engineers of opposing trains.

Back in the "Sixties"

Illustrated with this article is a watch then used in railroad service. It is of the same vintage as the old locomotive "C. P. Huntington." Those at the top of the seniority list will have no difficulty in connecting this key-wind watch, our grandfather's watch, with their early experiences in railroading.

The pressure upon the railroads for faster train schedules coupled with the accidents referred to, compelled the managements to take definite action toward obtaining expert advice on the subject of time. Webb C. Ball of Cleveland was called into consultation by an eastern road, as an expert. He investigated closely and diagnosed the malady. His recommendations contemplated standardization in watches and clocks, and a system of watch inspection for those engaged in the movement of trains which would make train meets, resulting in time variations, impossible. Those recommendations were adopted. Thirty years have elapsed and now all railroads maintain some system of watch inspection.

With the advent of railroad watch inspection, the American watch factories were compelled to introduce into their products those mechanical improvements which have made the Standard Watch the dependable safety appliance as we value it today. The double roller escapement, lever setting, and adjustments to temperature and positions have all been per-

History of the Watch

PREHISTORIC Cave Man used a grass rope, which he knotted, dampened and set afire. As it slowly burned it recorded the passing of the day.

First water clock attributed to Chinese Emperor, 2656 B. C.

First sun-dial, 742 B. C.

Twelve-hour sand glass of Charlmagne, 807 A. D.

Time candle of wax, King Alfred, 900 A. D.

Oldest clock; placed in a turret opposite Westminster Hall in London, 1288, during reign of Edward 1.

Tower clock built by Dutch inventor, De Vick, in 1364; real ancestor of the modern timepieces.

First portable clock or "watch" was constructed by Peter Henlein, young locksmith of Nuremberg, in 1504.

fected during the last fifty years. A few weeks ago it was my privilege to spend a day in one of our largest watch factories, employing 3,500 people. This factory is turning out 3,500 finished watch movements every eight hour working day. Automatic machinery plays a very important part, some of them almost human in the intricacy and exactness of their operations. With the exception of jewels (which come from Switzerland) this concern manufactures every part used in their watches.

"Box of Wonders"

Let us not brag of our "On Time" performance without giving due credit to that little "box of wonders" in the pockets of Engineers, Conductors, Brakemen, Operators, and other workers whose duties involve them in train operation. Neither should we forget the Watch Inspectors, of whom we have 105 on the Pacific System. Collectively these Inspectors care for 12,000 watches carried by our fellow employes who come under the Time Service Rules. Their business is to see that watches of these employes do not vary to exceed thirty seconds a week.

Employes subject to Time Service rules are required to visit a Watch Inspector semi-monthly, and have their watches cleaned and oiled every eighteen months.

One of the Southern Pacific Company's oldest Watch Inspectors in point of service is F. M. French of Albany, Oregon. I asked Mr. French recently for some reminiscences of his early experiences with the Company and here is what he said:

"The slogan 'safety first' should have originated with the Watch Inspectors, for after serving thirty-one years as a Watch Inspector, I am fully convinced that without his help,

the Dispatcher would have the greatest difficulty in the operating of trains. Where would the Dispatcher be without his standard clock and telegraph wire? Where would the Engineer and his train land if his watch and the Dispatcher's clock did not agree? Who is responsible for the two timepieces agreeing?—The Watch Inspector.

"Thirty years ago there were but fourteen trains a day over the railroads entering our town. Today we have fifty-eight. With this additional traffic there must be additional safety, each train maintaining its proper position on the track. The Dispatcher, Conductor and Engineer must work as a unit in maintaining this safety and it is only by each having the correct time that it can be accomplished.

"Is there an Inspector who has not felt the cold chills run up and down his back as he has stood on the depot platform and watched these fast trains arrive and depart on the time he has given them? These trains go thundering down the track to make a meet at some point designated by the dispatcher. Did the Inspector at the other end of the line give the oncoming train the same correct standard time that was given the departing train to make the meet?

"Going back thirty years when I first entered the service, it was all new, and we did not have the forms and systems in use today. Men were not checked up on Watch Inspection by the operating officials as they are now. At that early day any ordinary 15 jewel watch which the factory saw fit to stamp "adjusted," was considered good enough.

"The first watch I ever inspected belonged to an engineer on the Southern Pacific lines, an engineer with a record. I want to say something of this engineer as it illustrates what can be accomplished by a faithful and careful man. He presented for inspection a 15-jewel adjusted watch. This engineer soon found the need of a more perfect timepiece and thereafter he carried the best watch he could buy. In the forty years that he handled an engine he never had an accident that caused the loss of a single life. Surely a record to be proud of, and he did not have ninety pound rails all these long years either."

Veteran S. P. Safety Man

Inspector French refers to Engineer F. Craw, who was 41 years in the engine service on the Portland Division and who died in July 1920. Mr. French has loaned us one of the first watch inspection forms, (reproduced in this article) which was issued to Engineer Craw dated June 15th, 1892, and signed by L. R. Fields, now retired, who was the Superintendent of what were then known as the Southern Pacific Company Lines in Oregon.

Let us all follow Engineer Craw's

F. M. French, veteran watch inspector of the Southern Pacific, is shown in his
jewelry store at Albany, Ore., inspecting the watch of Robt. B. McCalley,
who retired as engineer of the Southern Pacific after 41 years 7 months service.

example, bearing in mind our slogan, "A careful man and an accurate watch are the two best safety devices."

Although I have emphasized the importance of time in railroading, we must not overlook the fact that through all human activity runs the measuring of time. The rails on which our trains run, the locomotives and the cars, everything we use in fact is manufactured on a time basis. The indication of a clock dial controls the whistle or bell on the factory, proclaiming the commencement or close of the working day. The workmen punch their time on an automatic recorder. Telegrams show the hour and minute the message is filed and the time received at destination. It matters not what our position in life may be; time is the essential factor in our success or failure. We can only take out of life in proportion to what we put into it, so it behooves us to make the best of every minute, every second, and in the process we will find that an accurate watch—time's meter —is an indispensable requisite. Rudyard Kipling must have had some such thought in mind when he wrote:

'If you can fill the unforgiving
  minute
With sixty seconds worth of distance run.
Yours is the earth and everything
  that's in it,
And, which is more, you'll be a
  man my son!"

There are 12,000 employees on the Pacific System whose watches are inspected by authorized company inspectors semi-monthly, and 105 Watch Inspectors of proven ability as watch mechanics are entrusted with the care of these watches.

Care of the Watch

To preserve the life of a watch, it should be cleaned and oiled once each year, in no case to exceed 18 months. It is a popular illusion with the public that a watch should be allowed to run until it stops for lack of attention. The average watch owner will boastfully tell you that his watch has run ten years or more without cleaning. But the same man will take infinite care to see that his automobile is regularly oiled. Damage is bound to result to a watch movement after a run of 18 months without cleaning

and oiling. The tiny pivots upon which the wheels revolve require but an infinitesimal amount of oil, but they must have this at regular periods. Neglect this cleaning and oiling process and the pivots wear, especially those of the balance wheel, and the watch owner will receive a correspondingly heavy bill for repairs.

So to protect the safety of the service a constant check is kept in Superintendent's offices of this very feature, the dates on which employes' watches are cleaned being properly recorded and followed up, to insure against watches running over the maximum time allowed.

Company Watch Inspectors issue standard Loaner Watches to employes while their own are under-going this necessary mechanical attention. No charge is made for the use of these watches.

Standard of the S. P.

Now a word about watches acceptable in the Company service:

The product of all representative American watch factories are acceptable, if the watches meet certain specifications which are briefly described as follows:

Sixteen size.

Adjusted to 5 positions, viz: Dial up, dial down, pendant up, pendant right, pendant left.

Adjusted to temperature.

A double roller escapement; steel escape wheel; 19, 21 or 23 jewels, lever set, and last but not least, a plain Arabic dial; plainness and clear reading of the dial being paramount where lights are uncertain such as in the cab or the caboose or by the light of the trainman's lamp.

Watches in railroad service may not vary to exceed 30 seconds per week, and it is the Watch Inspector's business to see that they don't.

Standard clocks are located at all terminal points, turn-around points and registering stations, for the use of train and engine-men. There are 144 of these on the Pacific System. First trick operators check the time of Standard clocks each day as it is received over the wire from the Naval Observatory at Mare Island, and they record the variation on standard forms and these records are closely inspected by this office at the close of each month. Standard clocks are not permitted to show a greater variation from standard time than 10 seconds.

Perhaps the most reliable timepieces in existence are the Riefler clocks installed in Observatories. They are anchored to concrete piers and encased in air tight glass cylinders, the clocks operating in a vacuum. The room in which they are contained is maintained at an even temperature at all times by thermostat control. They are wound every 30 seconds automatically.

Set Time From Stars

And yet, with all the precautions science can devise, these observatory clocks are not absolutely accurate.

2

(1) Conductor W. H. Banks gets ready to "highball" the "Cascade" out of Portland. ● (2) Engineer Chas. C. Benbow notes his watch as time nears for "Daylight" to leave.

ON THE DOT: *What's the Right Time? Just Ask Any Railroader Whose Job Has to do With Train Operations*

THE "railroad watch" has long enjoyed a unique reputation as the accepted standard of accurate time, especially with the traveling public. Anyone who has traveled on trains has noted how frequently passengers ask railroad men for the exact time and how those passengers, regardless of the excellent make or timekeeping record of their watches, are outwardly gratified if they correspond with the railroad standard.

If a railroad man tells you that your watch is 10, 15 or 20 seconds off, you don't argue with him because you *know* he's right. Leading American watch manufacturers, recognizing the railroad watch as a symbol of accurate time, often advertise their product by picturing it in the hand of a trainman or engineman, well aware that no other endorsement could be more convincing.

But how many folks, other than those in the branches of railroad service which require accurate time measurement, know the story of how railroad watches and clocks are kept in such close-to-perfect running order; how often they are inspected and adjusted, cleaned and oiled; and something of the department which does the job?

On Southern Pacific, Pacific System, and its affiliated lines, 16,000 railroad watches are ticking away regularly with never a variation, on the average, of

more than a few seconds. These watches are the heartbeat of the railroad, and the responsibility of keeping them in a healthy condition rests with our Time Service Bureau, under management of Stanley A. Pope.

The necessity for time service and watch inspection on a railroad is well illustrated in the following demonstration which was given by Pope during an address before the Pacific Railway Club in San Francisco on the subject of "Time Service as a Safety Measure":

Asking permission from members of his audience to compare their watches with the correct time, Pope noted these discrepancies: "One minute fast . . . three and a half minutes fast . . . two minutes slow . . . three minutes fast . . . two and a half minutes slow . . ." And as the demonstration continued, it showed that a majority of the watches varied from correct time.

No "Hit or Miss" Timing

Only a "few minutes." But picture the confusion on a railroad if its standard clocks and employes' watches were a few minutes off. Crack passenger trains, required to negotiate numerous meets and passings, would be unable to fulfill their schedules if those meets and passings were delayed because of time inaccuracies; arrivals, departures and connections would be made on a "hit or miss" basis and in a very short time a smooth-running transportation system would become disorganized. It is the business of the Time Service Bureau to prevent such interruptions.

The beginning of railroad time service dates back about 50 years to an accident which occurred on an eastern road when two trains met head-on because of an engineer's faulty watch. In those days trains were operated on a catch-as-catch-can plan, in which meets and passings were more often negotiated by chance than by timetable schedule. Train crews were more likely to get the correct time by glancing at the sun or the stars rather than their unreliable timepieces, which in some instances consisted of cheap alarm clocks dangling from the wall of the engine cab. One could scarcely blame the train crews for carelessness in keep-

◄ PICTURES: (3) All trains move under orders from the dispatcher. Here is H. E. Hood of Bakersfield, who is one of the staff that charts operations on San Joaquin Division. A Standard Watch and Clock are his constant guide. ● (4) The dispatcher's immediate contact is the telegrapher at on-line stations. C. D. Robertson, agent-telegrapher at Morganhill, is shown repeating an order back to his Coast Division dispatcher. He has, of course, a Standard Watch. ● (5) Crew Dispatcher Lee Breland (seated) of Taylor Roundhouse, Los Angeles, gives Caller F. S. Schrader the names of engineers and firemen to be called for passenger or freight runs. ● (6) Yuma is one of the division points where Standard Time moves up or back an hour. Crew Dispatcher Ed. Anaya is posting names of crews assigned to runs, while Engineer J. W. Beals signs train register. ● (7) Traveling Watch Inspector E. E. Dahlin of Ogden checks the watch of Section Foreman S. Kaneko, for maintenance jobs are among those requiring accurate timepieces.

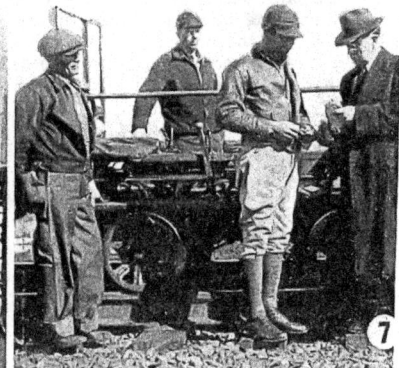

3

ing time, however, since practically every city and town through which they passed had its own method of time measurement (the present system of Standard Time throughout the world became effective November 18, 1883). Ringing of the school bell or factory whistle, for example, often regulated the habits of the populace, and at one railroad center five different times were in evidence—local time and one each for arrival and departure of trains north, south, east and west.

Supervision of watch inspection on Southern Pacific and affiliated lines was given by an outside concern prior to 1923, when Southern Pacific established its own Time Service Bureau, with Pope as manager, to maintain general supervision of all standard railroad timepieces, maintain a record of their performances under uniform rules and regulations, and appoint local watch inspectors to assure that they are kept in good order. Other railroads asked to become members, and today there are twelve transportation companies protected through the work of this bureau.

Included on the bureau's staff are 165 watch inspectors, located at all terminal points on Pacific Lines, also there are traveling inspectors who cover their territories at regular periods, inspecting the watches of employes who are not accessible to local inspectors, such as section and extra gang

SOUTHERN PACIFIC BULLETIN

Vol. 23   April, 1939   No. 4

ERLE HEATH, *Editor*

EMMETT G. FITZPATRICK, *Asst. Editor*

Published monthly by Southern Pacific Company, Bureau of News, in the interest of, and distributed free to, active and retired employes. Address of editor, 65 Market Street, San Francisco.

foremen, agents, operators, and others. All inspectors are thoroughly grounded in the mechanics of timepieces.

From the moment a student railroader enters service in the Operating Department, he becomes watch-conscious. One of his first acts upon passing his examination is to visit an authorized watch inspector for comparison of his watch with standard time and for any corrections or adjustments as may be necessary. He quickly learns that his railroad watch is a constant and valued companion.

Watches Closely Checked

Visits to the watch inspector are made semi-monthly by most of the employes affected, and should a watch show a variation of 20 or more seconds it must be set to correct time. Every employe who is subject to time service provisions carries a certificate on which a record of the performance of his watch is entered at each visit to the inspector. Thus, by glancing at one of these cards, the expert watchmaker is able to determine the "habits" of a particular watch, just as a physician knows the condition of a patient by reading his chart.

Watch mechanisms are inspected semi-annually, and if any unsafe conditions are noted the watch is taken out of service until the irregularities have been corrected. While his watch is undergoing repair, the employe is provided with a Standard Loaner watch which meets specifications of the Time Service Bureau.

Activities of the Time Service Bureau are not confined solely to supervision of railroad watches. There are about 5,000 clocks which are used for every purpose in railroad operation.

Most important of these are the 285 Standard Clocks which are located at all terminals and used by trainmen and enginemen in checking their watches— a necessity before every run. Embodying the highest type of workmanship, these clocks receive close attention. Telegraph operators check them daily with Naval Observatory Time, which

☞ **PICTURES:** (1) Time Service staff at S. F. (l. to r.): C. D. Fabrin, clock inspector; S. A. Pope, manager; H. A. Stabler, time service inspector; L. D. Blake, steno-clerk. Standard clock in window is used by many passersby to compare their watches. It is corrected every hour by means of wire connection with ● (2) Bureau workshop clock whose telechron motor keeps mainspring wound. Right: Pope points out features of 150-year-old English skeleton clock which was exhibited at recent Watch Inspectors convention in S. F. to Watch Inspector H. T. Harger of Sacramento. ● (3) Watch Inspector H. Bullard, Oakland, compares watch of Brakeman J. H. Childress, Western Div., with accurate time as latter signs comparison register. Rows of watches on wall belong to railroaders who left them to be serviced. ● (4) Fabrin at his bench where all railroad clocks are repaired. ● (5) Repaired and accurately set, this array of clocks is ready to leave the workshop and return to service.

is sounded over the telegraph wires, and if there is a variation of more than 5 seconds they must be corrected. A daily record of performance is maintained, indicating whether the Standard Clocks are correct or how many seconds off, and forwarded to the Time Service Bureau at the close of each month for careful scrutiny.

A fully-equipped clock repair shop, located in the Time Service Bureau's headquarters in San Francisco, periodically services and repairs all of the railroad's timekeeping mechanisms, which, in addition to Standard Clocks, include watchman clocks, employes time recorders, marine clocks, office clocks, "conductors' clocks" whose function at the new San Francisco Terminal was explained in the January *Bulletin*, and recording meters which include anemometers and thermographs. Visitors to the shop never fail to be impressed by the number and variety of clocks—old and new, large and small—which literally cover the walls, with dozens of swinging pendulums giving life to the room as the intricate machinery counts off the seconds in a noisy, but strangely soothing, rhythmic chorus.

"Some of these clocks arrive at the shop in the final stages of dilapidation and probably would now be spending their last days on some scrapheap if sent elsewhere for attention," Pope states. "But through the skill of a cabinet-maker and watch-maker they are reconditioned at moderate cost."

Southern Pacific's Time Service Bureau is nationally known. Manager Pope is vice president of the Horological Institute of America, and a recognized authority on railroad time service. He has devoted more than 25 years to the care of watches and clocks.

Rules for Watch Care

"So many people treat their watches with utter disregard for the delicate movement," he says, "yet wonder why they fail to keep accurate time. Often you hear persons say: 'I haven't taken my watch to a jeweler in a number of years, but it still keeps comparatively good time.' Such a remark makes any railroad operating man shudder."

Here are a few simple suggestions which Pope passes along for those who would increase the life of their watch:

Never allow your watch to suffer unnecessary shocks or bumps.
Wind it to the full extent of the mainspring as close to every 24 hours as possible.
Form the habit of placing it in the same position during rest hours, preferably dial up and resting on a flat surface. Lay it down with care.
Have it oiled and cleaned at regular periods no longer than 18 months apart.
When your watch needs regulation or any mechanical adjustment, under no circumstances attempt it yourself. Never open the movement.
If correction is necessary, take your watch to a watchmaker, whose training and profession it is to do such work.

"Abide by these rules," he adds, "and you'll take a railroader's pride in the performance of your watch."

NEW TRAINS: *Two More "Daylights" for San Francisco-Los Angeles Coast Line Run*

ORDERS for two additional 14-car Daylight streamliners for the San Francisco-Los Angeles coast line run have been placed by Southern Pacific Company with the Pullman Standard Car Manufacturing Company, President A. D. McDonald announced in San Francisco March 30.

The new trains, costing approximately $1,000,000 each, exclusive of locomotives, will be delivered for service some time late this year. They will follow the basic design of the present Daylights, among the most popular trains in the world, but will embody a number of improvements and refinements.

Major advancement will be a three-car articulated dining unit, made up of a full-length kitchen car in the center, serving both a full-length diner and a full-length coffee shop car. The new diner will seat 72 persons as compared with 40 in the present diner-kitchen car, and the new coffee shop car 80 as compared with 56 in the present coffee shop-kitchen car. This is the first dining unit of its type in America.

Typical of other refinements is a new type of luggage compartment in each chair and parlor car, with an outside door through which baggage may be loaded and unloaded at terminals without congesting the vestibules. Shelves in the new compartments will be operated electrically on an elevator principle to bring any shelf to the level of the outside door for easy access.

Although outside appearance of the train and interior appointments such as sponge-rubber seats will follow the design of the present trains, additional roominess will be provided by increasing the length of cars from 79 feet 2 inches to 81 feet.

The new trains will be powered, as are the present Daylights, by 5,000-horsepower locomotives, among the largest and most powerful streamlined steam motive power units in the world.

RECOVERY: *Stop Special Privileges in National Transportation, S. O. Dunn Urges*

"NO program for business recovery will accomplish its purpose unless it includes measures that will effectively promote financial and physical rehabilitation of the railroads," declared Samuel O. Dunn, chairman of the Simmons-Boardman Publishing Corp. and editor of *Railway Age*, in an address March 24 before the Commonwealth Club of San Francisco. "This is a vitally important fact which political considerations and the short-sighted selfishness of certain big business interests are causing to be disregarded in most current discussions of needed cooperation by Government and business.

"There is only one thing necessary to a solution of the railroad problem in the public interest, but that one thing is absolutely essential. This is application to all carriers of the sound American principle: 'Equal rights for all; special privileges for none.' The present plight of the railroads is mainly due to almost every conceivable violation of this principle by the transportation policies of both the federal and state governments, although also largely due to the depression. The purpose and effect of these policies is *government control and restriction* of the railways and *government promotion* of other carriers.

"Because all railway rates are controlled by the Interstate Commerce Commission, the managements of the railways are not free promptly to change rates to meet the competition of other carriers, while other carriers are free promptly to make any rates necessary to meet railway competition, the railways provide and maintain their own highways at a cost of about 40 per cent of their total expenses, while government charges nothing for use of the inland waterways that it provides, and charges much less than cost for the use of its roads by commercial buses and trucks.

"Thus the railroads are in competition, not only with the carriers using inland waterways and roads, but *also with government itself* as the owner and maintainer of these waterways and roads. No private industry can successfully meet such government competition; and the railroads are being bled to death by it. Their freight earnings in 1937 were 1,400 million dollars less than in 1926; and at least 1,000 million dollars of the loss was due to this government competition.

"A solution of the railroad problem by equalizing government treatment of all carriers is a necessary means of promoting recovery by largely increasing railroad employment and railroad buying from the durable goods industries. It is essential to stopping the unnecessary and enormous increase occurring in the total cost of transportation borne by the public due to artificial expansion of transportation facilities. And, most important, it is essential to arresting the rapid drift of the railroads toward government ownership, the adoption of which would be the longest single advance that could be made in this country toward state socialism."

Chapter Fourteen
Cool the Produce

Consider some of these major developments in agriculture: Aquaducts for moving water, the first iron-cast plow, horses—instead of hand labor—for field work, the cotton gin to remove the seeds from the fiber, the thresher for separating the kernels and straw in grains, the use of steam engines for powering mills, tractors, soil amendments and fertilizer.

Consider that all of these developments helped produce more food per acre or the same amount of food cheaper. Only indirectly did they increase the number of customers. Now that they had increased production, they needed to increase their scope of distribution. Trains provided the most obvious way to do this, but even though they could increase the speed of delivery, perishability was still a concern. Innovations for a cooled railroad car began in the mid-1840s when the Western Railroad of Massachusetts used a car that was designed to not only keep perishables cool in the summer but insulated in winter. Inside the voids of the car's walls, ice was used in the summer and charcoal in winter.

Around 1860, a new design called for ice to be put in the car through roof openings at both ends. The walls were filled with cork and the floors were slanted to drain water from melted ice.

Other innovators thought the ice tanks in the walls should be thinner, believing that the bin's surface area is more important than its volume. So, the ice needed to be crushed to pieces a bit bigger than a golf ball. Yet another practice included adding salt to the ice bin and thereby increasing the cooling properties of the melting ice.

The idea of adding the ice to tanks at the end of the cars was discarded by Joel Tiffany in 1877 when he designed a car with ice storage in an area on the roof that extended the length of the car. Yet another method was used by the Atlantic Coast Line in about 1878 when they placed ice-filled containers on flat cars to carry vegetables. The containers were moved by train to Charleston where they were put on ships. This idea of containerization would come to dominate the railroad business over 135 years later.

Despite the great opportunity to ship cooled goods, the railroads of the 1870s were hesitant to build cars that could handle enough ice for shipping meat. The railroad companies had large commitments to stockyards and already-built cattle cars that were useless for hauling ice and other kinds of cargo. The packers liked the idea of shipping just the edible carcass instead of the live cow given the fact that there was more edible meat in a ton of carcasses than in a ton of cows. The math was simple for the packers. It was also simple for the railroads—which raised the rates for shipping the carcasses. The packers responded by purchasing their own cars. From 1880 to 1900, the number of refrigerator cars not owned by railroads increased fifty times to about 55,000. During this period, however, the private owners realized that the cost of building a refrigerated car was prohibitive. The railroads, realizing that the customers needed cooled cars, were then more willing to re-enter the business of providing refrigerated cars.

One of the early packers that pushed the idea of shipping the dressed meat instead of the whole cow was the Armour Company. The line of business for Armour included operation of the Fruit Growers Express Company. The scope of this company was too broad and controlling according to the Federal Trade Commission in 1919 when they ordered a diversification of the company, which was eventually taken over by a group of eastern and southern railroads in 1920.

State-of-the-art cars and customers for the veggies are two important parts of a lucrative produce shipping business. But this business also required ice. Prior to 1869, most of the ice used in California was shipped from Alaska—an odd fact given the cold temperatures and lakes in the Sierra. But, as with any business, getting the product to the customer was the challenge. So, until the railroad allowed the transport of ice, the Sierra ice business could not grow. In Alaska, the Sitka Ice Company sent ice down to San Francisco at five cents per pound. During the years 1852-53, three large warehouses in Sitka and Kodiak were connected to the wharves by rail. It is believed that these rails were the first rails laid in Alaska. In California, the Alaskan ice was stored in warehouses in Sacramento, San Jose, Marysville, and Stockton. Each storage area could handle four hundred to thirteen hundred tons of ice. From the moment the railroad provided a link over the Sierra to the central valley and a conduit for the delivery of ice from the Sierra, ice companies in the summit area grew and the Sitka Ice Company didn't.

The early runs of trains over the summit were in 1868 and, as one could guess, the Summit Ice Company was incorporated in the same year. The company's operations were at Ice Lakes about two miles south of Southern Pacific's Norden station.

Donner Summit, the location of Ice Lakes, can have viciously cold temperatures, and large amounts of snow affected the business. So, after three years at Ice Lakes, the operation was moved to Prosser Lake about three and a half miles north-northeast of Truckee. In 1877, the Prosser business had a building 450 feet in length by 50 feet where seventy-five to eighty men,

working for about two dollars per day, produced eight thousand tons of ice per year. Ice was sent to the mines in Virginia City and to the bars and stores of San Francisco and the Sacramento Valley.

At Boca, a few miles east of Truckee, another ice production facility existed: The Boca Mill Ice Company, which began its business in 1868. On the Little Truckee River, a dam was built that formed a pond area of about 180 acres. Ice harvested from the pond was moved two hundred yards to the south where it was stored in sawdust inside the icehouses. By 1872, Boca was one of the largest shippers of ice, wood, and shingles of any town between San Francisco and Omaha. One advantage to working in Boca was the ability to purchase beer from the award-winning brewery on the south side of the river. The last harvest of Boca ice was in the mid-1920s when the Union Ice Company began making ice at its plant in Reno.

The story of transporting perishables grown in California is not complete unless there is some reference to the Pacific Fruit Express Company (PFE). Formed in 1907, this company purchased and maintained a fleet of refrigerator cars and built and operated a series of icing plants throughout the west. On its first day, it had no equipment, a manager, one clerk, and one stenographer. During the first year, 6,600 new cars were built. From 1908 to 1920, carloadings increased from 40,000 to 142,000.

In October of 1909 at Roseville, PFE began operating a very large icing facility. Made of brick and reinforced concrete, it was the largest building in the yard when completed. It featured a new process that was able to cool a group of twenty-four cars from near eighty degrees to forty degrees in two hours. It also had a daily ice making capacity of 250 tons.

By 1925, the Roseville plant was icing between seven hundred and eight hundred refrigerator cars a day during the harvest season. The abundance of produce shipped in these cars included vegetables, grapes from

Napa, pears from Oregon, and melons, oranges, peaches, plums, apples, and cherries from the central valley.

In Carlin, Nevada, the Pacific Fruit Express Company blocked a small river and made a fifty-acre artificial lake to hold water that would freeze in the winter. Added to this volume, was water pumped from the Humboldt River. The resulting lake was about thirty-two hundred feet long and sixty feet wide.

After nature provided the cold temperatures and the staff found that the ice was frozen to the desired thickness, from one hundred sixty to two hundred men began the harvest. The first step was to mark the ice in straight lines twenty-two inches apart to a depth of about two inches. This scoring, in the early days of ice harvesting, was accomplished by a horse pulling a device that looked like a cultivator. Next, a power saw was guided along the markings and cut the ice to a depth of about nine or ten inches. The visual effect of this chore was described by H.T. Whyte of the Pacific Fruit Express Company as being "a stream of ice chips spurting out behind the saw for thirty feet, much like a shower of sparks."

Following the cutter was the "Face-Spudder" who had the job of using a heavy two-pronged steel bar to separate the cut pieces so they could be floated along a channel toward the storage building. The ice eventually reached an inclined chain conveyor where the ice was brushed free of debris, caught on wooden bars, and moved upward to the top of the conveyor. From there, the ice was moved toward the opening of various rooms where the chunks were shunted into the room. Gravity helped again at this location when the ice pieces were moved down a chute to the proper location in the room. On a good day's operation, three thousand tons of ice were cut and stored.

In 1927, this facility was the only location where cars containing perishables were re-stocked with ice. The extent of the operation becomes clear when one considers that nearly sixty thousand refrigerator cars were filled with ice each year and each car needed fifteen hundred to two thousand pounds of ice. The eye-catching building on the property was a covered icing platform over one-half mile long with space to ice sixty-seven cars on each side at one time. The adjoining storage houses were large enough to hold fifty-four thousand tons of ice.

The business of harvesting ice from the Sierra began about 1870 and continued past 1925. Various designs of cars came and went as demand for California's produce increased. There was an interesting shift in the railroad's operation when, finally, ice was not part of the cooling process. Instead, electrical generators and refrigerators were the emphasis, and a new labor skill was needed to ensure that peaches grown near Fresno were fresh on arrival in Omaha.

The development of equipment and infrastructure needed to transport ice-cooled produce to eastern cities has a history of about fifty five years.

The six articles on the next pages depict the Southern Pacific's emphasis on the delivery of cooled produce. These articles from *The Bulletin* are:

"Southern Pacific Hauls First Fruit Special," December 1922

"Pacific Fruit Express Take Great Strides in 14 Years," 1921

"The Refrigerator Car Supply – Meeting the Demand When the Crops Are Ready," June 1920

"Handling Western Perishable Products," June 1924

"Ventilating Perishable Freight," February 1928

"Fast and Cold," June 1937

THE BULLETIN

S. P. HAULS FIRST FRUIT SPECIAL IN 1886

Hundreds of people gathered at the Southern Pacific station in Sacramento on August 5, 1886, to see California's first fruit special pull out of the yards bound for eastern markets. "Matt" Rudech was the Engineer of No. 18 on the first stretch of this notable trip and A. K. Prather, veteran Engineer on the Los Angeles Division, who was a Fireman then, is shown standing in the cab. Martin Halloran, then Yardmaster is in shirt sleeves standing by the tender. It is interesting to note the cord wood piled high in the tender for that was in the days of the historical wood burners.

THIRTY-SIX years ago, on June 24, 1886, the first solid train of fruit cars ever shipped from California to eastern markets was pulled out of the Sacramento yards of the Southern Pacific.

The train was chartered by W. R. Strong & Co. and Edwin T. Earl, who were later prominently identified in the fruit industry of California, and was composed of sixteen cars. The cars were not of the present day refrigerator type but were just the ordinary box cars of the type which would now be used for manifest freight.

As shown in the picture above, the event was recognized as a notable one in the annals of California and hundreds of Sacramento residents and visitors were on hand to see the train pull out of the yards. From cowcatcher to caboose the train was decorated with flags and large placards on the sides of the cars broadcasted the news that this was California's first special fruit train.

The picture and information used herewith was furnished the Bulletin by A. K. Prather, Engineer on the Los Angeles Division, who was the Fireman on Engine No. 18, which hauled the train from Sacramento to Truckee. "Matt" Rudech, one of the best known engineers at that time, was the Engineer in charge of the train.

Mr. Prather has remained continuously in the service of the Southern Pacific since August 5, 1884 and says that he is good for several more years at his seat in the cab before the seventy-year limit forces him to accept a pension retirement. He has been on the Los Angeles Division since 1890.

LINING OF LONG TUNNEL IS OVER HALF FINISHED

Work of lining with concrete the long San Fernando Tunnel (Tunnel No. 25) is now over half done. The tunnel, one of the longest on the Company's Pacific System, is 6978 feet long and the entire job calls for an expenditure of well over a million dollars. Work of placing the concrete lining in the Chatsworth Tunnel (Tunnel No. 26) has also started and over 450 feet have been lined with concrete to date. The latter tunnel is 7370 feet in length and this job will entail an expenditure of nearly a million and a half dollars.

The lining of the tunnels with concrete obviates any fire hazard and will do away with the necessity of trains slowing down while passing through them. In order to place the concrete lining the tunnels have to be enlarged, the timbers being renewed, set back and then cemented in. The concrete is reinforced with steel rods.

One of the most interesting features of the work of lining the San Fernando tunnel is that the concrete is mixed outside the tunnel and then blown by compressed air through a six inch wrought iron pipe directly into the forms. As the work progresses the concrete has to be carried a greater distance and recently the concrete has been blown by compressed air a distance of 4000 feet. This method has been used before in engineering operations but this is believed to be the longest distance concrete has ever been "shot" in this manner. The compressed air is obtained from two air compressors run by two 200 horse power motors.

The work is being carried on "under traffic," trains proceeding through the tunnels without delay. George W. Corrigan, Division Engineer for the Los Angeles Division is in direct charge of the work.

A force of about fifty men is at work on Tunnel 25 and, according to Mr. Corrigan, fine team work has been developed on the job, causing an excellent record of accomplishment. A number of good suggestions have been made by various members of the force which have resulted in facilitating the work to a great extent.

BANDITS ATTACK OPERATOR IN RAID ON STATION

Two bandits were unsuccessful in an attempted robbery of the station at Mecca, Los Angeles Division, on October 24th after having felled third-trick operator Geo. H. Reid, with blow on his head with a pistol when he refused to hold up his hands. The men searched the station and went through Mr. Reid's pockets, but did not find anything of value. They evidently operated very hurriedly for they overlooked a considerable sum in the cash drawer.

Mr. Reid's injury was not serious and after attention from the Company doctor he was again at his key. Chief Special Agent O'Connell's office is working with local authorities in attempting to trace the bandits.

Dolly—"They say she spends twice as much money as any other woman for complexion powder."
Polly—"Of course she does. She is two-faced."—N. Y. C. Lines Magazine.

P. F. E. Co. Takes Great Stride in 14 Years

Work of Handling Perishables from Producer to Markets Devolves on Organization Which Developed from Small Beginning

By C. M. SECRIST
Vice-President, Pacific Fruit Express Co.

In the following article Mr. Secrist discusses some of the salient features in connection with history and development of the Pacific Fruit Express Company, owned jointly by the Southern Pacific and Union Pacific companies.

In subsequent issues the operation of icing plants and other branches of this interesting service will be discussed.

EMPLOYEES of the Southern Pacific Company are perhaps, generally speaking, more or less familiar with the functions of the Pacific Fruit Express Company, which furnishes equipment and handles refrigerator traffic from the Pacific Coast, but perhaps a brief recital covering the cause of its origin, its growth, and manner of operation, will be of interest.

As is generally known, the production of fruit and vegetables for shipment to distant markets, started on the Pacific coast years ago in a small way, and it was some time before it reached proportions making it an important factor in the railroad's tonnage.

Prior to 1907, equipment was furnished, as was refrigerator service, by various private companies under contract with the railroad, but the rapid growth of the industry and increasing importance of the traffic justified better general control of the business by the railroad and in 1907 authority was given to organize a railroad controlled company to furnish cars for and handle the perishable business of the Southern Pacific and the Union Pacific System lines. This resulted in organization of the Pacific Fruit Express Company

When the company was organized it had no equipment, but with the first year 6600 new cars were built, icing facilities acquired at strategic points and active operations began in 1908.

At the start the company was obliged to depend almost entirely on commercial concerns for its supply of ice, which is used in large quantities for the refrigeration of perishable commodities in transit, but shortly afterward started erection of its own ice plants. Today the company owns and operates some of the largest and most modern ice plants on the Pacific Coast and at numerous other points on Southern Pacific and Union Pacific System Lines.

It was also necessary to provide car repair shops, for the company does its own building and repair work, and at present operates a number of up-to-date plants for this purpose.

The growth of the Pacific Fruit Express Company is interesting to those who have watched its progress. Starting in 1907 with a manager, one clerk and one stenographer, it has on its payrolls today 2300 employes.

During the first year's operations (1908) total business handled amounted to approximately 40,000 carloads. Last year (1920) it amounted to 142,000 carloads.

The original equipment consisted of 6600 cars. Today it is 19,500.

During the year 1920, the company used the enormous amount of 1,388,409 tons of ice for refrigerating shipments moving under its jurisdiction.

The general offices are located at San Francisco, with superintendent of transportation and traffic manager located at Chicago. Important district offices are maintained at San Francisco, Sacramento, Los Angeles, Portland, Ogden, Omaha and Houston; also agencies are maintained at important receiving points in the East, such as New York, Boston, Philadelphia, Pittsburgh, Baltimore, Buffalo, St. Louis, Kansas City and St. Paul, and agents or inspectors are located at important junction or railroad division points, and at regular icing stations between the Pacific Coast and the Atlantic Seaboard.

Facilities for Tracing

The operation of the Car Service Department of the Pacific Fruit Express Company is one of the most interesting branches of this service. The movement of P. F. E. cars is closely traced by car numbers on all foreign lines in order to prevent delays to or mis-

Upper: Ice manufacturing and storage plant at Roseville, California. This plant has a storage capacity of 45,000 tons and is capable of manufacturing 500 tons daily. Center: Last word in modern refrigerator cars, and one of 4,000 acquired by Company during last year. Lower: Ice storage and manufacturing plant at North Platte, Nebraska.

THE BULLETIN

Fire Losses and Prevention

1920 Figures Show Necessity for Greater Care— Results Will Follow Watchfulness

By H. POLLARD,
General Fire Inspector.

A FEW figures showing the fire loss on the Pacific System for the year 1920 will be of interest to readers of the Bulletin. We had a total of 316 fires, ranging from $5.00 up to $148,000.00 for the largest. Our gross loss was $284,329.48. Insurance collected was $14,568.55. Net loss amounted to $269,760.93, and the premiums on outside insurance were $181,266.93. The fire loss was $20.12 for every mile of railroad operated by the Pacific System and 22.64 cents for every $100.00 of insurance valuation, not including the field fires.

We have four classes of fires which claim the greatest share of the fire loss —field, car, building and bridge and trestle fires. In the tabulation herewith the amount of fire loss by divisions is shown.

stacks of all locomotives and that they are in good order. Care should be taken in removing carbon, so there is no danger of it breaking loose and flying through stack red-hot. Assurance should be had that burners are working in good order. This will save forming carbon and throwing it out when flues are sanded.

Precautions

The 32 fires set by section men were from several causes: Trying to burn when it is too windy, setting more fire than can be properly watched, burning late in the evening and fire not all out before going home. Foremen should see that the burners do not get ahead of the men who are heading the fire, and should quit burning in time to have all fire out before going home,

Division	FIELD		CAR		BUILDING		B. & T.	
	No.	Loss	No.	Loss	No.	Loss	No.	Loss
Western	53	$ 3,542	13	$ 1,625	9	$ 497	3	$3,568
Sacramento	23	5,015	8	11,171	4	10,939	2	260
Salt Lake	1	No value	10	624	8	1,095	1	37
Shasta	14	912	8	21,643	2	971	1	50
Portland	11	378	10	11,785	5	935	4	891
Stockton	28	1,153	4	6,309	3	205	2	230
Coast	18	2,159	9	2,602	2	100	..	....
San Joaquin	6	2,265	7	19,901	2	577	..	....
Los Angeles	2	100	11	13,515	5	1,567	3	1,503
Tucson	..	....	10			1,501	1	435
	156	$15,524	90	$94,094	45	$18,387	17	$6,974

In the tabulation of fires by divisions a total of 308 fires are recorded. The remaining eight are miscellaneous, three being rails and ties, amounting to $1,168.00, one a tunnel amounting to $148,000.00, and four locomotives, amounting to $180.00.

Causes Plentiful

It has been established that there were twenty-six causes for fires, during 1920, namely from adjacent property, cigar or cigarettes, candle, caused by children, carelessness of employees, fusees, friction, gasoline, gas jet, hot box, locomotive, lamp, matches, overflow of oil burner, Pintsch gas, section men burning right of way, stoves and ranges, spontaneous, sparks from brake shoes, short circuit, tramps, unknown, waste against steam pipe, wreck or derailment, and weed burner.

Of the 156 field fires, 98 were set by locomotives; 32 were set by section men; 26 were unknown. Western Division had 35 set by locomotives; Stockton second with 21; Coast third with 12; Shasta and Sacramento ten each. Portland and San Joaquin four each; Los Angeles and Salt Lake one each; Tucson none.

Master Mechanics should see that there are spark arrestors in the smoke-

and should not try to burn when it is too windy.

Car fires are the next to mention —90 car fires amounting to $94,094.40. Agents and others in charge of loading cars should see first that they are properly cleaned. Spontaneous ignition causes many car fires. Great care should be taken in loading merchandise subject to heat from moisture, or being packed close together, or becoming mixed with oil or acid. No shipment should be received that is not properly packed, thereby lessening the risk of packages being broken open and various materials being mixed together.

Friction sometimes causes fire. Merchandise easy to ignite should not be loaded with heavy boxes or packages that are likely to rub together. If in the same car, they should be well separated.

A word about buildings and bridges —keep them clean. Those in charge of such structures should not allow unnecessary trash, brush or grass and weeds near them.

Let us do everything possible to make our fire loss in 1921 much less than last year. We can do it if we all put our minds to it and work to accomplish it.

---

use of equipment, agents and inspectors being utilized for this service in addition to their other duties. This permits of the prompt location of any car loaded or empty wherever it may be.

In addition to the regular supervision given to shipments moving under its supervision, the P. F. E. Co. also handles all passing reports and diversion orders covering perishable freight for the railroad. This service extends from the Pacific coast to the Atlantic seaboard, and contemplates that shippers and receivers of perishable freight will be kept fully informed from day to day of the whereabouts or movement of cars consigned by or to them. This also permits prompt action in the accomplishment of diversion orders due to the very close check by the P. F. E. Co. on the movement of its cars.

In order to keep pace with this rapidly growing business approximately $16,000,000 was spent for new equipment and $4,000,000 for new and enlargement of existing icing facilities by the company during the last year.

HOUSE MOVED ON FLAT CAR SHOWS S. P. SERVICE

Moving a house on a Southern Pacific flat car without disturbing the tenant was an unusual stunt performed the other day at Crockett, Cal. The location of the residence on the bay front made it possible to handle the house otherwise without taking it apart.

J. R. Spurgeon, agent at Crockett, tells of the job as follows:

"With instructions from the Superintendent's office, H. L. Kent, General Yard Master, Port Costa, and I picked out a time right after train 49 leaves here at 12:37 p. m. We had a switch engine hook onto the flat car, upon which the house had been placed by house movers, and it was easily brought up along the bay on a track used to store cars for about 400 yards to a point opposite the station. Here it was handed to local freight 302, the engine pulling the car over the main track and placing it upon our team track near the station. It was unloaded at once. The movement took about 20 minutes, during which time a flagman was out each way to see that nothing ran into us.

"Except for the size of the building, 14 by 24 feet, it was an ordinary switch, arrangement merely being made by me in order to please our customers. Such a little extra exertion, I believe, might mean we would get additional passengers or freight shipments later."

Another dashing young fellow with a speedy flivver attempted to beat a locomotive across the grade crossing. The undertaker was called upon to finish up the job.—Exchange.

"It should be our daily aim to eliminate carelessness ourselves, demand the same from others and use Carefulness as our watchword."

The Refrigerator Car Supply

What the carriers are doing and have done to aid shippers

By C. M. SECRIST, Vice-Pres. and General Manager, Pac. Freight Express

The need for every available refrigerator was fully appreciated early in the season by our owning lines, the Southern Pacific Company and the Union Pacific Railroad, which authorized the purchase of 4000 refrigerators in January, at which time contracts were placed, providing for the delivery at the rate of 108 cars daily, commencing May 1.

United States Railroad Administration instructions in effect at the close of Federal control provided for the diversion of a very considerable number of P. F. E. cars into Southeastern States for the protection of citrus and vegetable movement, also into the States of Wisconsin and Minnesota for protection of the potato movement. These movements were due to the refrigerator car shortage existing in those sections.

On March 1 the Pacific Fruit Express Company cancelled out all United States Railroad Administration orders and issued instructions to all lines to turn its equipment to the Pacific Coast. These instructions were followed up by personal representation and the taking of men from our Chicago office and placing them in all large terminals to see that our instructions were carried out. After about two weeks it was evident that our instructions were without desired effect and the condition was brought to the attention of the executive officers of the Southern Pacific and the Union Pacific Railroad, which resulted in conference with the Interstate Commerce Commission, March 15 and 16.

In attendance at the conference were the following: Messrs. W. D. Tidwell, secretary West Fruit Jobbers' Association; R. F. French, secretary National League of Commerce Merchants; R. G. Phillips, secretary International Apple Growers' Association; F. B. Dow, attorney for the above concerns: W. C. Kendall and G. L. Barnes, representing Commission on Car Service; C. R. Gray, president Union Pacific Railroad; L. J. Spence, director of traffic, Southern Pacific; L. V. Baldwin, vice-president Illinois Central; J. F. Porterfield, superintendent transportation Illinois Central; Mr. Spencer, vice-president Southern Railway, representing Fruit Growers' Express Company; J. H. Young, president Norfolk & Western; A. P. Thom, representing American Association of Railway Executives, and C. B. Strohm, superintendent transportation A. T. & S. F. R. R.

It was recognized at this conference that the production of fruit and vegetables during the war period had largely increased and that their mar-

Shippers of perishable products along the lines of the Southern Pacific are being regularly informed as to the general efforts being made by carriers to furnish an adequate supply of refrigerator equipment, and the results accomplished, as well as the changing conditions on Eastern lines, a service that is admittedly of very much assistance and is greatly appreciated.

Agents and others whose duties bring them in contact with the general public will find this article by Mr. Secrist, giving a general outline of the situation, helpful in enabling them to properly present the facts when opportunity arises.— THE EDITOR.

keting was entirely dependent upon transportation to obtain proper distribution, also that the means of transportation—refrigerator cars—had not been added to in proportion to the increased production. Further, inasmuch as the shortage of refrigerator cars was then apparent and would continue throughout the present year, it was felt that there should be the fullest co-operation between representatives of the railroad-owned refrigeration equipment and those of private car lines, who would assist in meeting emergencies by authorized use of their cars when this could be done without detriment to the interests for which they were provided and that in return for this consideration there should be reciprocity.

To this end and to insure maximum use of refrigerator cars it was felt that the American Railway Association should establish, at Chicago, a bureau of the Commission on Car Service, under the direction of Mr. W. L. Barnes, commissioner, who during Federal control had supervision of refrigerator cars and tank cars.

On April 5, Mr. W. L. Barnes opened his office at Chicago.

During the interim between the conference and the opening of Mr. Barnes' office, due to a very active campaign being conducted by executives of Southern Pacific and Union Pacific, as well as Pacific Fruit Express Company, considerable improvement in the movement of refrigerator cars was effected, but the switchmen's strike, occurring the first week in April, retarded further progress in this direction and tended to materially impair the service that normally would have resulted from the appointment of Mr. Barnes.

This, of course, was quickly apparent to the railroads serving this section and the campaign was continued

by executives and general officers of the Southern Pacific Company and the Union Pacific System, as well as by the Pacific Fruit Express Company, by making known to all Eastern lines the disastrous condition that would result here if refrigerator cars were not promptly turned in this direction. The Pacific Fruit Express Company made heavy additions to its forces on Eastern lines, checking the detail movement of its cars and assisting in every way possible to effect prompt release and return of its equipment.

Perishable shippers here, as well as all receivers in the East, were kept informed of the situation and they also opened a campaign with the Eastern railroads, the Interstate Commerce Commission and their respective Senators and Congressmen, urging upon all the absolute necessity of prompt handling of refrigerator cars to avoid ruin to producers.

Through the efforts of our railroad executives and the Pacific Fruit Express Company, Mr. Casey, who succeeded Mr. Barnes about May 10, has ordered to California all surplus railroad owned refrigerators, which cars are now beginning to reach our lines at the rate of forty to fifty cars daily.

The out-turn of our new cars has been seriously affected by the switchmen's strike, our latest advice being that delivery will not commence until the second week in June. Executives of the owning companies, as well as the Pacific Fruit Express Company, are personally active in trying to get the car builders the necessary material for the construction of these cars, and everything possible is being done to secure the relief that would be afforded by such cars.

Formal meetings were held with representative shippers at Los Angeles, May 5; El Centro, May 6; Sacramento, May 10, at which the general situation was reviewed and it was uniformly agreed that nothing possible was being left undone by carriers here to meet the needs of this section. It was agreed further that shippers and carriers alike should continue their respective campaigns.

HERE'S ONE FOR THE BOOK.

Assistant General Passenger Agent F. C. Lathrop has heard of lots of ways of interrupting passenger service, but he ran across a brand new one the other day. Something happened to a semaphore signal near Folsom, Cal., on our Placerville branch. Trains approaching in the evening could not distinguish the familiar light. Signal Maintainer W. J. Berriesford rushed to investigate. He found that a swarm of honey bees had elected to convene over the face of the semaphore glass, shutting out the light completely. The bees were given a free pass elsewhere. Traffic proceeded.

July, 1924   THE BULLETIN   5

How Western Perishable Products are Handled

Methods Used to Transport Fruits and Vegetables to Distant Markets Illustrates Efficiency of American Railroad Transportation

By C. J. McDONALD
Superintendent Perishable Freight Service

The following is taken from an address delivered before the Pacific Railway Club at San Francisco, June 12.

NOTHING gives better evidence of the advances made in American railroading than the efficiency shown in transporting perishable commodities. We may be inclined to take this great activity as a matter of course, just as the public takes railroading in general as a matter of course, believing that since the railroads were here yesterday and are here today, they will be here tomorrow no matter what is done to them or whether they make any money or not. But when we stop to think of the way the railroads carry the products of orchard and field to markets thousands of miles away, and of the tremendous organization involved, we can regard the whole activity as a remarkable achievement.

Within the last twenty years the greatest strides have been taken in developing methods of handling this traffic so that all perishable commodities might be carried to destination without loss. This development in efficiency has not only improved the methods of handling, but has also resulted in reducing the cost of transportation with consequent benefit to the shipper.

First Shipments East

Fresh fruit shipments were first sent East from California in 1869, totaling in that year 35 tons, and it is said that more fruit was grown in California that year than could be consumed. In 1884 only 6,000 tons of fresh fruit were shipped East. Then the markets east of Chicago were opened to the California shipper and in 1886 Boston received its first car of fruit direct from California.

In the late eighties or early nineties it was realized that special equipment was needed for the transportation of fruit. Ventilator cars were built to handle this traffic. A few years later the railroads were asked to build refrigerator cars so that time in transit might not be such a vital factor. Refrigerator cars were furnished, but the fruit growers claimed they were unsatisfactory and recommended that they be taken out of service. However, from time to time the cars were improved, so that in 1907 the present type of refrigerator car came into being.

As compared with the meager shipments of the eighties, it is interesting to note that 218.526 cars of grapes alone have been shipped out of California within the last eight years. In 1923 the Southern Pacific Company handled more than 176,000 cars of perishable freight. This means an average for every day in the year of more than 480 cars.

It is difficult to visualize the enormous volume of perishable freight that is represented by 176,000 cars. This number of cars would require 3,520 trains of 50 cars each. These trains would cover 1,753 miles of railroad, with no spacing between trains. Try to think of one solid train of perishable freight, the caboose at San Francisco, occupying the entire main track down the Coast route to Los Angeles and that same train likewise occupying the entire main track from Los Angeles to Ogden via Fresno and Roseville. And then the engine and head end of this train would be 146 miles East of Ogden. The Southern Pacific Company handles more perishable freight than any railroad in the world.

With the development of the refrigerator car and the growth of perishable traffic, it was seen that this business was of such a specialized nature that special organizations were needed to handle it. A number of companies were organized to provide the refrigerator cars and icing facilities.

In 1907 the Pacific Fruit Express Company, owned jointly by the Southern Pacific and Union Pacific, was organized by those railroads as the medium through which to provide refrigerator cars and look after the icing of fruit shipments originating on those roads. When the company was organized it had no equipment, but 6,600 cars were built the first year, icing facilities were acquired and active operations began in 1908. At first the company had to depend almost entirely on commercial concerns for ice, but soon started to build ice plants. In 1907 the company was composed of a manager, one clerk, and one stenographer. Today it is the largest organization of its kind in the world. The general offices are at San Francisco, with an eastern office at Chicago, important district offices throughout

the West, and agencies at principal receiving points in the East.

As compared with the 6,600 cars built shortly after its organization, the company will have this season more than 33,000 cars. Of these, nearly 16,000 have been built since the end of Federal control, March 1, 1920, an increase in equipment of more than 100% in four years. Under a recent agreement, Pacific Fruit Express Company acquired, by lease, the right to operate with its own, all Western Pacific refrigerator cars, under conditions that make for equitable treatment to the shippers of perishable products on Southern Pacific, Union Pacific, and Western Pacific lines. This is intended to increase efficiency in the handling of western perishable shipments because the difficulty of car distribution in times of heavy shipments is decreased when the number of agencies making such distribution is lessened.

In addition to providing cars, the P. F. E. also handles, for the railroads, all passing reports and diversion orders, this service extending from the Pacific Coast to the Atlantic Seaboard. Very close check is kept on the movement of cars, which permits prompt action in accomplishment of diversions. These diversion privileges, and the prompt execution of them, has been a very important factor in the marketing of perishable crops, as western growers, by keeping an attentive eye on market conditions, have been able to divert their shipments from flooded markets where prices were low, to other markets where there was a demand and better prices. This ability of shippers by rail to divert their cars at any time is one reason why the great bulk of perishable products of the west must always be marketed by rail.

Garden of the U. S.

The Pacific Coast, and especially California, has been rightly called the orchard and garden of the United States. The bulk of the crops from this section has to be carried great distances in order to find a market. About 70 per cent of California's products are marketed East of Chicago, the average haul being about 2,600 miles. So we can see that the fruit and vegetable industry is very important to the railroads, and likewise the fruit and vegetable producers are very dependent on the rail carriers.

Each variety of products has its own particular season which, as a rule, is of limited duration. In the case of grapes, for instance, the great bulk of the crop has to be moved in a period of ninety days, and the heaviest shipments come within a much shorter period. The difficulty is greatly increased when the various crop periods overlap. The months of greatest strain come from August to November when perishable commodities in a great flood must be cared for. We have the same situation as exists during the peak hour in the street car service of great cities such as New York and Chicago when everyone wants to go somewhere at once and crowding and congestion unavoidably results.

In the handling of perishable commodities, with the greatest load coming in a brief period, the furnishing of cars is, indeed, a grave problem. As it is now, a great number of refrigerator cars are idle a large part of the year. The investment in these cars is enormous, for each refriegrator car costs around $3,500.

60,000 Cars This Season

This season the Pacific Fruit Express Company will have more than 33,000 cars in service and the Santa Fe nearly 16,000. Each year about 10,000 foreign line cars are sent to California to tide over the peak movement, making a total of nearly 60,000 refrigerator cars availabe to take care of California shipments.

If enough refrigerator cars were provided to maintain a surplus of cars at all times, the investment in such equipment would be so great that the rates necessary to furnish any return on the investment would be too burdensome to allow the products to move, and if this investment brought no return, the roads would quickly be reduced to such financial straits that the service would break down.

Think of the millions of dollars these refrigerator cars represent and of the millions more that have been expended in other facilities to handle the ever-swelling tonnage of fruits and vegetables from this region.

This reference to the staggering sums that have been invested in order that western perishable products may be marketed successfully brings the observation that if the roads are to continue to provide the facilities needed to keep pace with the constantly increasing production, they must be allowed to receive a fair wage for providing this service. Cheap freight rates won't pay for refrigerator cars and icing plants. What the producer saves in freight rates will be swallowed up in his losses through inability to get his grapes, his apricots, his oranges or his cantaloupes to the markets across the continent.

The only way the heavy perishable business can be handled with the cars available is by an endless chain system whereby refrigerator cars, as soon as unloaded in the East, are rushed back to be reloaded and sent on their way East again. Much difficulty has been encountered in the past in getting the cars back from the East, through failure of consignee to unload cars promptly, some of them preferring to pay demurrage and use the cars for storage purposes until such time as they can sell their contents to advantage. This has been an important factor in car shortage and it seems to me that a thorough investigation by shippers as to the possibility of improving the distributing system in the East might bring worth-while results. Each year at the peak of the perishable season we have men in the East trying to speed up the release and return of the cars, but it is a discouraging experience to try to get the cars unloaded and started home again.

One difficult feature about the perishable business is the necessity of hauling from 60 to 65 per cent, and sometimes more, of the cars used in this service to the point of production as empties. The situation is aggravated by the inability of the roads, under existing governmental regulation, to make such rates as will enable them to compete with water rates through the Panama Canal and so get some business which would help fill the empty refrigerator cars coming West.

In addition to the car supply, the icing facilities of the roads have also been greatly increased in recent years. In California alone the Pacific Fruit Express Company has an annual production at its own plants of about 425,-000 tons, all of which is used for refrigeration of perishables. Icing facilities have been so perfected that it is possible to re-ice a train of refrigerator cars in one hour, which is slightly less than one car a minute.

World's Largest Ice Plant

The largest ice plant in the world from the standpoint of daily ice production is the railroad icing plant at Blue Island, Illinois, which is capable of producing 1,000 tons of ice daily. Within the year, this plant will be exceeded in size by the Pacific Fruit Express Company's plant at Roseville. which will have a daily production of 1,200 tons. From the standpoint of annual production, the Roseville plant is now the largest. The production at the Roseville plant was 131,570 tons in 1923 and 200,000 tons is the mark set for 1924. The Pacific Fruit Express Company used in 1923, 1,900,000 tons of ice.

The transportation of perishable freight is on a thoroughly scientific basis. One of the important phases of the work is that of preparing for a crop movement. We must know beforehand what tonnage is to move from each section so that there can be a proper distribution of cars. Surveys are made before the movement starts, and information from every available source is obtained. Government, state, county, and district authorities are consulted and figures are obtained from shippers' organizations and individual shippers and growers. This information is supplemented by our own observations. Figures on tonnage to move are kept up to date throughout the season and every effort is made to distribute cars impartially.

Co-operative Meetings

Much credit for the continual improvement in the handling of perishable traffic must, however, be given to the increased cooperation between railroads and shippers. In January, 1921, and again in 1923, meetings of shippers and railroad executives from the West to the East Coast were held in Chicago and at these meetings a number of common problems were brought up and harmoniously worked out. In this way the present transcontinental schedules were arranged and have given satisfaction generally since that time. Every year prior to crop movement, meetings of shippers and railroad men are held in the producing districts and here questions of service are brought up and settled after full discussion by all parties concerned. Definite hours are agreed upon for the placing of cars at various stations and the hours such cars will be picked up when loaded, and moved to concentration points where they are assembled into blocks for movement East. These conferences and the railroads' efforts to give the best possible service have brought good results.

I have told you that the California railroads cooperate with the fruit interests at every opportunity. Let me give you just one instance. New York City is the largest market for California fruits and nearly all of our fruits are sold there by auction. For a num-

This group of employees at the San Jose Freight Station are setting good marks in the efficient and careful handling of freight. The picture was taken by Clerk F. M. Beckett following a station OS&D meeting on May 15 when the subject of preventing loss and damage to freight and other station matters were discussed. The meeting was conducted by Freight Agent H. A. Stiver.

ber of years the city piers used for auctions have been under lease to the Erie Railroad. The lease was to expire in May, 1923, and prior to that time it seemed apparent that the Erie Railroad preferred not to continue their lease on these piers. At the time the outlook for marketing California fruits in New York City was very gloomy and some action was necessary to keep that market open. In response to an appeal from the fruit interests, the Southern Pacific Company offered to take over New York City piers 20 and 21 and operate them for fruit auction purposes, open to all railroads entering New York, on the basis of actual cost with no profit. We did not lease the piers because the Erie Railroad officers apparently had a change of heart and exerted their influence with Port authorities to keep the piers, with the result that New York City auction market continued to function for California fruits. However, the fact that a California railroad, handling this fruit but 800 miles, would go more than 3,000 miles away from California and offer to lease two very expensive piers in New York City to keep open a market for California fruits is evidence of our desire to foster that industry.

While the handling of perishables has been developed to a high stage of efficiency, there are two great economic wastes involved in this business. One of these is a problem of the shippers and growers and the other a problem of the railroads. The shippers' problem is to devise some means of packing fruit that will do away with the enormous waste of lumber and the expense of boxes, and the loss of space in cars through the use of this material. The railroads' problem is to devise a refrigerator car that will transport and refrigerate its full cubical capacity instead of only a part of it. These problems are being actively attacked by all concerned and I anticipate that, in time, they will be solved, just as other probems encountered in the development of our transportation system have been overcome.

PLAIN TALK ON TAXATION IN MELLON'S BOOK

Copies of a book, "Taxation, The People's Business," by Mr. Andrew Mellon, Secretary of the United States Treasury, have been placed in the library at the General Office. McMillian and Company, the publishers of the volume, have also sent copies to public libraries so that this book is generally available.

The book is a very clear discussion of the tax question, dealing with the problem from an economic and not a political point of view.

Grabbing a Bargain
An ancient car chugged painfully up to the gates of the races. The gatekeeper, demanding the usual fee for automobiles called, "A dollar for the car!"

The owner looked up with a pathetic smile of relief. "Sold," he said. —McKendree Review.

S. P. Finds New Pets

The little daughters of section foreman Griffith find their new pets, given them by the Southern Pacific, pleasing substitutes for the ones killed by a passing train.

TWO little sisters are happily engaged in gaining the complete confidence of four fluffy kittens presented them as pets by officials of a large railroad system who are supposed to think only of traffic problems and kindred subjects.

An approving mother cat usually may be seen watching the antics of kittens and children—characters in a real life story that has developed as happy an ending as that of any fairy book ever read by the little girls.

The children of this story are Mary Griffith, aged 11, and her 5 year old sister, the daughters of the section foreman at Bloomington, near Colton, Calif.

Recently Superintendent W. H. Whalen of the Los Angeles Division, received a letter painstakingly written with pencil in a childish hand. It read as follows:

"About two months ago I had a pretty angora kitty which got killed by a train. My sister had a plain yellow cat but a good cat, and it got killed last week by No. 109. My sister cried, and I am going to ask you to ask the man who pays for things the trains kill if he will send us a kitty. My dad is section foreman of Bloomington. I live in the section house. I am going to plant flowers in the park. I am 11 years old and my sister is only 5 years. I thank you."

The letter was signed by Mary. It was turned over to S. C. McClung of the claims department, who promptly wrote to the little girl. He pointed out that "it is a pretty hard proposition, as your father will tell you, to fence against cats," but added that "however, I am looking around for a

nice kitten and hope to be able to send it to you and your sister this week."

There was some difficulty in locating kittens deemed suitable as replacements for the original pets, but through the co-operation of J. E. Newman, claims attorney at San Francisco, Mr. McClung and E. T. McNeill, agent at Colton, a mother cat and her four little kittens were obtained. McNeill and his wife went to Bloomington to formally present the new kittens.

"We are so happy to have the new pets," said Mary. "And we are trying to make them know they must not play on the tracks where the trains go by."

S. D. & A. BUSINESS GETTERS SHOW REAL RESULTS

Co-operative efforts of employes on the San Diego & Arizona Railway in business solicitation resulted in San Diego's first solid "Golden State Limited" through Chicago train to leave that city May 22 with a capacity load of "back east" excursionists

Early in May S. D. & A. employes held a meeting in the San Diego machine shop for the purpose of stimulating business getting. Ernest Ebersole, of the master mechanic's office, presided at the meeting. General railroad matters, as well as business getting, were discussed with the idea of bringing out all possible suggestions for maintaining efficient service. Master Mechanic Thomas F. O'Connell had the machine shop decorated and arranged an interesting program for the occasion.

Among those called on for short talks were: F. B. Dorsey, traffic manager; Dave Sutor, car foreman; Robert Rennie, blacksmith shop; Glenn Crawford, warehouse: Al Carney, conductor; Dr. Harry Wegeforth, company physician; S. A. Lamey, yardmaster; A. J. Mello, T. D. Hood, Charles McLachlan, H. G. Burke, E. Christiansen, A. D. Hagaman, C. A. Vincent, R. J. Schussler, Joe Hild, and William Barker.

NEW NAMES ARE ADDED TO SAFETY COMMITTEES

During May the following changes were made in safety committeemen account terms expired or leaving the service:

Western Division: J. Prendiville, Section Foreman, vice John Kenny; A. Dobbie, Carpenter, vice R. Rouse.

East Bay Electric Division: H. Dieckman, Inspector, vice J. H. Johnson.

Portland Division: N. Rose, Supervisor B&B, vice O. V. Chesney; R. Stein, Fireman, vice R. N. Lowry.

Surely a Celtic Star
A man was walking down the street one evening with a friend, and pausing to look up at the starry sky, remarked with enthusiasm:

"How bright Orion is tonight!"

"So that is O'Ryan, is it?" replied Pat. "Well, thank goodness, there's one Irishman in heaven, anyhow'"—Exchange.

Before the railroad in Sacramento, ice was shipped to central California from Alaska by boat. Once the railroad provided a means to deliver ice from the Sierra and western Nevada, the boat owners were faced with competition they could not match. For a long time thereafter, business was good for suppliers of ice from the mountains. But, technology progressed, and with the advent of electricity and refrigeration, providers of ice for railroad cars were faced with too much competition. Today, it is easy to find a similar pattern in retail. Before the Internet, buyers went to stores, but now that the Internet provides the opportunity to buy from home on a computer, brick and mortar retailers are facing potentially unmatchable competition. Railroads changed the culture then, much as the Internet is changing ours here in the twenty-first century. - Courtesy California State Railroad Museum

How to Ventilate Perishable Freight

By R. G. FAGAN
Superintendent of Freight Protection

AS an aid to the shipper of fruits, melons and vegetables in placing his consignments on the market in sound condition, the governing tariffs provide for accessorial or "protective" service such as heat, refrigeration and ventilation to retard or hasten, as may be desired, the degree of ripeness.

Each class of protective service presents its difficulties from an operating standpoint, aggravated somewhat from lack of time due to rapidity with which perishable freight must be transported. Only a short time is allowed at terminals, icing stations and inspection points for the numerous operations which must be performed and so it becomes important that each function be timed that they may be performed in sequence. Compliance with the tariff definition of "Standard Ventilation" and maintenance of an affirmative record of the performance are problems which have been very hard to solve without creating an undue operating expense.

Adjust the "Vents"

When cars are billed "vents open" or "vents closed" to destination, a compliance with instructions is comparatively an easy matter. The process is to adjust ventilators to the proper position at shipping point, seal them accordingly, and make record based on inspection at designated inspection stations.

Definition of "Standard Ventilation" as provided by tariff is: "Standard ventilation means that on all perishable freight, except pineapples, tomatoes and sweet potatoes, all ventilating devices will be closed as soon as practicable after the outside temperature falls to 32 degrees above zero, and all ventilating devices will be opened as soon as practicable after the outside temperature rises above 32 degrees above zero." The same rule applies to pineapples, tomatoes and sweet potatoes except that the degree of temperature at which ventilating devices will be regulated is 40 instead of 32.

Watch Temperature

To comply with the rule, the party in charge of the car, that is the yard clerk, agent or conductor, whether it be moving or at rest—night or day—must know what the prevailing temperature is. For that reason producing stations are supplied with thermometers and record is made showing position of ventilators and prevailing temperature. Designated inspection stations are provided with self registering thermographs, which, being delicate and expensive instruments, must be located, cared for and adjusted as required by chief engineer's instructions of February 13, 1925. Weekly charts are removed from thermograph and sent to chief engineer, who has blue print copy made and furnished Freight Protection Department so that temperature records may be furnished for claim purpose without the necessity of directing inquiry to inspection stations. Position of ventilators is determined by inspection, adjustment made if necessary, and record posted.

Conductor of each train containing one or more cars of perishable freight under ventilation is furnished a form containing a list of such cars with ventilation instructions. Cabooses in perishable freight service are equipped with thermometers. Position of ventilators is checked by conductor and record of changes when required and prevailing temperature is recorded on the form, which is forwarded by the terminal agent to Freight Protection Department for check and record purposes in event of claim, thus facilitating claim investigations and eliminating necessity of writing conductors for record.

In view of the carriers' obligation to afford the service provided by the contract, which is the bill of lading, and the severe claim penalties which result from failures, the very best care and judgment must be and has been used in handling perishable freight under ventilation. The record which the Southern Pacific Co., Pacific Lines, has made is evidence of that fact. During the year 1927, investigation of claims on perishable freight under ventilation required use of 13,799 records. 13,741, or 99.58 per cent, proved that the ventilators had been regulated as required. Incidentally, the small percentage of error, 0.42 per cent, resulted in claim penalties of $5,926.37.

B. & B. Man's Urging Gets Some Nice Passenger Business

Fred S. Hawley, of Bridge and Building Gang No. 14 on Coast Division, has been getting some nice passenger business for the Company, unassisted and unaided by traffic solicitors. Some friends of his living in Peoria, Ill., have been accustomed to making trips to California over a competing line. Hawley finally convinced them that Southern Pacific had the best service to offer. One trial trip was sufficient, and now they intend to make their usual periodical trips over Southern Pacific Lines.

Only One Train in 188 Late on Pacific Lines in Day

An "on time" record to be proud of was performed one day recently when there was only one train late in arriving at destination out of the 48 through and 140 local passenger trains operating on Pacific Lines that day. A local train, No. 79, lost twenty minutes on the Stockton and Western divisions because of unavoidable delays.

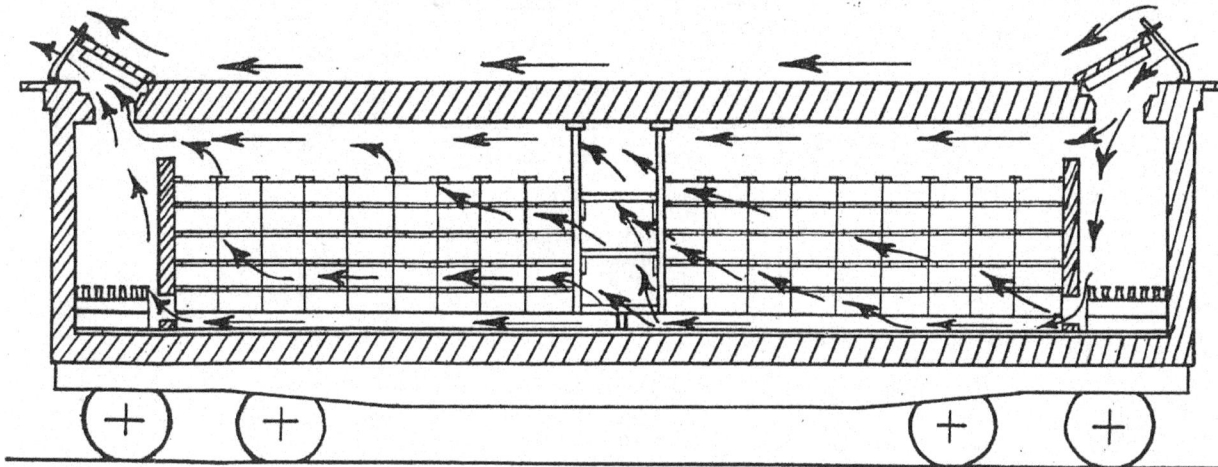

Refrigerator car loaded with fruit ventilators open permitting air to circulate through car

More than 5000 freight trains were required to move the Pacific Coast's fruit and vegetable crops to the nation's markets last year. Of course, only a few of the trains comprised 99 cars and 5 locomotives as does this one shown winding around the famous "horseshoe curve" near San Luis Obispo on the Coast Division. The picture was taken by the Aston Photo Shop of San Luis Obispo.

FAST & COLD: *Service of S. P. and P. F. E. Carries Fruits and Vegetables of West to Nation's Dining Tables*

IT is nearing midnight at Salinas, California, "lettuce capital" of the world!

The last crate of firm, crisp "Iceberg" lettuce for this day's pack has been shunted along the conveyor into a Pacific Fruit Express "reefer."

As workers scurry from the expansive packing house — one of 40 in the Salinas district—a Southern Pacific switch crew stands by ready to couple onto the car as soon as "top icing" has been blown over the crates and the car door is sealed.

Then the car is switched over to the "yard" to be assembled into a solid block along with the cleanup of cars from other packing sheds, for this will be the last lettuce train of the day—a day that has seen more than 300 cars loaded and sent to markets of the nation.

From Salinas the train is moved 18 miles to Watsonville Junction, rail concentration point for this lettuce growing region. The Watsonville area itself is second only to Salinas territory in the production of lettuce, and has 16 packing houses in active production. At the Junction cars gathered from all sections are quickly made up into full trains for the run to San Jose and Roseville—first lap of the long transcontinental journey to eastern cities.

Ten days later Mrs. Housewife, going about her shopping in a New York City market, will very likely exclaim:

"My, how lovely and firm the lettuce looks!"

The salesman, nodding graciously, will probably add: "Yes, madam, it arrived just this morning from the Pacific Coast."

Rail Delivery Boy
From the Pacific Coast!

Chances are neither customer nor salesman give a thought to the astonishing fact that this highly perishable lettuce has been shipped by freight train nearly 3,000 miles; that only a little over a week before it was still growing in a California or Arizona field; that in its movement across the continent to reach distant markets in perfect edible condition the railroads have performed what is perhaps their greatest transportation achievement—an achievement that has become a vital factor in the nation's prosperity and has contributed much to the welfare of its people.

Mrs. Housewife might then make the

Cover Illustration

This picture of Lee Bemiro icing the bunker of a Pacific Fruit Express "reefer" at the Watsonville Junction plant of the Union Ice Company, is typical of night and day operations at many points on Pacific Lines as this season's vegetable and fruit crops move to markets. At this time of the year, Watsonville and Salinas are key points in the handling of heavy lettuce shipments from California.

same comment about the cantaloupes, the cherries, or other fruits and vegetables she sees in the market; and the salesman makes the same response. For during this time of the year eastern markets are bountifully supplied with early produce from sunny, fertile valleys of the western states—made possible by a highly specialized and carefully organized phase of railroad operation that has long since ceased to be a source of wonderment to the average person. It is taken for granted, just as is the mystery of electricity and its many human blessings.

Winter and summer; during blizzards or sweltering weather, Mr. and Mrs. Public in the far corners of America expect their markets to display all fruits and vegetables in season. It gives them little concern where their favorite delicacies are grown, or how the products of fields and orchards reach their home town. They simply demand, and they are rarely disappointed; for the railroads are on the job to deliver.

The story behind these perishable fruit and vegetable deliveries is not news. It has been told many times during the years the famed Pacific Coast industry has developed and flourished. But it is a story that never grows old, and is one that railroad people should keep constantly in mind; for the entire operation from coast to coast is a striking example of the cooperation existing between the nation's rail carriers in serving the public, and repre-

SOUTHERN PACIFIC BULLETIN—Volume 21. Number 6—Published monthly by Southern Pacific Company, Bureau of News, in the interest of, and distributed free to, active and retired employes. Signed contributions of news items regarding employes or activities of the railroad are invited, and should reach the editor before the 15th of the month. Poetry can not be used due to small size of magazine. ERLE HEATH, editor. EMMETT G. FITZPATRICK, assistant editor. 65 Market Street, San Francisco, California.

sents an accomplishment in which all railroaders may take great pride.

Just how important is this perishable freight business to Southern Pacific and to employe payrolls? Take a look at these figures:

Money taken in last year for all classes of freight moved on Pacific Lines amounted to $122,087,867. Of this total gross revenue, $28,538,521, or more than 23%, came from the fresh fruit and vegetable business originating in the Pacific Coast states.

When it is remembered that about 80% of all the money the Company took in last year came from the freight it hauled, the importance of perishable traffic can be readily understood.

In no section of the country are more fruits and vegetables grown than in the western states of California, Oregon and Arizona. As a consequence more perishable shipments originate on the Pacific Lines of Southern Pacific than on the lines of any other railroad.

Nation's Garden Spot

Last year lettuce shipments from these three states amounted to 95% of the nation's total; of the grapes, 98%; 57% of the citrus fruits, including all the lemons; 79% of the cantaloupes and melons of that type; 26% of all the vegetables, including most of the asparagus, carrots, cauliflower, and about half of the celery and peas; and 25% of the principal varieties of deciduous fruits, including more than half the pears and cherries, and 70% of the shipments of fresh prunes and plums.

Truly, this region is the greatest source of supply for the "salad bowls" of the nation; and truly, a great responsibility rests with the railroads in getting these perishable products to markets in condition for use in these same salad bowls.

Sharing this responsibility with Southern Pacific, is the Pacific Fruit

Express Company—a company organized by the S. P. and Union Pacific in 1906 during the infancy of the Pacific Coast fruit and vegetable industry, and which company is today the largest operator of refrigerator freight cars in the world.

Growth of the P. F. E.

When Pacific Fruit Express began operations in October 1907, it owned 6,600 cars and during its first year handled 48,902 carloads of business. Today it has 39,573 freight and 298 express cars in refrigerator service and last year originated 339,336 carloads of perishable commodities for its contracting railroads—Southern Pacific, Union Pacific and Western Pacific. Its record year was 1930 with 378,422 carloads originated. This was during the Prohibition era when great quantities of grapes were shipped from California.

Starting with a manager, clerk and stenographer, Pacific Fruit Express now employs more than 3,000 persons during peak seasons. Its general office is in San Francisco, with eastern headquarters in Chicago. Also there are important offices in many other cities, and agencies at principal receiving points in the East and in loading territories of the West.

Pacific Fruit Express is owned jointly by Southern Pacific and Union Pacific, on whose lines it furnishes refrigerator cars for perishable loadings and provides the protective service—refrigeration or heat—for such shipments while enroute to markets. For the construction, rebuilding or repair of cars, P.F.E. has the most modern shops of their kind in the country, located at Roseville, Los Angeles, Colton, Tucson, Nampa and Pocatello, Idaho, and near New Orleans.

In its early years P.F.E. had to depend almost entirely on commercial ice companies for its ice supply. It now owns 18 ice manufacturing plants with combined production capacity of 5,560

Citrus

Lettuce

Grapes

Potatoes

Major perishable shipments over Pacific Lines during 1936 ranked in the following order: Lettuce, 47,000 cars; Citrus, 42,000 cars; Deciduous tree fruit, 25,000 cars; Grapes, 16,500 cars; Cantaloupes and kindred melons, 14,600 cars; Potatoes, 7000 cars.

Deciduous

Melons

First special train of fruit ever shipped from California was given a rousing sendoff when it pulled out of Sacramento June 24, 1886. The train of 16 cars was chartered by W. R. Strong & Co. and Edwin T. Earl. Single car shipments of fruit had gone east from California before that date, but never a full train. There were no refrigerator cars in those days. Several blocks of ice were piled in each end of the car, but when this had melted the shippers trusted to luck that the fruit would reach a market before it spoiled.

tons daily, and combined storage capacity of 330,900 tons. Largest of these plants is at Roseville, which is also the largest of its kind in the world, having daily manufacturing capacity of 1,300 tons, and storage capacity of 53,000 tons.

Harvesting Ice

P.F.E. also has three natural ice plants at Carlin, Nev., Laramie, Wyo., and North Powder, Ore., where (during the winter months) ice is harvested from artificial lakes adjacent to these plants and stored for summer use. These three plants have combined storage capacity of 141,800 tons.

In addition to the manufacturing and natural plants to supply its cars with ice, the P.F.E. owns 45 icing platforms where ice is purchased from commercial ice companies or shipped from some nearby P.F.E. plant. Combined length of these platforms and those adjacent to P.F.E. plants is 97,394 feet, or about 18.5 miles, and would accommodate 3,758 cars at one spot. Most of these platforms are of the island type permitting icing of cars from both sides at the same time. Car icing is also done at about 190 other points on S.P. and U. P. lines.

Quantity of ice used is not now as tremendous as it was a few years ago when practically all perishables moved eastward under standard refrigeration (icing at every regular icing station enroute). In one year (1929) the bunkers on the P.F.E. "reefers" consumed 2,283,294 tons of ice. In recent years icing services have been modified in various manners, and the ice consumed by P.F.E. last year was 1,229,965 tons.

The perishable produce business offers a year around job for the railroad and the P.F.E. Some crop of major proportions is ready for shipment during every month (as has been described in recent issues of the *Bulletin*), and there is not a day in the year but that at some point on Pacific Lines carloads of fruits or vegetables may be seen on their way to eastern markets.

To cope with this situation and to insure ample cars being on hand at all times to handle the crops, regardless of how fast the produce may ripen or what other emergencies may arise, S.P. and P.F.E. observers are constantly in the field making surveys with shippers, government and association experts to determine when the crop will start to move, when the peak will be reached, how long shipping will continue, and how the supply of cars must be spaced. Sudden changes in weather or market conditions often disrupt the most carefully made plans.

Last year 210,928 carloads of perishables were moved over Pacific Lines. Fresh vegetable shipments were slightly greater than in 1930, the previous record year. This was principally due to increased production of lettuce. The citrus movement was under that of 1935 when record crops were harvested. Deciduous shipments remained much smaller than a few years back when the parade of grape trains to the east was a high spot in the perishable freight business.

Of these total carload shipments, nearly 170,000 cars were handled in some 5,100 solid fruit and vegetable trains which moved north and eastbound to the nation's markets from the five Pacific Coast rail concentration points at Roseville, Colton, Yuma, Tucson and Medford.

Fast and On Time

All these trains move on fast schedules—schedules that are kept just as religiously as are those of the extra fare streamliners.

Principal perishable-shipping areas originating traffic for movement through the California concentration points are zoned. Perishables that are loaded in these zone areas by established closing hours will be provided first-day departure from the Colton or Roseville concentration points, and will be handled on fast schedule to give 7th morning arrival at Chicago and 9th evening arrival at New York from the loading stations.

Of the nearly 5000 trains moved over Pacific Lines in 1935, more than 97% reached their destinations on time or ahead of schedule!

If the business of the railroad and the Pacific Fruit Express was simply to take these carloads of perishables from point of origin and deliver them direct to a given point of destination upon a given track, the transportation problem would not be so great.

Not only do these crates of vegetables and lugs of fruit require fast rides and

Harvesting ice on the Pacific Fruit Express Company's artificial lake at Carlin, Nevada. The ice is stored for use during the summer. P.F.E. also harvests ice in this manner at Laramie, Wyo., and North Powder, Ore.

constant protection against hot or cold weather, but they are quite often loaded into cars and started on their way without the shipper knowing for sure where they are going or to whom they will ultimately be consigned. In fact, the majority of them are on their way to market before they are actually sold.

How Diversion Handled

For instance, during the Imperial Valley cantaloupe "deal," a local shipper or one of the small "army" of buyers on the ground representing national produce distributors, routes a car of melons out of El Centro. The car is most likely 'consigned to the shipper himself and is billed to, say, El Paso. Before the car has gone very far conditions change in the eastern markets, possibly due to weather conditions affecting some competitive production area. The shipper then requests the P.F.E. office to change the billing on the car to another consignee and move it to Chicago. Possibly before the car goes much further, market conditions are reversed, and the routing and billing again changed.

These services performed by the P.F.E. are known as "diversions," and are most vital to the successful marketing of western perishables. It is in the functioning of this service that the railroad's telegraph facilities are brought into play. Through an elaborate record of train consists, the P.F.E. offices can determine almost instantly the location of any car of perishables.

Last year the P.F.E. forces handled 969,746 diversions with almost a perfect score, a remarkable demonstration of efficiency and accuracy. Of these diversions, 463,643 originated in the Pacific Coast territory. More than half of them came from the district comprising Imperial Valley, Los Angeles, Colton and Arizona points. Of the 506,103 balance, although originated in the eastern offices of the P.F.E., the greater number affected perishable shipments originating on the Pacific Lines of S.P.

Lettuce Heads the List

As previously mentioned, the movement of fruits and vegetables is an all-year activity for the railroad. Lettuce and citrus shipments both fall into this category. Last year more cars of lettuce were moved over Pacific Lines than any other perishable crop—a total of nearly 47,000 cars, to set a new shipping record for an industry that only produced five carloads in 1915. Starting with the winter and spring crops of lettuce in the Salt River and Yuma valleys of Arizona and Imperial Valley, during December-April, harvest of the green "Icebergs" moves northward during March to the two-crop production areas of Salinas-Watsonville and Santa Maria-Guadalupe where shipping continues on into December.

The citrus industry was second in

number of carloads moved on Pacific Lines last year. Nearly 42,000 cars being the S.P. share of the estimated 84,000 cars shipped from California and Arizona during the year by all rail and steamship lines. Navel oranges are shipped November to May, and the Valencias follow during the summer and fall months.

Deciduous tree fruit shipments totaled nearly 25,000 cars in 1936 on Pacific Lines, beginning with cherries in April and continuing through succeeding months with plums, apricots, peaches, pears, and winding up with apples in the fall.

Grapes figured fourth in Pacific Lines shipments last year with 16,521 cars of California grapes (and a few from Arizona) handled as compared with the heyday of an all-time high that hit 43,079 cars in 1927. Shipping season extends from June until December.

Cantaloupes and kindred melons contributed 14,595 carloads to perishable shipments last year, with watermelons adding 3,500 cars. This is the most spectacular of the perishable movements on Pacific Lines, particularly in Imperial Valley where half the cantaloupes grown in the United States ripen quickly to create the heaviest concen-

MR. STORK: *New Arrival on train; Conductor Acts Fast*

In looking out for the comfort of their passengers, trainmen are occasionally called upon to render services that test the limits of their resourcefulness and quick-thinking. Caring for ill or other-

wise upset travelers, aiding mothers with fretful children, preparation of special diets, locating misplaced baggage, finding lost pocketbooks are but a few of their experiences.

No exception was recent experience of conductor W. L. Scruggs on the "Shasta" enroute Portland to San Francisco, when old Doc Stork overtook the train at Harrisburg to pay a visit to Mrs. Augustina Bradillo, traveling with her husband and their five children. With no time to remove Mrs. Bradillo to a hospital Scruggs put in a hurried call for Dr. D. G. Clark of Harrisburg who delivered the new passenger. Mrs. Bradillo and her latest offspring were taken to a hospital at Eugene while the rest of the family continued their journey.

Mother and baby later departed for home with a lovely layette presented to them by the wives of some of the S. P. people at Eugene.

trated freight movement handled during the year on the entire railroad. Shipments from this valley begin about the third week in April, reach the peak in June and windup about the middle of August. The record movement of cantaloupes for any one year on S.P. lines was 21,504 cars in 1931, and the record day was June 27, 1922, with 647 cars moved out of Imperial Valley.

Potatoes and onions, cauliflower, celery, asparagus, cabbage and mixed vegetables, all fit into the annual merry-go-round of perishable crops that get individual and special attention from Southern Pacific and Pacific Fruit Express companies; so that every day of the year housewives and restaurateurs throughout the nation are assured that Pacific Coast fruits and vegetables in season will be found awaiting them at their favorite market.

OUTINGS: *S. P. Clubs Select June 13 for Annual Picnics*

Preparations are being made at Alameda and Santa Cruz to welcome hundreds of railroad folks when Southern Pacific Clubs of Sacramento and San Francisco hold their annual outings, Sunday June 13.

Picnickers from the Bay district will travel by special trains from Oakland and San Francisco to enjoy a day of frolicking on the sands at Santa Cruz. Special rates have been secured for swimming, boating and rides on the numerous concessions, and it is anticipated that last year's record-breaking attendance of 4,500 persons will be exceeded. The Club will also hold a dance in the Casino ballroom Saturday evening preceding the outing. Charles Kleupfer is general chairman of the committee making arrangements.

Expectations are that most of the Sacramento employes will be present when their Club festivities start at Neptune Beach in Alameda. A full program of sports including a baseball game between the Club team and Oakland News Service, and entertainment has been arranged by the committee in charge, headed by O. E. Chappell, Club president.

SPOTLESS: *Health Experts Praise Housekeeping of S. P.*

Cleanliness of commissaries and dining cars, always a subject of pride with Southern Pacific, was recently paid a high tribute by a most qualified critic. Following an inspection of Oakland and San Francisco commissaries and dining cars by the United States Public Health service, H. B. Hommon, senior sanitary engineer in charge, wrote H. A. Butler: "We wish to congratulate you and the men in charge of the commissaries for the satisfactory sanitary conditions and generally good housekeeping found during the inspections."

Chapter Fifteen
Shining the Dining Car Silver
and the Elegance of Dining

In the mid-1880s, a family on a trip from Omaha to Sacramento expected to spend about three and a quarter days traveling. The temperature inside a train car could be ninety-eight degrees, and air-conditioning technology was about 150 years away. Likewise, refrigeration; it was impossible to quench one's thirst with a cool drink between Lovelock and Reno. An 1876 edition of the *Pacific Tourist* suggested carrying "a little lunch basket nicely stowed with sweet and substantial bits of food" and warned, "In packing your little lunch basket, avoid tongue, by all means, for it will not keep over a day or two and its fumes in a sleeping car are anything but like those from Araby the Blest."

If one did not carry their own food or had food that spoiled, eating was limited to restaurants at stations. (The only exception was first-class service to California that provided canned goods, coffee, and other liquid refreshments on the train.) There was on-board food service on eastern trains as early as the 1870s. But, trains through Sacramento had the trackside services for about another twenty years. Consequently, in 1880, most travelers that had come from east of Omaha may have been especially frustrated being part of a jostling group of hungry passengers scrambling to eat at the same time.

The restaurants, also called station houses, were operated by private individuals under contract and subject to the oversight by the railroad. Historian Oscar Lewis, on a transcontinental trip, described one meal experience:

Long before the towns were reached, passengers aroused themselves to activity and crowded on the steps and open platforms, the less encumbered swinging to the ground while the cars were still in motion and leading a headlong race toward the eating houses. Frame structures with the depressing aspect common to nineteenth century railroad architecture, they were filled with long tables, laden with thick crockery and steaming platters of food newly rushed in from the kitchen. The trains remained twenty minutes, the meals were table d'hote, and the price, whether for breakfast, dinner, or supper was uniformly a dollar greenback or, in California and Nevada, seventy-five cents in silver. (Loveland 1996, 28)

The oversight by the railroad was not thorough enough to ensure similar quality meals at all stations.

The profitability of the restaurants and the railroad's distaste for developing a fleet of diners led to an agreement signed on November 11, 1881 by the Union Pacific, Chicago, Burlington and Quincy, and the Atchison, Topeka and Santa Fe. The agreement between these three railroads stated that trains between the Missouri River and Denver would not contain dining cars without first giving six months' notice. It can be assumed that the same principle applied to through trains heading west of Denver. In May 1884, the Chicago, Burlington, and Quincy announced their withdrawal from the agreement.

Two years later, Union Pacific, with its connection to the Southern Pacific, stated its intention to pull out of the agreement also. The announcement included the advisement that the railroad did not have the intention of commencing dining car service. Was this a sincere statement of their plans?

During the following three years, the business philosophy of the Union Pacific changed completely and an era during which dinner in the diner was anticipated for its refinement and elegance began. The Union Pacific finally gave the industry notice of their intention to inaugurate dining car service beginning May 12, 1889. The early reluctance of western railroads to add dining cars because of the profit generated by track-side restaurants was gone.

Southern Pacific fell in line with competition and customer expectations when it introduced three Pullman diners (not owned by the Southern Pacific) in 1890 that operated between Ogden and Truckee. The first diners owned by the railroad were converted from business cars in 1892. One of these cars, the *T. W. Pierce*, was used between Oakland and Sacramento. Dining car service was gradually added to other lines on the Southern Pacific, including the *Del Monte* which served the affluent clientele of the San Francisco-Monterey area. In 1895, for the first time, dining car service was offered along the entire route between Oakland and Ogden.

The participation of the Sacramento Shops in the new boom of rolling restaurants occurred in 1912 when Southern Pacific built the company's first lunch car. It seated sixteen passengers, had a kitchen at one end, and operated between Oakland and Bakersfield. The idea was successful and was followed by a design change: a lower counter, which Southern Pacific described as better accommodating women. The railroad added that the less-formal food service was for the convenience of passengers who "lack the time to sit down to an elaborate dinner in the regular dining cars...." (Loveland, 61)

In 1914, Allan Pollok, Manager of Dining Cars, Hotels, Restaurants and Railway Clubs, announced:

> The Southern Pacific Company this month gained the distinction of owning and operating more dining cars than any other railroad in America, and the record probably holds good for the world. The company now owns 105 diners, five more than a Canadian railroad which is its nearest rival in North America, and forty more than an Eastern railroad, its nearest rival for the same distinction in the United States. (Loveland, 59)

In 1926, there were 3.3 million meals served in Southern Pacific dining cars.

When a railroad shopped for silver-plated dinnerware in the mid-1920s, there were three classifications of silver plate. The lowest grade was called "A1 Plate" or "Standard Plate." "Extra Plate" or "Hotel Quality" was the next higher quality flatware and had a plating thickness 25 percent to 33 percent thicker than the "Standard Plate." The highest quality plating classification called "Triple Plate Quality" was 240 percent thicker than the "Extra Plate," and was the most commonly used due to its durability. The dinnerware did not last forever. Each item eventually had worn areas that were unacceptable and led to the item being withdrawn from service. Wear was not limited to thinning silver plate. Repairing dents and hinges were also part of the maintenance program.

Even though the Southern Pacific was reluctant to serve meals on the train early in its history, fine dining eventually became an anticipated experience. To provide this service, the Southern Pacific had to have a large staff that involved employees in Sacramento, where the work included maintenance of silverware. Food preparation, dining car operations, and plating work in Sacramento and on trains passing through the city are subjects in the next four articles of *The Bulletin*:

"New Passenger Car Tin Shop Makes Everything Used in Passenger Cars," June 1919

"Food Preparation-How Does Southern Pacific Serve 4 Million Meals Per Year?" December 1935

"Polishing and Plating-From Silver Bullion to Cutlery," December 1935

"Behind the Scenes in the Dining Car Department," November 1927

COAST COMPLIMENTED ON LOAN DRIVE.

By H.R. HICKS
(Superintendent's Office Coast Division)

Many of our Coast Division readers as well as those on the other divisions will no doubt be glad to know that the Coast Division went over the top on the Liberty Loan drive. In fact, it ranked No. 1 among the divisions of the system. We want to take this means of expressing our appreciation of the work of the officers and employees of that division. As regards San Francisco, the latest word is that she subscribed for the full quota of the bonds assigned and that they are still counting. Let the good work go on. San Francisco is always "there."

The old timers will be interested to hear that T. L. O'Neil, conductor, and Messrs. E. A. Kelsey and J. A. Leavitt have placed their applications for pensions. We are all sorry to see these old timers leave the active list, and hope they will enjoy the rest they deserve.

No doubt many of the boys on the Coast Division will be glad to know that Jack Brennan has recuperated sufficiently to enable him to enjoy short walks in the ward to which he is confined in the Southern Pacific Hospital. We all hope that Jack will soon be "in the thick of it" again.

Genial Bill Hunt, smiling Eddie McDevitt and plodding Harry Roadknight have received their discharges from Uncle Sam and are back at their respective desks shoving the quill.

The suggestion made through these columns to our agents that they read carefully the pamphlet, "Rules and Instructions Relative to the Handling of United States Mail and Railroad Business Mail," effective September 1, 1910, also pass it around among their subordinates to read, has brought some improvement but there still seems too many mail failures occurring due to lack of full knowledge of those important instructions.

We find in the file a copy of a letter addressed to our agent at Santa Cruz from the local exemption board for the county of Santa Cruz. The letter reads:

"Before finally closing up the business of the local exemption board of Santa Cruz County we desire to express to you our appreciation of the uniform courtesy and helpfulness of yourself and your associates in connection with the entrainment of inducted men sent to the various mobilization camps. You have at all times not only performed the duties required of you by the regulations but have gone out of your way whenever necessary to smooth out difficulties and prevent complications. The result was that all our men got through without trouble, and so far as we know entrainments from this point were up to all the requirements in every respect."

We know Jim Doig was glad to get that.

LOOK OUT FOR J. BARLEYCORN!

All railroad men who have the interests of their Government and their own reputation at heart will be impressed with the following circular issued by Regional Director Hale Holden:

"A representative of the War Department advises that many reports are being received of intoxication of soldiers and sailors on railroad trains and the drinking of intoxicating liquors upon such trains by civilians in violation of local laws.

"Will you please bring the matter to the attention of operating officials, passenger train conductors and brakemen and Pullman car employees, making clear the fact that the United States Railroad Administration cannot tolerate the violation of Federal and State laws relating to liquor upon its trains. This is especially true in view of the extraordinary duty of all Government departments to safeguard the armed forces of the United States until demobilization has been accomplished. It is thought that railroad employees will be glad to co-operate in this matter both from a sense of patriotism, decency and propriety.

NEW PASSENGER CAR TIN-SHOP.

By J. HALL (Sacramento Shops)

The new passenger car tinshop at Sacramento is complete, with modern tools, enabling us to perform all work in copper, tin or sheet metal. This plays quite a part in our dining car equipment. The amount of work for this department found in dining cars, buffet cars and private cars has grown beyond all anticipation, requiring a force now of eight sheet metal workers, eight tinsmiths and five helpers.

We manufacture everything used in this line on passenger cars. Dining car ranges are rebuilt, using nothing but the old heavy iron trimmings. Ranges are built that weigh as high as 950 pounds, and steam stables that are used in depot restaurants.

This department contains a complete tinning plant, where all kitchenware is retined and repaired. We also manufacture and repair water coolers, supply tanks, refrigerators, ice boxes, ventilators, in fact, everything used on cars.

The shop is located about 100 feet east of the car paint shop, making a most convenient center for passenger car work, and avoiding loss of effort in transferring material from place to place.

G. M. Crocker, who is in charge of this shop, began his service with this company as an apprentice boy in the old shop, known as the copper shop, in August, 1890. He learned his trade and step by step has advanced to his present position.

During the past month six of our soldier boys have returned to their old positions with the company. They are: H. C. F. Fowler, machinist; B. J. Buccellia, apprentice upholsterer; F. S. J. Rath, trimmer; M. R. Wilson, apprentice machinist; A. Lascari, inside finisher, and Samuel Petta, painter.

We feel very proud of these boys, as they have returned full of ambition to become good employees and better Americans than ever.

D. S. MEDAL AWARDED LATE GEORGE HODGES.

Announcement is made by Walker D. Hines, Director General of Railroads, that the Distinguished Service Medal has been posthumously awarded to Mr. George Hodges, former manager of the Troop Movement Section of the Division of Operations of the United States Railroad Administration.

The information was contained in a letter from Newton D. Baker, Secretary of War, to the Director General, under date of May 13, 1919, reading as follows:

"My Dear Mr. Hines—It is with sincere regret that I learn of the death of Mr. George Hodges, manager of the Troop Movement Section of the Division of Operations of the United States Railroad Administration.

"The services of the railroads during the great war are gratefully remembered, and I thank you for the suggestion that some suitable recognition of the work of the man who was largely responsible for their success might be made by the War Department. I take pleasure in advising you that, by direction of the President, and under the provisions of the Act of Congress of July 9, 1917, the Distinguished Service Medal has been posthumously awarded to Mr. George Hodges 'for especially meritorious and conspicuous service as manager of the Troop Movement Section of the Division of Operations of the United States Railroad Administration. Mr. Hodges arranged all the details of the movement of troops from local draft boards to mobilization camps, between camps, or from mobilization camps to the ports of embarkation for shipment overseas. Troops in large numbers were moved on short notice and he was responsible for the successful coordination and carrying out of these movements.' "

Remember that it is our patriotic duty to prevent fire losses by preaching the doctrine of carefulness, and in the same way that we met the important demands arising in the winning of the war we may meet them in the reconstruction and readjustment of national and international affairs.

FOOD: *How S. P. Serves 4,500,000 Meals Annually on A Moving Dining Table That Extends Over Eight States*

Picture, if you can, an extension dining table long enough to reach across mountains and valleys, deserts and plains in eight western states!

This may sound too fanciful for serious consideration. But such a table is not as fabulous as it seems. As a matter of fact, one of this very kind is represented by the dining cars of the Southern Pacific Company.

Operating on 16 established runs over 4800 miles of line, the Company's diners, in effect, extend from Portland, in the north, to El Paso and Tucumcari, in the south, and from Oakland to Green River, Wyoming, in the east.

Just to provide the regular daily service, with no accounting for special trains or extra sections, 67 dining cars are required. These cars are put to use during 96 meal periods each day. And since a train travels approximately 100 miles during a meal period, it develops that meals are being served while trains move 9600 miles a day.

Presiding at the head of the table, and charged with the responsibility of serving it, is Harry A. Butler, manager of Southern Pacific's dining car, hotel, restaurant and news service activities. From headquarters in the Company's general office at San Francisco, he calmly surveys, as his principal field of action, the full extent of Southern Pacific's Pacific Lines.

This brings into immediate view the states of California, Oregon, Nevada, Utah, Arizona, New Mexico and bits of

Spick and span in fresh linen and shining silver, the spacious air-conditioned diner presents an inviting appearance.

Texas and Wyoming. But that isn't all. He must look even farther and include Oklahoma, Kansas, Iowa, Louisiana, and Illinois, to which states his activities, other than dining car service, are extended over connecting railroads.

Always Ready

Wherever meal time finds a passenger train on Pacific Lines, then and there must the Company's highly trained organization be ready to lay snowy linen upon the tables, set the places with shining silverware and china, and serve the finest food and drink—for that and nothing less will do for Southern Pacific patrons.

In the territory reached by connecting lines, Southern Pacific's catering operations are confined to serving refreshments in club cars. But this is no small task in itself, since it reaches

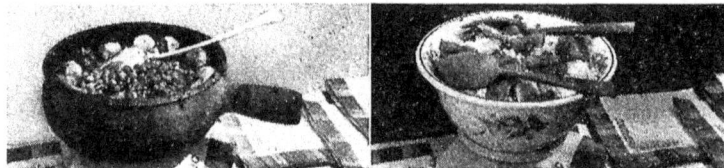

Twin favorites with our dining car patrons. The Casserole, left, contains choice cuts of lamb cooked with vegetables and served piping hot in an earthenware bowl. The Salad Bowl is made of crisp lettuce and vegetables, with distinctive French dressing.

so far from the Company's main base of control. Yet the work is carried on as if it were near at hand.

Viewed in its entirety, this department operates in dining cars, club cars, observation cars, all-day lunch cars, ferryboat restaurants and soda fountains, station restaurants and hotels, station soda fountains and employes' railway clubs, in addition to the handling of the news service on trains and boats and at stations.

These combined activities are conducted over 7300 miles of line in 14 states. A force of 1250 employes is required for this extensive enterprise, distinguished as one of the largest of its kind.

On account of its intimate nature, the service performed by the Company's dining car department is of major importance in the maintenance of friendly relations between the railroad and its patrons, and the success of the Company, in no small part, is due to the high standard of efficiency maintained in its catering branch, which strives constantly to render prompt, courteous and complete service.

Harry A. Butler, manager of dining cars, hotels, restaurants and news service, Pacific Lines.

Until quite recently, the commonest criticism of dining car service, as a whole, was that dining car prices were beyond the average traveler's budget.

The inauguration of what is termed "Meals Select" service was the solution of the problem. This forward step in dining car catering was taken in January 1933. It introduced the serving of full course meals with the price of the entree determining the price of the entire service. Immediate success attended the venture, with the result that it was not only extended over Southern Pacific's vast system, but has since been widely copied by railroads throughout the United States.

Meals at Moderate Cost

A la carte selections are still available on all Southern Pacific diners and the patron has the option of choosing his meal from this source or taking advantage of the Meals Select feature which enables him to obtain a complete meal at moderate cost. Breakfasts of this type range in price from 50 cents to 90 cents, while luncheons and dinners range from 80 cents to $1.25, the price of the entree, as has been pointed out, determing the cost of the entire meal, which includes soup, salad, bread and butter, entree, vegetables, dessert and beverage.

On certain trains Southern Pacific also serves what is known as the "Continental Dinner," a meal of elaborate proportions, consisting of fruit or avocado cocktail, olives, celery, soup, fish, entree (with choice of tenderloin or sirloin

In this tiny kitchen, 14 feet 9 inches long by 6½ feet wide, the dining car chef and his three assistants prepare meals for as many as 150 persons during a meal period. Left to right—Wm. Bartlow, Suthern J. Conley, Joe Alonzo, and Anderson Fields.

steak, lamb chops or broiled chicken), vegetables, salad, bread and butter, dessert, the "cheese crock" and demi-tasse. The charge is $1.50, and the dinner is unsurpassed for quality and price.

The great majority of passengers, however, prefer the Meals Select. Immediately following the adoption of this type of service, the number of meals served increased 14 per cent. This was based on the relation of the patronage to the number of passengers carried on trains where diners were operated, and the increase in popularity took place at a time when the financial condition of the country was becoming more unfavorable from month to month, with a tendency toward a diminishing patronage for the diners. It would therefore be reasonable to assume that popularity had increased 20 per cent as a result of the innovation.

Million and Half Meals

In 1933, Southern Pacific diners served a total of 1,068,061 meals; 1,261,814 in 1934, and it is estimated that 1,480,000 will be served in 1935.

During the 10 months ended October 31, 1935, 85,300 meals were served in diners in connection with special train movements. For instance, the Rotarian convention in Mexico City called for the assignment of five diners, and these cars served 14,237 revenue meals in a period of 13 days. In connection with the Boy Scout excursions to eastern points this last summer, 29,683 revenue meals were served by seven diners.

The results of the dining car service are all the more remarkable when it is

pointed out that the average diner is approximately 75 feet long, and nine feet, 10 inches wide. Since by far the greater part of the floor space is given over to the tables, the kitchen crew has little room in which to work. The pantry measures 6½ feet by 11 feet over all, and has standing space measuring 4 feet by 8 feet. The kitchen measures 14 feet, nine inches by 6½ feet, and has standing space measuring 2½ feet by 14 feet, 9 inches.

In these tiny quarters meals are prepared for as many as 150 or more persons during a meal period. This is remarkable when compared with the performance of a commercial restaurant where quarters are ample and all other conditions are favorable to the expeditious handling of patrons.

In the diner kitchen are a chef and three assistants working in a compartment which allows each man a space approximately 4 feet square. But this seemingly impossible condition is compensated for by the skillful arrangement of equipment, so that each man is enabled to perform his duties within the scope of small space, and without interfering with the movements of his co-workers.

On Land and Water

So far, this account of Southern Pacific's dining car department activities has been the story of a land-going enterprise, for railroads are attached to solid ground and seldom, of course, run over water. But Southern Pacific is different. Not only does its trains "go to sea by rail" over Great Salt Lake; the Company also engages in extensive ferryboat operations on San Francisco bay, both in the passenger and automobile carrying fields. It follows, therefore, that Southern Pacific's dining car department takes to the water like a true amphibian.

Restaurants are maintained on all ferryboats of the Southern Pacific and the Southern Pacific Golden Gate Ferries, Ltd. There are seven established runs and during normal times a total of 19 boats are in service. But on days when football games are played at the University of California in Berkeley, and on holidays when motor travel in and out of the San Francisco bay dis-

trict is heavy, the number of boats is increased greatly, sometimes to as many as 32. For the conduct of this service, the department has a force of 300 employes.

In 1933, a total of 1,946,519 revenue meals were served on the boats; in 1934 the number rose to 2,010,985, and it is estimated that 2,120,000 patrons will be handled in 1935.

Ferry Soda Fountains

Soda fountain service also plays an important part in connection with the ferryboat operations. The popular trend toward this type of service was clearly recognized a number of years ago, and in 1928 a soda fountain was opened, as an experiment, on one of the Company's boats. It met with immediate success, and additional units were soon installed on other boats. A total of 926,000 persons were served at the soda stands in 1934, and it is believed the 1935 patronage will increase to 1,080,000. Nearly all of this patronage represents business that otherwise would have been lost.

When the activities of the dining car department were listed in the fore part of this article, News Service was included. This function may properly be coupled with Coach Luncheon Service in the following descriptive matter, for the two are closely allied.

Southern Pacific took over operation of news service on all trains on Pacific Lines in 1927, and the ferryboat and station news stand service was acquired at the start of 1929. Previously the service had been conducted by a private company, in the role of concessionaire.

News stands are now operated by the Company at 15 principal stations, and train news agents operate on eight runs out of Oakland Pier, eight out of Los Angeles and four out of El Paso. A total of 70 news agents and 27 news agents' helpers are employed.

Effective August 1, 1935, Southern Pacific established what was termed Coach Luncheon Service on the Cascade. It was in the nature of an experiment, which proved successful, and was extended progressively to all trains, the installation being completed September 17. This service provides for the dispensing of food and beverages in coaches and tourist cars, along with the rental of pillows, and represents a popularization of the "Off the Tray" service which had been conducted from dining cars.

POLISH: *Thousands of Pieces of Dining Car Silverware Made to Shine Like New in Plating Room at Sacramento*

If you placed end-to-end each piece of silverware owned by the Southern Pacific Company and used on its dining cars and in its restaurants, it would reach to —ask George Lester, foreman of the Plating Room, Sacramento General Shops, and he'll probably answer "To Sacramento."

For eventually practically every piece of silverware on Pacific Lines finds its way to Lester's department, where it is reconditioned, replated, polished and turned back to service brand new.

The Plating Department is one of the activities of a railroad about which the average employe and the general public know very little, writes Lloyd Phillips, *The Bulletin* correspondent and secretary to superintendent of motive power at Sacramento. One does not usually think of railroading in terms of silver-plating or polishing, yet these operations are of utmost importance in the dining car service of our Company.

The average patron of a dining car probably estimates that about eight or ten dozen pieces of silver are used in the car. If you told him, as George Lester would, that there are between eight hundred and a thousand pieces of silver on the average dining car, he might be a little dubious. Yet these figures are approximately correct, and so it is obvious that some means must be provided for taking care of this great quantity of silver.

Forty-five years ago the Company operated only five dining cars, and when one was shopped for repairs, it created considerable interest and was considered a big job. But times have changed. Right now there are thirteen diners in Sacramento Shops being air-conditioned and otherwise generally overhauled. On each of these diners there is much silverware that must be reconditioned, ready for service when the car is released from the shop.

These items, after being checked from the car, are sent to the plating department. Here every piece is inspected and where necessary, repairs are made, dents are removed, hinges replaced and the silverware put in proper condition by a silversmith.

When the silversmith has finished his job, every piece is buffed and passed along to the platers, who clean the articles thoroughly. Then, after passing through several preliminary solutions, they are suspended in the silver plating solution. The silver comes in six by six inch squares a quarter inch thick, and weighs about a pound. One of these squares is dissolved in the cyanide solution. The silver is electro-deposited, copper wire being used for that purpose. The articles remain in the solution from three and a half to four hours, being constantly agitated, until the proper plating is obtained. They are then removed to be finished and polished with a fine wire power brush.

When the proper luster is obtained, the articles are carefully dried and stored until again called for service in the dining car to which they belong, each diner receiving the identical silverware removed from it when the car was shopped.

This routine of keeping silverware in proper condition is followed about every two years. The Pacific Lines has 120 dining cars, 67 being in constant operation on regular trains, so it is easy to understand why George Lester's plating department at Sacramento usually has plenty of work on hand.

His department also takes care of copper and brass plating, and the reconditioning of aluminiumware.

PICTURES: In the Plating Room at Sacramento Shops. Top to Bottom—George Lester, foreman, getting duplicate keys to a diner. There are duplicate keys here for every dining car, with silverware to be cleaned; Paul Poissant, plater, cleaning silver with pumice; L. L. Shearer, polisher, who also checks silver in and out of the plant; Anthony Cvitan, apprentice plater, removing silverware from plating solution.

Behind the Scenes in the Dining Car Dept.

By ALLAN POLLOK
Manager, Dining Cars, Hotels, Restaurants and Railway Clubs

WHAT is the Dining Car Department? On the pay roll there are 2100 employes. We have altogether a total of 140 dining cars, 26 all-day-lunch cars, 20 steamer restaurants, 7 hotels and station restaurants, 10 railway clubs, 36 club cars, and 73 news agents employed on our trains. We have a General Store at Kirkham Street, West Oakland, with an average

Allan Pollok

stock of over $100,000.

During the year 1926 we served 3,260,000 meals on our dining cars, 2,243,000 meals in our steamer restaurants, 1,000,000 meals in our station restaurants, and 679,000 meals at our railway clubs.

I wish you would take a little time some day and go through the very fine Commissary we have right here in San Francisco at the Ferry Landing. I am sure it would be worth your time to see for yourselves just what we have to offer in this fine and modern plant, with all the latest appliances of all kinds. We also have commissaries at Oakland, Los Angeles, Portland, El Paso, Houston, San Antonio and New Orleans. The last three, however, are operated by the Atlantic Lines and are not under our supervision.

We also draw supplies from commissaries at Ogden, Omaha and Chicago, and I may say, without criticism, that these foreign commissaries only serve to show the very high excellence of our own, and in particular our Steamer Commissary, which is the last word in commissary efficiency. Any time you are on the line you can judge as to this for yourselves. Just drop off at any of these

THIS article is taken from Mr. Pollok's address before the Chief Clerk's Council at San Francisco, October 4, 1927. Mr. Pollok gives many interesting sidelights on the many important details of dining car operation. The attention given these details is an important reason for the high standard of our dining car service.

places and go in the commissary and inspect it.

Must Have Good Men

The difficulty in getting men for catering departments everywhere is getting greater and greater each year. Men are not being brought in from Europe, and we are not training stewards, cooks and waiters in America. American boys don't take to this profession. Most of our stewards are foreign born and foreign trained, and more and more the Dining Car Departments of the railroads throughout this country are coming to realize the importance of training their own stewards.

I would rather get a young fellow without any experience, take him through the various branches of our commissaries, from the commissaries to our dining cars, and thus teach him our system. In this regard, I don't know of any profession in the country today that offers the opportunity to young men of ability and ambition, who want to get into a business that will make them independent, as quickly as the catering business. You may smile at this, but, after a close study of the situation, I make that statement. I don't mean as a cook or a steward, but as a proprietor. The difficulty in getting trained men is evidenced by the fact that when the St. Francis Hotel wanted a Maitre d'Hotel, after making inquiries throughout California, the management finally had to go to New York and get a man from the Plaza Hotel at

$10,000 a year. They are paying that amount today. Mr. Mainwaring at the "Palace" had the same experience, and also pays that much.

Of course, these are the high marks, such as for positions at Biltmore and Ambassador hotels in Los Angeles, and at other fine hotels throughout the country, and these places must have competent help.

As to the training of stewards, we have to pick our men carefully for many reasons, for a steward, without more than a turn of his hand in laying down a menu card, can affront a patron.

We have very severe discipline and inspection. The severest discipline we have is of course from our passengers. It is a peculiar thing, but a man will go for a year to the Palace Hotel and take luncheon every day and be well satisfied. If he does see some little inattention or lack of service, or perhaps receive a tough steak, he will pass it off as all right and not think, because of the tough steak, that the "service in the Palace Hotel is the worst of any hotel in the country."

Diplomacy Needed

As manager of hotels I have been in, I don't recall such an incident. Nor would we ask the assistant manager of the hotel to ask the guests: "Well, how was the dining room service?" Nor say that any complaint the guest might make "would be taken up for correction." The guest in a hotel would, of course, say everything was all right, and be surprised at being approached in the matter. Unfortunately many a passenger who gets on one of our dining cars is seemingly a proprietor of that dining car, and if he sees any chance to complain he does not hesitate in calling attention to it. However, that is something we have to put up with.

Our stewards must be in the dining car ready to serve breakfast to passengers at six-thirty, that is to say,

East end of Southern Pacific's fine Steamer Commissary Kitchen, San Francisco.

they have to get up at five o'clock. The cooks have to start breakfast at five in order to be prepared to serve the train crew at six o'clock. The dining car has to be cleaned, swept up, tables set up and everything ready at six-thirty to commence serving breakfast to passengers.

Breakfast lasts until ten o'clock. In the winter time passengers get up later, and breakfast drags along until ten o'clock. At ten o'clock, with nobody else to come in, the dining car is closed. The crew then sets to work to clean the silver, clean the pantries, wash the dishes, glasses, etc., get the soiled

Our Instructing Forces

Then we have an Instructing Chef and an Instructing Waiter operating out of Los Angeles, and an Instructing Chef and Instructing Waiter covering the northern lines from Oakland Commissary. They report to the Supervising Chef, Mr. Reiss, who is a European, a Frenchman who received all his education in the "Old Country," and who is a past master in the art of catering. He gets up our menus; writes our instruction book for guidance of dining car chefs and cooks; is responsible for the recipes

tion of the cook on that dining car. If the cooking of that particular cook is found to be deficient in any manner, he is sent for further instruction to our commissary kitchen at West Oakland or Los Angeles, where schools are maintained for instruction of cooks and waiters.

As to instructing waiters: Their duty is to work in cooperation with the commissaries and inspectors. If any dining car is deficient in service, one of the instructing waiters is assigned to that particular car to make a round trip, or two if necessary, and if any waiters are not up to standard they are sent to the "school car" for further instruction. If an employe shows he is unsuitable as a waiter, we discharge him, but we take great pains in instructing our employes.

The School Car

The "school car" for waiters is an obsolete dining car, completely set up in the yard for instruction of waiters, where "make believe" meals are served. Everything is exactly as it would be on a dining car in service, only food, of course, is not cooked or served. We have one of these cars at Los Angeles and another at West Oakland.

The service on a dining car is quite different from that in a hotel or restaurant. Owing to the swaying of the car, waiters have to have "sea legs," and yet it is the rarest kind of

linen stored away, and the clean linen out. By that time it is eleven o'clock and they have to dress themselves and get prepared for luncheon at twelve o'clock. The dining car must be in perfect condition, everything ship-shape, clean, without a single thing out of place. Luncheon goes on until nearly two o'clock. The men are on the floor all the time working.

At two or two-fifteen p. m. the diner is closed down again and the same preparations made for serving the evening meal at five-thirty. The trainmen have to be served between meals. I mention these things to show that the duty of a steward is arduous work. He must be up and coming every moment, on his toes, and yet be diplomatic and courteous to all kinds of passengers. A traveler going to Chicago, for instance, eats heartily the first meal out and after that is tired and cross; sleeps most of the time and is loggy. The steward has to take all these things into consideration and endeavor to please the most exacting patrons.

Our inspectors are selected and promoted from our best stewards. They must know everything about dining car service; how to make up menus; how every item should be served; must watch all delinquencies in cooking and service and lecture and instruct the dining car crew in all details of good service.

These scenes show new cooks and waiters being trained in the school car at Oakland Commissary. Max Hall, instructing waiter, is demonstrating the proper way to serve patrons, while Jefferson Davis, instructing chef, is showing how various dishes should be arranged.

and for proper cooking and instruction of the chefs.

Under him are the Instructing Chefs, colored men, whom we select from the very cream of our colored cooks. And let me say some of our colored cooks are able to go anywhere in the catering field. We have some exceptionally fine colored cooks. The Supervising Chef goes over all the new recipes with these Traveling and Instructing Chefs, or, if he has any complaint regarding the cooking on any dining car, it is called to atten-

a thing for one of them to meet with an accident. I have never seen a tray overturned or anything spilled on a guest in a dining car, and yet it is a most difficult thing to handle a full tray under such conditions. Of course I know that accidents occasionally occur, but I have never seen one.

As soon as a dining car gets into the yard, an inspector immediately boards the car and makes a thorough inspection of the kitchen, pantries, lockers and ice boxes. All perishable

SOUTHERN PACIFIC BULLETIN

supplies are removed and a report is made as to their condition. The kitchen is washed and all woodwork in the dining room is thoroughly cleaned. All kitchen utensils are stripped from the car, taken to the Commissary, where they are washed and thoroughly sterilized in a big steam sterilizer, so that once a trip every utensil is thoroughly cleaned. Once a trip the faucet and all parts of the coffee urn are cleaned to take away caffeine.

Menu planning is very important. We have a man at each commissary who gives careful attention to this. Most of our stewards have been many years in the service and are well posted as to preparation and "balancing" of menus. That is, the various foods such as vegetables, fruits, salads, meats, fish, etc., are all properly balanced. We can't have all expensive articles, nor all cheap ones, as we must cater to every class of patronage.

At the Palace Hotel they cater to one class, which can afford to pay whatever is asked. At other places they cater to people who want good service but can't afford such high prices. The Dining Car caters to all kinds of people mixed together. The man who comes from the Palace Hotel goes into the dining car and expects to find the best, while another expects to get a meal at a price to fit a modest purse, so that the menu builders have to study all these things out and make arrangements accordingly.

In the purchase of supplies, we are very big buyers in this market. For instance, we make the following purchases daily: Eggs, 900 dozen; beef, 2150 lbs.; lamb, 640 lbs.; poultry, 750 lbs.; pork, 690 lbs.; fish, 685 lbs.; ham and bacon, 825 lbs.; butter, 710 lbs.; milk, 530 gals.; coffee, 570 lbs.; potatoes, 2500 lbs.; flour, 1800 lbs.; vegetables, not including potatoes, 5750 lbs; shortening, lard, etc., 600 lbs.; sugar, 1900 lbs.; apples, oranges, grapefruit, 70 boxes.

A Big Market Basket

In buying in such volume, a wide field is open to us, and has to be watched. Take cantaloupes, for instance. About the first of June we commence drawing cantaloupes from the Imperial Valley and Arizona. Around the middle of July from the Turlock district and the central part of the state. Later, in September and October, from the Fallon district in Nevada.

Lettuce is secured in similar manner throughout the entire year; first from the Imperial Valley and Arizona, and later from the Watsonville district. The latter district has developed a tremendous volume of new business in the past few years.

Tomatoes are first secured in Mexico, later from southern California, and then throughout central California.

Formerly we drew a great many things from other states, but the western states are coming to be self-supporting and we now need to go out of this territory for very few supplies. We used to secure rice from Japan, but now California is a large

---

S. P. Aviators!

SOUTHERN PACIFIC aviators are to be given a special article in an early issue of the BULLETIN. If you are an aviator or a student of aviation write the editor about your flying career and about your most interesting experiences. Photograph of yourself with your plane is particularly desired.

A number of articles with photographs have already been received from division correspondents of the BULLETIN, and this final opportunity is given for the benefit of those not reached by the correspondents.

Material for this special feature must be in the BULLETIN office by December 1.

---

producer of this commodity and we can fill our requirements right at home. The same with sugar; and with flour—another great industry in California.

Avocados formerly cost a great deal of money, but thousands of acres have been developed around Los Angeles, a tremendous business has been established, and the demand is spreading all over the country.

The Best in the Market

We have four buyers. In San Francisco and Los Angeles each morning at 5:00 o'clock they go into the markets for the very finest perishable supplies, vegetables, fish, etc. These are brought into the markets very early in the morning, and the first man into the markets gets the cream of the lot.

We are very insistent upon quality. Above everything, quality first—then price. We can get some of the best fish in the country at Portland, San Francisco and Los Angeles, and as for fruits and vegetables, California is the leading state in the Union. Our buyers go down to the packing houses for our meats, where they inspect hundreds of cattle and select and stamp only the very primest for our service. When the meat reaches the Commissary, the man who receives it is responsible for its reception. He thoroughly inspects and places it under refrigeration in our ice boxes, so that if there is any inferior quality we know exactly who is to blame.

Eggs are one of the most difficult commodities to handle in a business like ours. We give great attention to quality, and draw our principal supply from farms throughout the state. The majority of eggs served on our diners are received the evening before or early that morning and immediately put under refrigeration on the dining cars. There is no place in the country that you can get eggs of better quality than on our dining cars. We send samples of eggs from various sources to the Nutrition Department of the University of California, and Professor Jaffa, who is in charge of that department, has made a thorough study of our supplies. Formerly there was no way we knew of to tell the age of an egg, but Professor Jaffa, through long study, has devised a means of telling the age of an egg almost to within a day. It

was found that, as the egg grows older, the air space beneath the shell enlarges; that air enters through the shell, which is porous. Eggs are also candled to show condition of the whites and yolks.

Handling of Cream

Samples of milk, butter and cream also are sent periodically to the University for test. We pay great attention to our cream. The state requirements call for 18 per cent of butterfat, but we will not use any cream having a butterfat content of less than 24 per cent. Our cream is put up at the dairies in non-refillable 3-quart tins, sent to us in refrigerated containers, placed in ice boxes at our commissaries, and from there issued to the dining cars and kept under constant refrigeration. These tins cost us about 10 cents a piece, and we don't use them again for cream, distributing them over the line for use as water or oil cans. It is a very expensive thing for the Dining Car Department, but in the course of a year we have had almost no sour cream, perhaps not more than two or three tins during a very hot spell.

We purchase our meats in wholesale quantities. In the butcher shop at West Oakland, these meats are cut into standard sizes for our dining cars. We handled last year, through this butcher shop, 400,000 pounds of beef, 100,000 pounds of lamb, 175,000 pounds of pork.

From the pork we made over 177,000 pounds of our famous pork sausage. Formerly we had difficulty in securing good pork sausage. We decided to make our own, and I don't know where you can get as fine pork sausage, with nothing in it but the finest ingredients, as we now serve on our dining cars.

We cure and smoke our own bacon. And don't be afraid to eat a "hot dog" on our dining cars—last year we produced more than 50,000 pounds of Frankfurters and other sausages, and we know the quality is the very best that can be had.

At the Steamer Commissary we have the best equipment that I know of anywhere. We had great difficulty in getting a really good, "home-made" doughnut, so we equipped this commissary with a doughnut department and now make over 250 dozen doughnuts a day.

At this commissary we have two electric bake ovens of the latest type, and bake 200 loaves of bread, 300 pies and 855 cuts of coffee-cake per day.

Get Uniform Quality

We used to experience difficulty in getting uniform hot cakes, bran muffins, corn bread and biscuits, etc., on our dining cars, as on nearly every car there was a difference. So we experimented and now have a Blending Department, where we blend the ingredients for all of these articles.

Chef Reiss, with his great experience, studied and perfected his recipes, so that the blends contain so much flour, so much dried milk and dried eggs, so much shortening, so much sugar, so much baking powder, etc. The different blends are put in

7-pound packages and issued to the dining cars. Instead of preparing his dough in the old manner, all the chef has to do is to take a cup of one of the blends, pour it into his mixer, add sufficient water or milk, and in five minutes he has a pan of fine hot biscuits, or whatever is wanted. And the quality is always uniform.

This also has resulted in considerable saving in money, as it enables us to know exactly what these things cost. We know just how many packages are issued to the cars, and how many pans a 7-pound package can make up.

At the Steamer Commissary, we also put up our own mayonnaise, Thousand Island, and other dressings. Some years ago I noticed a cook on one of the dining cars making mayonnaise. It curdled, as very often happens in heated atmosphere, and he threw it out, commencing to mix a fresh lot.

We talked this over at the Commissary and commenced to make our own dressings. The first month we saved $500 on olive oil alone. We put up our mayonnaise under refrigeration and issue it to the dining cars in quart Mason jars. This not only relieves the chefs of this work, but enables us to keep a line on the expense, and also gives us a check on the number of salads served.

A Careful Check

We have a check sheet which shows every item on the menu, the cost price and the number of orders served on each dining car—so many salads, so many jars of mayonnaise, etc. The same with steaks. We have a machine which cuts all our hams and bacons, and the number of slices issued and the number of orders sold should tally, as should the number of eggs issued and the number of orders sold. If these do not tally, an investigation is made.

We buy the finest prunes in the market. We used to boil them, and half of the lot would break and turn to mush. So we studied this out and now we never boil or stew our prunes, but bake them and put them up in quart Mason jars. We are very careful in purchasing prunes and only get the best sun-dried quality, then select them carefully so that not one damaged prune goes into the jars. A prune broken in baking is used for other purposes. Last year I believe we sold over a ton of prunes a month.

In 1926 we got an average of ninety-five cents for every meal served on our dining cars. Against this are many expenses. Food costs us forty-six cents per meal served— just about 48 per cent of the amount received goes into raw materials. Crews' wages amount to forty-four cents; fuel, ice and water, six cents; breakage of crockery and glassware, two cents; laundry and upkeep of linen, six cents. Other expenses amount to sixteen cents, making a total commissary expense of $1.20 for each meal served, for which we receive 95c. That is to say, we lose twenty-five cents for every meal served in our dining cars. These figures are for commissary expense only, and do not include car repairs, interior or exterior car cleaning, etc., which are charged to the Operating Department.

We buy very fine linen, "Old Country" linen of the finest flax. We must have linen of good wearing quality. When we buy linen, our specifications call for so many threads to the square inch; such and such a count of threads for warp and weft; then anybody can bid on our requirements. This new theory the Southern Pacific has worked out and it is the method now followed by the General Purchasing Department and other branches of the Company. Also, it is being put into effect by nearly every railroad

in the country. We don't care from whom we purchase, so long as our specifications are lived up to and the price is right. When we receive linen on these bids, samples are sent to the Testing Bureau in Washington for analysis. Thorough tests are made and the Testing Bureau renders detailed reports, which are compared with the specifications stated in our bids.

A Modern Laundry

At West Oakland we have a fine laundry, and last year laundered over 12,000,000 pieces. This is a modern plant, with all the latest machinery, and it employs an average of seventy people. Over the laundry a repair unit is maintained to mend and keep the stock of linen in good shape. Nineteen seamstresses devote their entire time to this work, and make repairs on 200,000 pieces of linen annually, cut and hem towels of all kinds and prepare other linen for use. They use electric sewing machines and the new articles thus prepared amount to 120,000 pieces a year.

One of the difficulties we have is the handling of big conventions, which invariably are held during the season of year when travel is heaviest. During a recent convention over 56,000 extra meals were served within a period of 10 days—with 100 per cent efficiency. This involved 105 distinct special dining car trips, and could only have been accomplished through the greatest of cooperation, not only on the part of our own department, but of the Passenger and Operating departments as well.

When you go into a dining car and find some little deficiency, speak to the steward about it in a nice way. If everything is all right, walk up to him and congratulate him. Give the steward the benefit of a doubt if possible—he needs it. The same with the waiters and the chefs. If the food and service is good, tell the waiter so, and, as you pass the kitchen on your way out, tell the chef that everything was first rate. That's the sort of thing the dining car crew likes to hear, and I can assure you it's appreciated more than your tips.

NEW ORLEANS RACING SEASON

Much attention is being attracted this season to the horse-racing program to be held at New Orleans. The 1927-28 dates are as follows: Jefferson Race Track—Thanksgiving Day, November 24 to January 1. Fair Grounds—January 1, 1928, to February 21, inclusive.

His Party

A matron of determined character was encountered by a young woman reporter on a country paper, who was sent out to interview leading citizens as to their politics.

"May I see Mr. ————?" she asked of a stern looking woman who opened the door at one house.

"No, you can't," answered the matron decisively.

"But I want to know what party he belongs to," pleaded the girl.

The woman drew up her tall figure. "Well, take a good look at me," she said, "I'm the party he belongs to!"—Capper's Weekly.

This is one of the yard crews that keeps things on the move at Mission Bay yard near San Francisco. Front row, left to right—Thomas Lacey, engineer; P. Fergon, yardman; H. D. Miner, engine foreman; J. C. Johnson, fireman; N. L. Sudduth, yardman; J. H. Stepp, engine foreman; S. Lanzi, trackwalker; C. H. Kason, yardman; E. Broberg, yardmaster; E. G. Mailloux, yardman. Back row—J. K. Baccich car repairer; and T. Leonard, yardman.

Appendix One

Timeline of Events related to the Central Pacific Railroad and the Southern Pacific Railroad from 1861 to 1930

1861

Abe Lincoln was elected president and the Civil War began. June 28 - The Central Pacific Railroad of California was incorporated. The directors were: James Bailey, Theodore Judah, L.A. Booth, Dr. Daniel W. Strong, Charles Marsh, and the "Big Four" (also known as "The Associates"), Charles Crocker, Mark Hopkins, Collis Huntington, and Leland Stanford.

The first transcontinental telegraph message was transmitted to the Pioneer Telegraph Building at 1015 Second Street.

Service began on the Sacramento Pioneers Railroad Company's horse-drawn streetcars.

1862

Abraham Lincoln signed the Pacific Railway Act.

1863

The battles of Gettysburg and Vicksburg were fought.

Theodore Judah, the initial supporter and planner of a railroad line over the Sierra died in New York. He had crossed the Isthmus of Panama in an attempt to visit Washington D.C. The trip was to secure the financial support of Cornelius Vanderbilt and buyout the Big Four for $100,000 per person.

January 8 - Ground was broken for the Central Pacific Railroad (CPRR) at Front and K Streets and signified the beginning of construction of the CPRR.

October 26 - The first rail was laid on the CPRR.

November 10 - The first locomotive of the CPRR, the *Governor Stanford*, was entered into service.

1864

The CPRR purchased Goss and Lambard Machine Shop located between Front and Second on the north side of I Street for use as a temporary shop.

March 25 - The *Governor Stanford* pulled the CPRR's first revenue freight train.

1865

General Lee met Grant at Appomattox bringing an end to the Civil War.

A plan for new CPRR shops was laid out in 1865-66.

1867

The laying of rail reached the summit.

1869

The roundhouse, planing mill, and the first segments of the car and machine shops were completed.

Twenty acres of Lake Sutter had been filled.

April 28 - Using more than 4,000 men, 25,800 ties and 3,520 rails, the CPRR laid ten miles of track in one day west of Promontory.

May 10 - The golden spike was inserted in Promontory, Utah. The two railroads, the Central Pacific and the Union Pacific, had passed each other while progressing in opposite directions as each company was paid by the mile.

May 13 - The Central Pacific and Union Pacific Railroads inaugurated regular service between Omaha and Sacramento.

September 6 - The Western Pacific Railroad was completed from Sacramento to San Jose via Stockton.

1870

The Central Pacific Railroad built an ice-cooled freight car at its Sacramento Shops.

1872

The second segment of the machine shop was opened.

1873

The first all new locomotive, a Central Pacific 4-4-0 number 55 was completed at the Shops.

1874

The California State Capitol building received the finishing touches.

The first train shipment of produce and salmon from Sacramento arrived on the East Coast.

1876

3,000 eucalyptus trees were planted around the Shops. Some people at the time thought they would cleanse the air of malaria.

1879

The boiler in the powerhouse exploded.

1883

The *Stanford*, a private car, was built at the Shops.

1886

The Statue of Liberty was dedicated in New York.

Three 4-4-0 locomotives, numbers 122, 123, and 125 were completed.

The first solid train of fruit cars ever shipped from California to eastern markets left the Sacramento yards.

1888

The third segment of the machine shop was opened. Also, the car machine shop, known as the car wheel shop was opened.

Battery operated streetcars operate for a brief period signaling an end to the era of horse-drawn streetcars.

Southern Pacific sent five-car trains called "California on Wheels" through the Midwest with exhibits of California products and agricultural displays.

1889

Electricity was used for the first time when a small steam driven generator supplied enough power for thirty-five lights.

1894

Employment at the Shops was in the range of 2,000-2,500 people.

1895

A new 70-foot turntable was installed.

The first successful installation of an oil-burning firebox was installed in a Southern Pacific locomotive (most locomotives were converted to burn oil instead of wood or coal between 1900 and 1915).

The Pullman Strike shut down the railroads across the country and led to troops being stationed in Sacramento, shots being fired, and a fatal train derailment.

1896

The Shops were connected to the Folsom Powerhouse, which was the beginning of the shift from line shaft power (power gained from rotating overhead shafts connected to a centralized powerhouse) to electric motors.

1897

Gold was discovered in Alaska.

1898

A fire destroyed one of the pattern shop storehouses

and another fire destroyed the car machine shop and the planing mill.

Southern Pacific Company's Passenger Department published the first monthly issue of *Sunset Magazine* to promote settlement, travel, and investment in the states it served.

1904

The locomotive machine shop nearly doubled in size with completion of the new erecting shop. The old erecting shop was converted to a bay for large tools.

1906

A prototype of an all new steel Harriman passenger car was built.

The eucalyptus trees, many of which were planted in 1876, were removed.

1910

The last wooden passenger car was built.

Southern Pacific provided 33 percent of all jobs in Sacramento.

Western Pacific Railroad began through passenger service between San Francisco and Salt Lake City.

1915

The Panama Canal was built.

1916

Three separate fires damaged or completely destroyed the second floor of store building number 1, store building number 4, and car shop number 3.

1917 and 1918

Due to the inability to purchase new locomotives, construction of new locomotives began and included one 2-8-0, two 4-6-2's, five 2-6-0's, eleven 4-6-0's, and thirty-four 0-6-0's.

1919

The area of the Shop grounds reached 145 acres.

1922

During the last four years, an estimated 3,000 freight cars had been built at the Shops. At this time, 1,200 freight cars went through the Shops each month for servicing and repairs. The workforce reached 3,100.

1924

An Oxy-Acetylene plant was built to supply piped gas to the Shops.

A new planing mill was built and driven by individual electric motors.

1926

The new Southern Pacific Station was dedicated.

1930

A statue in honor of Theodore Judah was cast in the Shops.

Appendix Two

Local Newspaper Articles Describing Shop Activity and Growth, 1863-1887

The *Sacramento Daily Union*
March 21, 1863

PACIFIC RAILROAD IRON

The gratifying news was telegraphed yesterday to the Governor, that five thousand tons of iron had been purchased for the California Central Pacific Railroad, by the agent of the company, C.P. Huntington, and paid for by him from moneys furnished by the Board of Directors. This places at the disposal of the Directors iron enough to lay the first fifty miles, and insures the completion of the road across the Sierra Nevada within the time specified in the bill passed by Congress. That bill requires that the first fifty miles shall be completed within two years from the date of the acceptance of the conditions of the same by the company, and fifty miles a year thereafter. The acceptance of the terms, we believe, dates from last December. It is, we understand, the determination of the President and the Directors to have the eighteen miles now under contract completed and in running order by the first of next September. The contracts for the locomotives, cars, etc. were made some time since, and are in process of fulfillment. They will be here in time for service at the date named. Everything appears to be going on very encouragingly, and before the people of California and Nevada wake up to the importance of this great undertaking, the locomotives and trains will be crossing the mountains. As soon as possible now the balance of the fifty miles should be put under contract.

❧ ❧ ❧

The *Sacramento Daily Union*
August 4, 1863

STRIKE FOR WAGES

The work of grading the track of the Pacific Railroad from the Chinese Chapel to Front and K streets, which was to have commenced yesterday, was deferred in consequence of a strike for higher pay. Teams have been at work at $4 per day, but yesterday an advance of $1 per day was demanded

RAILROAD MATERIAL

Material in the shape of railroad iron and ties is now arriving daily at the levee, from San Francisco, for the Central Pacific Railroad, the Northern California Railroad, and the Freeport Railroad, on all of which the work of construction is progressing rapidly.

❧ ❧ ❧

The *Sacramento Daily Union*
September 30, 1863

RAILROAD AFFAIRS

The arrival of a portion of the iron for the Pacific Railroad opens a new chapter in the history of the enterprise. Although the shipment was not a large one by the ship which came into port, it is sufficient for a beginning. The hundred tons will soon be on the levee, to be immediately transferred from that to the track. As many as three ships sailed with iron before the Herald of the Morning, which has dropped her anchor in San Francisco. They are therefore overdue, and may be expected any hour in the day. It is understood that over three thousand tons – including the hundred which have arrived – have left Boston. The contract was for five hundred tons per month until the sixty thousand contracted for had been delivered. It may be anticipated that the iron will, from this time forward, come into port as fast as the contractors are ready to lay it down. The first subdivision of eighteen miles is now substantially ready for the ties and iron, and no delay will be experienced in putting it down as it arrives. The bridge over the American will be finished in time for the rails. The trestling on this side of the river is all up, and the timbers for the bridge all formed and prepared for raising. The line of trestling on this side of the river is twenty two hundred feet in length, and on the north side it is six hundred and fifteen feet long. The bridge, which is on the Howe principle, crosses the river in two spans of one hundred and ninety two feet each in the clear, with an elevation above the highest water ever known of six feet in the clear. For present purposes the piers are to be built of wood, and so arranged as to permit stone piers to be built within them. The subjoined dimensions of the bridge we take from a late report of the chief engineer to the Board of Directors: It contains 188,000 feet, board measure, of timber; 44 tons of iron bolts and washers, and weighs about 400 tons, which, with 375 tons (the greatest load to which it can ever be subjected), gives 775 tons as the whole weight of structure and load, one-half of which, or 388 tons, is supported by each span. The total length is 397 feet; height of truss, 24 feet; width of truss from out to out, 18 feet 8 inches; width of truss in clear, 13 feet. It has two bottom chords, consisting of six pieces each, two of which are 5 ½ by 16 inches, and four are 4 ½ by 16 inches, with an intermediate space of 1 inch, admitting the main tension bolts, giving an aggregate cross section of lower chord 896 square inches. There are also corbels at either end of each span, with a cross-section of 792 square inches, the corbels extending 22 feet from either end of chords and resting on four wall plates, 10 by 12 inches. The upper chords also consist of six pieces of the same width as the lower chords, 12 inches deep, with a center rib of 6 inches in depth, extending over six panels, and firmly bolted to the upper chord, giving an aggregate cross-section of 1008 square inches.

The abutments, the engineer says, are of redwood piling, driven at intervals of two feet from center to center. It was necessary, in the middle pier, to use iron shoes to drive the piles to the required depth, the hard bottom being composed of [word not readable in copy]. The piles are tenened and capped with timbers twelve inches square, upon which are held longitudinal timbers of the same dimensions as the caps, one foot apart, and secured by drift bolts. On these timbers a solid flooring of ten by twelve inch timbers is laid, projecting one foot beyond the footing center of the intended masonry. There are three hundred and seventy piles in the foundations, of about thirty feet in length.

The first subdivision extends eighteen miles, where the Pacific intersects with the California Central Railroad. Of the first division of fifty miles, which the law of Congress requires to be completed within two years from the date of the commencement of the work – thirty miles are yet to be graded. Contracts for grading thirteen have been let, and the contractors are working over three hundred men on the thirteen sections. The contracts for grading are to be completed by the first of January, and before the first of February, 1864, the iron will be down and trains running for a distance of thirty one miles on the Pacific Railroad, which will land passengers and freight within four miles of Auburn. The other nineteen miles of the first division have been bid upon, and will be let within a month. A slight delay has been caused by the factious opposition made to the issuing of county bonds voted the company by the people of Placer, Sacramento and San Francisco. The legal issues raised have all been decided by the Supreme Court, notwithstanding which the Supervisors of San Francisco seem determined not to issue the bonds until after a case from that county has been acted upon in the highest Court. But these questions will soon be finally disposed of, and the bonds voted placed at the disposal of the company. The nineteen miles of the first division yet to be put under contract, include some of the heaviest work to be encountered in crossing the mountain, and in the opinion of the engineer, cannot be completed much, if any, before the end of the two years given by Congress; those two years will expire on the 1st of December, 1864. The length, therefore, of the Pacific Railroad which will be in operation when the trade and travel fairly open next spring to Washoe will be thirty-one miles. This will place passengers and freight within a hundred and twenty miles of Virginia by the Dutch Flat Wagon Road. Had it been possible to complete the fifty miles by next June, it would have added wonderfully to the usefulness of the road and to the income of the company.

The law of Congress provides that the second division of fifty miles shall be completed by the company within one year after the first is in running order. To build fifty miles of railroad in the Sierra Nevada mountains in one year, in the opinion of the engineer, is not within the range of possibilities. It is physically impracticable to build it within that time, and hence the contract for those fifty miles must be let so as to begin work next spring, or the company will be forced to approach Congress with a petition for an extension of time. This would prove embarrassing, if not absolutely dangerous, and if possible should be avoided. To provide for this contingency we understand the Directors have under consideration a plan for letting the second fifty miles to some company [word not readable in copy] this or the other side which is able, from its own means and credit, to put the second division in running order by the 1st of December, 1865. Building that fifty miles is a giant undertaking of itself, the cost will be over five millions of dollars. It will require a company with immense resources at command to take the contract and successfully complete it. The estimated cost of the first division of fifty miles is three millions two hundred and twenty-one thousand four hundred and ninety-six dollars. The first and second divisions, when completed, will fall short a few miles of reaching the summit of the mountain. It will require another division of fifty miles to place the temporary terminus of the road within commanding distance of all the trade and travel of the State of Washoe and of Utah Territory. The under taking is one of immense magnitude, but the business of the road, when in a running condition, will exceed that of any railroad on the face of the globe.

❧ ❧ ❧

The *Sacramento Daily Union*
October 20, 1863

THE GOVERNOR STANFORD

The work of putting together the new locomotive "Governor Stanford", which was discharged at the foot of I Street several days ago, will be commenced today. It will be required, of course, for service in laying the track of the road which work will be commenced with but little delay.

❧ ❧ ❧

The *Sacramento Daily Union*
October 27, 1863

THE FIRST RAIL LAID

Yesterday morning the contractor to build a section of eighteen miles laid the first rail on the western end of the Pacific Railroad, as described in the bill passed by Congress. Quite a number of persons were present to witness the work, though no notice that it was to be done had been published. Those engaged in the enterprise did not choose to have any ceremony over the affair; they made a regular business matter of an event which in the eye of the public is the first certain step taken in building the great Pacific Railroad. Grading has been done, bridges built; but nothing looks to the public so much like making a railroad as the work of laying down the iron on the road bed. On the Atlantic side the contract for building the section through Kansas has been let two or three times, but up to this date we have seen no report of rails laid, though not long since we saw it stated that a shipment of iron had been made from New York for the Kansas section. But no iron has yet been laid. The credit, therefore, of having put down the first rail on the line must be awarded to the California Central Pacific Railroad Company. A few weeks since it was reported that all the stock of the Union Central Pacific Railroad had been subscribed, the ten per cent paid in, and the company organized. The company is to build the road from the western line of Kansas to the eastern boundary of California. The law, however, provides that the Central California Pacific Railroad Company may continue to build east through Nevada and Utah Territories, in the event of their building their road to the east line of California before the Union Pacific Company reach that point from the East with a railroad. The prospects now are that the California company will complete their road to the east line of the State before the Union Company finish theirs through Nebraska. In fact, the road must be built from the two ends; upon the center section little can be done until it can be reached by rail each way. Hence the vast importance of pushing the work at the east and west ends of the road as rapidly as possible. On this point we maintain that the Central California Pacific Railroad Company has accomplished more than could have been expected under the circumstances. It is but little over a year since the Pacific Railroad Bill was received in California. It was signed by the President on the 1st of July, 1862, and reached California in August following. Within the intervening time the company has obtained subscriptions to the stock for nearly a million of dollars; sent an agent to the East, who purchased the iron and rolling stock for seventy miles of the road, six hundred tons of which have arrived, while four thousand tons are known to be afloat; seventy-five miles of the road have been carefully surveyed and located, and thirty miles put under contract, eighteen of which is now ready for the iron; and, as before stated, a commencement to lay it

down was made yesterday. Unless delayed by the failure of the iron to arrive, the eighteen miles will be in running condition before the first of December. The twelve more to make the thirty are under contract, to be completed on the 1st of January, 1864. During the winter, the locomotives and trains of the company will be running to the Thirty Mile Station. The means, so far, have mostly been obtained from stockholders, who have paid the assessments levied by the Board with remarkable promptness. In addition to the private subscriptions obtained, the officers of the company, in conjunction with the friends of the enterprise, succeeded in getting bills passed authorizing the county of Sacramento to subscribe for stock to the amount of $800,000, Placer $150,000, and San Francisco $600,000- provided the people of said counties voted in favor of the proposition. At the special elections called for the purpose, the majority of the people of each county voted to subscribe for the stock. A factious opposition was made by a rival railroad interest, to prevent the bonds of each county from being issued to the company, but the Supreme Court has decided the Acts authorizing the counties to subscribe to be constitutional, and the bonds have been issued in all except San Francisco. In the latter city a malignant opposition has manifested itself, and every technical legal obstacle has been thrown in the way of issuing the bonds. The first step was to obtain an injunction which, after hearing testimony and argument, the Court dissolved, and now the parties resisting are causing further delay by appealing the Supreme Court in the face of the fact that the Court has decided every point which can be raised in favor of the right of the counties to subscribe for the stock and issue bonds for that purpose. The Board of Supervisors, though, might, if so disposed, issue the bonds, as all legal obstacles to their doing so have been removed. But, that body does not seem disposed to aid the Pacific Company so far as to issue them. Their plea in justification is that the case has yet to be heard in the Supreme Court, though it is known that in other cases that Court has decided every point in favor of the company which can possibly be raised in the

San Francisco case. The twenty miles required to make up the first fifty are surveyed and prepared to let, and as soon as the bonds of San Francisco are issued, bids will be asked for by advertisement. The Board of Directors make no contracts until they know definitely where the funds to meet the payments are to be obtained. The first fifty miles of the road are to be built by the means furnished by stockholders – the counties – named being classed as stockholders. The bonus to be obtained from Government cannot be realized until after the fifty miles are completed. The amount to be received from the State is in nearly the same condition. Hence the company is called upon to build the first fifty miles from its own resources and the subscriptions of the counties, and some of the heaviest work on the line is met in the twenty miles above Auburn. When the cars are running fifty miles the company will, besides the earnings of the road, receive the $48,000 per mile and the loan donated by Congress, as well as the $10,000 per mile granted by the State to aid the enterprise. Therefore, the completion of the first fifty miles solves the financial problem connected with building a railroad over the Sierra Nevadas. That fifty miles the company will have in running condition before the first of December, 1864, which is the date named in the bill for the completion of that section. Opposition to the issuing of bonds in San Francisco may cause delay, but it will only be temporary. The road will continue to advance. The work of laying the rails has begun, and it will continue until California and Washoe are united by iron bands, and until the iron rails are stretched across the continent. It is hardly twelve months since work was actually commenced, and yet within that short space of time the greatest obstacles in the way of building a Pacific Railroad have been met and surmounted. With fifty miles of road in operation, the company will have become an institution which will be recognized in the financial markets of the world. It will possess character and credit equal to millions, and will be enabled to proceed with full confidence in its resources and in the future.

The *Sacramento Daily Union*
November 6, 1863

NEW BUILDING

A new frame building was commenced on Wednesday by the Pacific Railroad Company on the west side of Sixth Street, north of "I". Opposition was made to the building on account of its being on the street. The workmen were stopped for an hour or two. Yesterday morning opposition was withdrawn and work on the building was resumed. The building, when finished, will be about twenty feet wide and one hundred and forty feet long. It is designed as a work shop, to be used for fitting up railroad cars, etc. The southern end of the building, some thirty feet in length, was erected yesterday.

The *Sacramento Daily Union*
November 7, 1863

STEAM UP

Yesterday afternoon steam was gotten up in the locomotive Governor Stanford, at the foot of "I" Street. The engine could not be set in motion, as it has not yet been placed upon the track. On Monday morning it is expected that it will go to work hauling iron for the purpose of laying the track.

The *Sacramento Daily Union*
November 7, 1863

THE FIRST RIDE ON THE RAILROAD

The locomotive Governor Stanford was so far prepared for service last week that steam was gotten up in her boiler. It was decided yesterday morning to place her on the track and make the trial trip on the road. The track had been laid from Front and K streets to about Twenty-first street, on the north levee-a distance of about two miles. The officers of the railroad company, about noon, gave verbal invitation to a dozen or two of citizens, including State officers, bankers, merchants, editors, etc. to be hand at Front and I Streets, at 2 1/2 o'clock, to take a ride. Everybody was, of course, anxious to take the first ride on the Pacific Railroad, and everybody was therefore on the spot. Considerable delay occurred in getting the locomotive on the track. It had been put together several rods from the permanent track, and rails were temporarily laid for the purpose of forming the connection. Steam was finally let on, and after several trials, the engine passed over the temporary and gained its position on the permanent track, and moved slowly forward. The invited guests took their position alongside the track, ready to jump on board, but before the locomotive halted, the tender- which constituted the only accommodation for passengers-was crowded to the extent of its capacity by men and boys, who climbed into it two or three deep while in motion. The engine moved slowly as far east as Fourth Street, with the design on the part of the engineer of returning to Front Street for the expectant passengers. It was found, however, that the engine was not in running order, and that some of her valves, pumps, etc., required overhauling. She returned to the starting point, and the trip was for the time being abandoned. Young America, however, ad enjoyed "the first ride on the Pacific Railroad." That was a fixed fact. The ride was not long nor fast, but that it was the "first" there is no denying. A basket or two of champagne which had been placed on the tender was withdrawn and stored away for future reference. Previous to the starting of the engine the new Union gun purchased by W. Siddons was brought on to the ground. As the locomotive started the event was heralded by the report of a salute of thirty-five guns. The gun, which is a twelve-pounder, spoke in more emphatic tones than any heretofore used in this city. After the crowd dispersed the engineer went to work on the locomotive, and by about half-past eight o'clock had the defect in the machinery remedied and steam up again. After that hour the engine made two or three trips to Fifteenth or Sixteenth streets, the tender being crowded with volunteer passengers who cheered lustily as they passed along the streets. The Governor Stanford ran finely after being overhauled, and will doubtless to-day be used in transporting iron over the road.

The *Sacramento Daily Union*
November 19, 1863

CAR BUILDING
The workshop of the Pacific Railroad Company, on Sixth Street, above "I", having been finished, is now occupied by workmen employed by the Company. Some eight or ten mechanics are there busily engaged in getting out the wood work for freight cars for the Company. Oregon lumber is currently use for the purpose. The iron work will of course be brought from the East.

The *Sacramento Daily Union*
December 3, 1863

RAILROAD IRON
There is now railroad iron enough in the State, belonging to the Pacific Railroad Company, to lay about four more miles of the track than is now constructed. There are now out from Boston or New York, for San Francisco, sixteen different vessels which are freighted in part with iron for this company. Of this number eight are already due, and either may be looked for at any time. These are the Winfield Scott, from Boston, out 180 days; the Courier, from Boston, out 159 days; the Cremorne, from New York, out 155 days; the George Peabody, from New York, out 155 days; the Otis Norcross, from Boston, out 150 days; the Thatcher Magoun, from New York, out 133 days; the Governor Langdon, from Boston, out 127 days; the Queen of the East, from New York, out 125 days.

The *Sacramento Daily Union*
December 13, 1863

PORTABLE HOUSES

Three cars have been constructed at the workshop of the Pacific Railroad Company on Sixth Street, to be used as houses for the workmen in laying the track. They are constructed like the ordinary platform car, except that the planking projects so as to make the floor considerably wider than that of the car. On this floor are constructed a commodious house. One of these will be used as a storehouse and for cooking; another for eating in, and the third is provided with bunks for lodging purposes. These houses on wheels will be placed on the track and moved along as the track advances toward the mountains.

❧ ❧ ❧

The *Sacramento Daily Union*
March 15, 1864

THE FIRST CAR

The first passenger car of the Pacific Railroad Company was completed last week, and placed yesterday on the track, on I Street. Another finished car remains in the workshop of the company, on Sixth Street, and will be brought out in a day or two.

❧ ❧ ❧

The *Sacramento Daily Union*
March 26, 1864

THE FIRST FREIGHT

The locomotive Gov. Stanford arrived at dusk last evening from the foothills, with three car loads of granite, about thirty tons, taken from Brigham's quarries, on the line of the Pacific Railroad. This is the first freight which has passed over the road. It is the first installment of a contract of 500 tons which Brigham is to furnish to P. Caduc, of San Francisco. It will be shipped to-day by the schooner J. A. Burr, Captain Johnson, now moored near the foot of J Street. At that point a large derrick will soon be put up for the purpose of handling granite. The Pacific Railroad Company has contracted to bring down the 500 tons at $1 per ton. This granite – which may be seen at the foot of J street – presents a fine appearance, but being taken from near the surface is not, of course, equal to what it will be when the quarry is fairly opened.

❧ ❧ ❧

The *Sacramento Daily Union*
April 26, 1864

THE FIRST TRIP

The first trip over the Pacific Railroad for the purpose of carrying passengers was made yesterday. The company had advertised to run at stated hours, and, accordingly, at 6 ¼ o'clock the locomotive Gov. Stanford, with a number of passengers, left the foot of J street. In consequence of the difference of gauge between the Pacific and the California Central roads the same engine and cars cannot run over the same track. Arrangements had, therefore, been made for the cars of the Central road to connect at Roseville, the junction, with the Sacramento train, and convey the passengers to Folsom. The time made by yesterday's train was, from Sacramento to Folsom, twenty-six miles, fifty-nine minutes. From Folsom to Sacramento, fifty minutes. From Sacramento to Roseville, eighteen miles, thirty-nine minutes. From Roseville to Sacramento, thirty-three minutes. Arrangements have been made to narrow the track of the California Central road to correspond with the gauge of the Pacific road. This change will be made, it is said, within a week, when the rolling stock of the Pacific road can run to either Folsom or Lincoln, as may be desired. The rolling stock of the Central road will of course be altered to suit the narrow gauge. The gauge of the Pacific road is four feet eight and a half inches – that of the Central road five feet.

❧ ❧ ❧

The *Sacramento Daily Union*
December 20, 1864

PACIFIC RAILROAD REPORT

We are indebted to S. S. Montague, Chief Engineer of the Central Pacific Railroad, for a copy of his report just made, upon the recent surveys, progress of construction, estimate of receipts, etc. of the road, together with the report of E. H. Miller, Jr., Secretary of the Company, showing the actual business transacted by it from the 26th of April last, at which time the cars commenced running, up to the 1st instant.

❧ ❧ ❧

The *Sacramento Daily Union*
May 18, 1865

JUBILANT

The passenger cars of the Pacific Railroad made their first trip to Auburn on Saturday last, the 18th inst. The California Stage Company have removed their stables from Newcastle to this place, and the change has made a material difference in the bustle and activity about Auburn. An Auburnite can now take ample time for a hearty breakfast-step aboard the cars-take his seat in the baggage car, light his meerschaum, and, by the time he has finished his smoke, find himself in the city of Sacramento. Two or three hours having been devoted to business, he once more enters the cars and is home for dinner. Ain't it nice? Some improvement on the old Concord coach style-don't you think so?

❧ ❧ ❧

The *Sacramento Daily Union*
May 18, 1865

THE TUNNEL COMMENCED

S. S. Montague, Engineer of the Pacific Railroad, has just returned from the summit of the Sierras, which point he visited for the purpose of starting the work on the summit tunnel. This tunnel will be 1,750 feet, or about the third of a mile long, twenty-six feet wide and twenty feet high. The excavation will be sufficiently wide for a double track. The entire work runs through solid granite. It is expected that a year and a half will be required to complete it. The work was started by Montague at both ends. This tunnel is not level, but descends to the east at the rate of ninety feet to the mile. The summit is about fifty miles east of Colfax. Thirteen miles of the road between Colfax and Dutch Flat will be graded by the first of January. It is expected that the summit will be reached before the tunnel is completed, and that a temporary track will be laid over it for the purpose of facilitating the work on the eastern slope.

❧ ❧ ❧

The *Sacramento Daily Union*
October 24, 1866

THE SNOW PLOW

The directors of the Pacific Railroad Company have had constructed at their workshops on Sixth street a novelty, at least in this section of country, in the shape of a snow plow. This apparatus runs on two ordinary car trucks, which are entirely concealed from view by the outer casing of the frame work. The machine, which is ten feet wide, eleven feet high and thirty feet long, resembles a huge wooden box, made wedge-shaped in front, with the sharp point resting as near the ground as practicable for working purposes. The slope of the wedged portion runs up at an angle of about forty-five degrees. On this portion of the apparatus, several feet from the ground, is a large double iron plow. The machine, when is use, will be placed in front of a locomotive. The wooden incline will raise the snow from the track, and the iron attachment will throw it off at either side. The iron attachment can be shifted to one side or the other, as necessity may require. Machines somewhat similar to this are said to be in use on the railroads of the Northern States.

❧ ❧ ❧

The *Sacramento Daily Union*
December 10, 1866

PACIFIC RAILROAD WORKSHOPS

The Directors of the Central Pacific Railroad Company are preparing to commence at once the erection of new and extensive workshops. These important works will be located on the line of the new track north of the slough, between Second and Fourth streets. They will consist of a machine shop, a car shop, an engine-house, a blacksmith shop, a foundry, a paint shop, a stationary engine-house, two storehouses, and two transfer tables. All these buildings will be built of granite from the Rocklyn quarries. The dimensions of the buildings are as follows: Machine Shop, 100 feet wide, 200 feet long and one story and a basement high. Car shop, 90 feet wide, 200 feet long, with 25 foot L, and two stories and a basement high. Engine-house – capacity 30 locomotives – semi-circle 380 feet diameter, 60 feet deep, one story high; turn-table in the center, with thirty tracks diverging to the building. Blacksmith shop, 60 feet wide, 80 feet long and one story high. Foundry, 40 feet wide, 80 feet long and one story high. Paint shop, 50 feet wide, 100 feet long, one story high; stationary engine house, 30 feet wide, 40 feet long, one story high; two storehouses, each 40 by 40 feet, one story high. For the machine shop and car shop the foundations will be piles. The base of the granite walls will be five feet wide. They will rest on three tiers of redwood piles, driven about two feet apart. The walls of the basement stories will be three feet wide and those of the second stories will be two feet wide. These buildings will all be erected and arranged with a view to their extension and enlargement whenever the business of the company may require it. The pile driver now in use on the ground will be employed about two weeks in driving piles for the bridge across the slough. It will then be set to work at once in driving piles for the foundation of the machine shop. A considerable quantity of sand has already been hauled upon the ground to be used in mortar for laying the granite. A full supply of machinery and tools, including engines, lathes, forges, shafting, gearing, etc., etc., for equipping these workshops, was received from the East several months ago. When complete, the company will be prepared to build their own passenger cars, freight cars and locomotives, including the casting of their own car wheels. It is expected that five hundred hands in all will be employed in the several departments of the establishment. The business and machinery of the shops now in operation on Sixth street will, of course, be transferred as soon as practicable to the new establishment.

❧ ❧ ❧

The *Sacramento Daily Union*
April 8, 1867

SACRAMENTO AND CISCO

The singular conjunction of climates in California is a constant cause of wonder to the tourist. The phenomenon is perhaps more remarkable at this season than at any other period of the year. Spring smiles in the valley of the Sacramento and wreathes her brow with floral beauties of richest hues and sweetest fragrances. The plains are covered with a luxury of verdure, and myriads of golden, blue, pink and purple wild flowers are floating on the ripples of an emerald sea. The youngest wheat outshines the grass. Orchards are adorned with white and pink promises of fruit. The full, liquid note of the lark and the twitter of the linnet make music in every field. The sun laughs through vails of April showers. The air is pure, genial and balmy. Yet the resident of the valley, listening to the chorus of the birds and arranging bouquets of the untamed beauties of the plains, can see the white summits ridging the eastern sky, and knows that within five hours he may ride in among the ice and snow of a Canadian Winter. Yes, the sleigh-bells are ringing in the mountains, and muffled people are speeding along beaten roads, walled with snow to the height of from ten to fifteen feet, while the bloom and beauty, odor and music of spring are around us here. For each glittering shower that gives a lusher growth to grass and flowers in the valley there is a cold and quiet shower of snow among those distant heights. For each warm and gentle breeze that calls fresh blossoms forth in these orchards the mountains have a chill and furious gust. The resident of the valley fears the flood; the worker on the mountain highway dreads the avalanche. The Sierra may be imagined sitting upon a throne of gold-veined rock, resting his feet on soft cushions of flowers, armed with the "thunderbolt of snow", lifting his silver brow to the blue dome of sky.

The influence of the severe climate of the mountains upon transportation is a matter of deep concern to Californians and of not trifling interest beyond our borders. Now that the Spring has returned and heavy snow-storms are not to be expected, it is important to inquire how the railroad men have fought the battle against the tremendous obstacles of Winter and what lessons they have learned to govern the operations of the future. The snowplows are still ready for service; but, except in the case of avalanches thundering down upon the track, they may be regarded as having completed their share of work for the season. The first great fall of snow was easily managed, though deeper than the snow which stopped all railroad travel in the Eastern States for two days, and in spite of the washing down of sections of the road below the wintry belt, which checked the running of trains through to Cisco. Before the line had been fairly restored, other storms broke upon the mountains, and from that time until the middle of March, the snowplows were kept in almost continual service. The winter was one of the fiercest ever known in the Sierra Nevada. If the report which some half-smothered miner sent from Meadow Lake to the effect that "it stormed for a month," was not exactly square with the truth, it is certain that the storms were unusually frequent and prolonged. Snow fell five days of one week. Yet the railroad men, though sometimes stopped at Colfax or Alta, persisted in grappling with the obstacle, determined to prove that even in this terrible season, without the instruction of experience and bothered by the settling of a new road-bed, they could run trains to Cisco.

For the greater portion of the distance traversed by the locomotive among the mountains, the line is on the side of a ridge, the height towering on one hand and a ravine yawning on the other. It was found that with the aid of the big, independent snow plows, where the track was not walled in by rock and earth, the result of the heaviest storm could be soon cleared away, the bulk of the snow being thrown over into the ravines. The greatest difficulty was caused by the accumulation of snow in the deep cuts. For a time the plows cleared the track, but then the snow was banked up against the rocks on either side, and there was no place to receive the heaps thrown from the railroad after succeeding storms. In such cases it was necessary to employ a force of laborers, armed with shovels, and to haul the obstacle away in the trains of freight cars returning empty from above. The freight cars which have come into Sacramento loaded with snow, to the infinite delight of snow-balling youth, have borne witness to the kind of work done in clearing the cuts. This costly experience has satisfied the railroad managers that to avoid such trouble hereafter they must either widen the cuts or cover them. The former remedy would be expensive and of limited effect. Between Alta and Emigrant Gap the cuts are numerous and most of them are deep. They could not be widened without an outlay of about one-third the original cost, and then it would be doubtful whether sufficient space could be gained for depositing the snow removed from the track. The railroad men have therefore decided in favor of covering the cuts, and have accepted a plan for roofing structures which seems capable of furnishing ample protection. Where the snow falls to the depth of ten or twelve feet on a level in the course of a winter, such roofs must be strong. We have seen, near the summit of Johnson's Pass, on the Placerville route, the roof of a new barn, with a sharp gable and supported by huge rafters, crushed in by the weight of the accumulations of many storms. According to the design adopted by the Railroad Company, the roofing of the cuts will have a double support of stout timbers at the eaves and half-way between the eaves and the peaked top, so as to be strong enough to sustain the weight of many tons of snow. Thus far no train has suffered from the fall of an avalanche, and the drifts have not proved to be such serious obstacles as they were supposed to be. Snow-slides are rare incidents along the line, and there is hardly one chance in a hundred that disaster will result. To sum up the wintry experience of the Railroad Company, it is satisfactory in proving that with strongly covered cuts, a well-settled road bed, the free employment of plows and frequent trains, such as the business of the road will require after the locomotive crosses the summit, the mountain road can be worked through in the severest seasons, carrying freight and passengers with more regularity and, of course, with more speed, than has hitherto been found practicable. Between Blue Canyon and the summit the snow is deeper now than at any previous period of the winter; yet the railroad trains and the sleighs are making regular trips and excellent time. There is snow on the hills above Alta, within seventy miles of Sacramento, where the gardens are in bloom and the fields are arrayed in a refreshing green. And the swift transition from the reign of spring to the Arctic realm of the Sierra is one of the most novel phases of experience in California.

The *Sacramento Daily Union*
August 9, 1867

RAILROAD WORKS

Several months since notice was made of the contemplated Central Pacific Railroad workshops, car and engine houses to be erected in this city. On yesterday we visited the grounds and, through the politeness of Woolaver, the draftsman, and Wilkinson, the engineer, obtained information relative to the extent and manner of the structures whose foundations are now being made. The buildings laid out for the present cover about ten acres of ground and are so planned that extensions can be conveniently made to each one of them. The extreme eastern line is occupied by a temporary car shop, which is 60 by 160.

This superstructure is entirely of wood. The floor is supported by wooden columns standing on blocks of granite. This building is immediately connected with the main trunk of the road, and though intended to answer the present purpose only, yet in all probability the company will find it both useful and convenient for some department of the immense work they will have to make. West and a little south of this temporary car shop is the permanent car shop, whose foundation is now being made. This building is 230 by 90, but with an ell 90 by 45 feet, and can be added to when required without interfering with or having to pull down any portion of the work. Adjoining on the southwest corner there is to be an engine house, a fuel apartment and water tank on top sufficiently high to flood water all over the entire works. Twenty-five feet from the carhouse is the building or machine shop, 205 by 100 feet. The east half of this building is to be appropriated to lathe machines and work benches. The west half will have eleven tracks or pits, one of which is a drop table. An engine needing repairs is run on this table, and by machinery the track is let down so that the frame of the engine can rest on bearings while the wheels and undergearing can be taken out. The tracks or pits extend westward to a transfer table which connects with small turn tables and then with the main turn table that intersects the various tracks in the round-house. The round or engine house is north of the railroad shop track and is semi-circular in form, being three hundred and seventy-eight feet in diameter by 64 feet deep, and can accommodate twenty-nine engines at once. In the center of this semi-circular building is to be the main turn table on the track that will connect with the various departments of the shops, and also with the main road. The foundation of the last mentioned building is not piled. The trenches are dug and large blocks of granite eight feet long, by all thicknesses, from two to four feet are being laid. The manner of masonry is known as rubble work, which is to extend about twelve feet from the present surface, upon which the superstructure is to be of brick. The car and machine shop foundation is first piled with sawed red wood twelve inches square, driven down twenty-five feet, and standing about twelve inches apart. Upon these piles a layer of sand is wet down, then a course of cobble, upon that granite broken in small pieces and capped with large blocks of granite laid in cement, and the whole graveled. The foundation is twelve feet wide, and will extend twelve or fifteen feet in height, on a level with the main track, which passes outside of the entire works on the north. Besides the buildings mentioned the paint and blacksmith shops are arranged for and properly connected with an eye to convenience and adaptability for their purposes. S. D. Smith has the contract to make the granite foundations of all the buildings and has men on the ground busily engaged. He has already three derricks up in working order to handle the massive blocks of granite and contemplates erecting five others in a few days.

❧ ❧ ❧

The *Sacramento Daily Union*
September 27, 1867

PACIFIC RAILROAD SHOPS
The workmen engaged at the foundations of the new shops of the Pacific Railroad, north of the slough, are progressing so rapidly that their work begins to show to advantage. Immense foundation walls of granite are being laid, which indicate the size and character of the building to be erected. Portions of these walls, which are from six to eight feet wide at the base, are now eight or nine feet high. It is the design of the builders to complete them to the high grade or the level of the railroad track before the rainy season fairly sets in. In that case the work of building can be continued with but little interruption during the winter. A large frame shop has already been erected on the ground and is being supplied with machinery, which will be in running order in about two weeks. This shop is 50 feet wide by 150 feet long. It contains a steam-engine, one of Daniels' planing machines, capable of planning timber 32 feet long; a morticing and boring machine; a tenoning machine; a surface planer;

two wood-lathes; three rip-saws, one circular saw; one scroll saw; one molding machine, etc. A portion of this machinery will be employed, when started, for car building, and a portion in making sash-doors, etc. for the new work-shops.

❧ ❧ ❧

The *Sacramento Daily Union*
December 18, 1867

NEW REPAIR SHOP
The Pacific Railroad Company have in process of construction on Sixth street, in the neighborhood of their machine shops, a frame building over two hundred feet long, designed apparently for use as a repair shop or car-house. Yesterday two fine locomotives were occupying a portion of it-the Idaho, which is having some little repairing done; and the Santa Clara, a new engine, which is being put together. Work at the shops, in manufacturing and repairing, appears very brisk.

❧ ❧ ❧

The *Sacramento Daily Union*
April 14, 1868

PACIFIC RAILROAD SHOPS
The contract for erecting the brick workshops of the Central Pacific Railroad Company has been let to Mc-Cants & Penman. It is expected that about four million bricks will be required to complete the work. Active operations will commence north of the slough in about ten days if additional rains do not render an additional postponement necessary.

❧ ❧ ❧

The *Sacramento Daily Union*
October 31, 1868

CENTRAL PACIFIC ROUNDHOUSE
The roundhouse of the Central Pacific Railroad is nearly

completed, and will be occupied by the locomotives of the company soon. The twenty-nine tracks diverging from the turn-table and leading into the roundhouse have all been laid, and the turn-table itself, which is fifty-six feet in diameter, is well advanced toward completion. A force of forty or fifty workmen are engaged in the roundhouse, putting in running order new engines, of which four are being operated upon at present-the White Bear, Buffalo and Mountaineer, from Rogers' Works, Patterson (N.J.) and the Growler, from the Danforth Locomotive and Machine Works, likewise of Patterson.

❦ ❦ ❦

The *Sacramento Daily Union*
February 2, 1869

THE SNOW TROUBLES

A correspondent of the Virginia *Enterprise*, writing from Reno, February 16th, has a long account of "snow life" in the Sierra Nevada on the line of the Central Pacific Railroad, which is decidedly interesting, both as showing the difficulties of that route in winter and the hard experience of the traveler. We make the following extract:

At Auburn an occasional flake of snow of huge proportions foretold the approach of the train to the mountains; at Colfax the earth was covered, and each revolution of the engine "drivers" was more difficult; and at Alta the train was brought to a stand-still to await the arrival of the train west. The air was thick with fleecy specks; the wind swept down the heavy canyons above; the two car-loads of passengers settled to a patient waiting; the youth at the telegraph office reported "wires down in the mountains," and right then "trouble began."

The long hours of the afternoon crept slowly on. Conductor Dennison gazed on his block of passengers pleasantly, and reported that there was no hope of progress until the coming day. And so it got dark. White Piners (there was a gay old crowd of them along) got rather impatient; several Virginians manifested regret at the detention, but at least the crowd dropped off to sleep, and gave an occasional snore, and all was quiet, except when a car seat got too hard to rest on, and the shifting (we had several ladies aboard) pilgrim muttered curses too profane to do good.

Right here we desire to lay down as a proposition not to be gainsayed-there is no such thing as a comfortable position for sleep in a rail car seat-it is impossible.

Morning came, and found us in the midst of a boundless waste of snow, at least two and one-half feet having fallen during the night, and more coming. Patience weakened considerably, and a large amount of growling was indulged in, mixed with some merriment, at the delay. Toilets were made of snow water and a finger comb, and then a breakfast of good quality enabled us to await orders with more composure. Eight, nine, ten, eleven o'clock came, and then the order was given to take the train, with such passengers as chose, back to Sacramento.

Five locomotives and a huge snow-plow had arrived, bound to the front, and so your correspondent, with a route agent of Wells, Fargo and Co., bearing two huge sacks of letters, got aboard the locomotive Auburn, engineer H. Spence, and were made comfortable by that quick-eyed, strong-hearted man, who managed the huge machine with consummate skill. Hank Lancaster conducted the train, and after half an hour consumed in preparation, and to see the passenger train on its return, the signal was given for a charge at the unbroken snow bank which covered the track to a depth of nearly three feet. Two sharp whistles from the front engine, echoed by ten similar shrieks from the followers, and off they went. The snow was turned in huge furrows from the track, and the bright sun came out and seemed to make partly cheerful the heavy stillness which hung around the line of road, all undisturbed except by the puffing engines. A short run and China Ranch was reached and passed, when suddenly the snow began to increase in volume, the cuts became filled, the locomotives wheezed and labored, and directly the train came to a dead stop in the middle of a huge bank of snow.

The signal came to reverse, and haul back; half a mile down the track went the train, the advance was sounded, and with the speed of the wind the immense plow swept to the snow bank. The men on the plow swung their arms in the air with frantic motion, the engineers each whistled a double-quick advance, each throttle valve was pulled wide open and every pound of steam let on the cylinder, and the bank was again struck. A move of a few rods, and then another halt. So on, for hours, during which the hind engine became disabled and put back for repairs. The force was thus diminished one-fifth, and the snow increasing every mile-there being at the Ranch, four miles from Alta, about five feet of snow-conductor Lancaster shook his head, said more power must be had, and started a man to telegraph to Sacramento for assistance. So went the hours, and not until dark did the engines force the plow to the wood shed at Blue Canyon.

The messenger with letters was not a very patient babe; in fact, the progress of the train was all too slow. What though the snow was deep, the express must go through, and go through it did; the messenger crawling and swimming almost to the station, and then taken in hand by two hardy men of the mountain, who took it to Cisco, thence by two more to Truckee, and thence to this place by rail, where letters were forwarded East, to Virginia, Carson and smaller places. The enterprise displayed by Wells, Fargo & Co. is certainly commendable.

All day Friday was passed by Lancaster in trying to force his train along, but not a mile was accomplished, and at dusk the train backed to Alta for reinforcement. At midnight two more engines arrived, and early Saturday morning seven locomotives, behind the immense plow, rushed again to the fray. Passing the station with great speed, the bank was reached and gave way; curve, cut and trestle were passed, and a halt was not made until two miles had been accomplished. Bravo, Lancaster! Bully, Hawkins! Hurrah, Spence! Go it, Masser Daily! Then came a lull; another back, two more runs, and at six o'clock the train made the sheds at Emigrant Gap, having accomplished four miles of track through snow that had drifted, and was from five to twenty feet in depth.

The recent storm has demonstrated just this: From Truckee to Alta, the Central Pacific Railroad must be shedded-nearly ever rod-to be rendered practicable in the winter. Wherever the sheds are, two engines with a plow can clear the way; in other places, ten are inadequate to the work. We predict the road will be entirely shedded before another winter.

Sunday morning the sun rose bright and beautiful. A large force of men had arrived from Sacramento under charge of roadmaster John Fogarty, and were set shoveling. The snow had increased in quantity, and above the Gap had packed every "fill and cut". The engines were set at work, but it was slow progress they made, and during the entire day but half a mile as accomplished.

About two o'clock am we concluded to walk to Cisco, a distance of nine miles, and in company with the inevitable messenger (how the fat rascal did puff and blow before he reached Cisco) started. Pits had been dug the first four miles eight feet apart and four feet deep, so it was into one, a scrambling out, into another, and so on, the hardest work possible. Four hours tedious walking and the bridge, a mile west of Cisco, was reached. Here a tremendous slide of snow from the mountain had come down and carried away four "bents" of the bridge, at which an hundred men were at work rebuilding-which will have been accomplished long before this is in type. The train we left behind us did not get through to the bridge until Tuesday at ten o'clock, making the time from Alta just one week, and every man on the road working his level best.

❦ ❦ ❦

The *Sacramento Daily Union*
April 23, 1869

CENTRAL PACIFIC RAILROAD HOSPITAL

We paid a visit yesterday to the Central Pacific Hospital, located in the large frame building at Thirteenth and D streets, formerly occupied by the Protestant Orphan Asylum, and were pleased with the aspect of cleanliness and comfort exhibited. The institution is conducted under the superintendence of Dr. S.P. Thomas, who appears to have been remarkably successful in its management, considering the crowded condition of the building. A door opening on the right side of the main hall leads the visitor into a large ward, every bed in which is occupied by sick and disabled men. Upstairs there are three smaller wards, each of which contains three or four patients. One of these wards is assigned to patients suffering from affections of the eye. Down stairs, the small front room on the left side of the main hall is used as a dispensary, reception-room and office. Back of this is the dining-room, and still further back the kitchen. A large wing built on to the southern end of the main building is used as a sick ward, and here lie many of the worst cases. All of the bedsteads used in the Hospital are of iron. The greater portion of them are supplied with excellent spring bed hair mattresses, clean sheets and white counterpanes. Lately, however, sponge mattresses have been introduced, with an under mattress of straw. These are thought to be superior to hair mattresses, as they do not pack; and in cases of amputation, broken limbs, etc., where the patient needs perfect quiet, they are preferable to the spring beds. The walls of the wards in the main building are bright from whitewash; in the addition spoken of they are papered. The floors are clean, and the ventilation good. In the Hospital, at present, there are about thirty patients, over two-thirds being surgical cases. All appear to be in good spirits, and in a fair way to recover. Among these cases are many that are singular and interesting to medical men, comprising injuries to almost every portion of the body, rheumatism, paralysis, aneurisms, abscesses, gunshot wounds, broken limbs, cuts, etc. One the most serious cases at present is that of Perry, the engineer of the Klamath, who, by the explosion of that engine, lost his left eye, had his left thigh broken, was quite severely scalded, and had a nail buried in his forehead the depth of an inch. About a week ago it was found necessary to amputate his leg to save his life. It was doubtful whether he would recover even then, but we were glad to find him yesterday doing well. Another patient, who was shot through the bowels by a pistol ball some time since, and injured in a manner that appeared almost necessarily fatal, is also recovering. A third, of the seriously injured, has had both thighs fractured. In addition to the patients at the Hospital there are about ten convalescents boarding a the Railroad Hotel at the company's expense; there is also a class of patients who reside at their homes in the city or board themselves, but who are attended by the Hospital physician or receive medicine from the dispensary. During the month of March one hundred and seventy of these railroad patients were treated by Dr. Thomas at the Hospital or at home. A record, short but comprehensive, is kept of each, which shows his age, birthplace, when he came to this coast and where from, how long he has worked for the company, and the cause and nature of his injury or disease.

The railroad company is now having constructed on the grounds a new building, the plans for which were drawn by John Woolaver and Benjamin Welch, from suggestions made by Dr. Thomas. This building is 35 x 60 feet, with wings on the northern and southern sides each 24 x 52 feet, making an extreme frontage on Thirteenth Street of 139 feet. Back of the main building will be a kitchen, 24 x 25 feet. The first story of the Hospital will be nine feet high, the second, fourteen; the third, twelve; and an attic, eight. The foundation will be of brick; the rest of the structure, frame, to be finished in a tasty and ornamental manner. There are to be four large wards, two in each wing, capable of accommodating eighty patients. Besides these, the new building will contain a commodious dining-room, office, dispensary, operating room, bath rooms, etc., the whole rendering it, to judge from the plan, a model institution of the kind. The cost of building, furniture, etc., is estimated at $25,000 to $30,000. E.F. Woodward, who has charge of the brick work, informs us that it will be done next week, when the carpenters will be set at work. The railroad company [will] pay the cost of constructing and fitting up the Hospital, and also the salary of the physician; the employees of the road pay the cost of running the institution by a voluntary contribution of fifty cents per month.

❧ ❧ ❧

The *Sacramento Daily Record*
April 9, 1875

THE FIRST CASTING OF THE NEW CAR WHEEL FOUNDRY-EXTENT OF THE WORKS-MEN EMPLOYED

Yesterday was an initial day at the new car wheel foundry of the Central Pacific Railroad shops in this city. The first casting of wheels was made, and all the new and improved machinery put to a thorough test.

The foundry has been built on the most southerly edge of the filled in railroad lots. It is a spacious wooden structure, flat roof, with a central terrace ventilator one third the width. At the westerly end, and in the center, stands the cupola in front of a base of masonry some six feet high, and flanked by walls of brick to the floor of the platform. On the left are two ovens for baking cores, each oven having six shelves, and being six by ten feet in size and seven feet high. Back of each oven is a furnace. The shelves being filled with cores, the ovens are closed and the furnaces heated. Flame and smoke alike pass into the ovens, through flues in the floor, and thence to a chimney between the ovens. This secures the passage of heat to the lower as well as the upper shelves. On the right of the cupola are two grinding mills, consisting of two huge cast iron globes pivoted, and revolving constantly. These are

for grinding charcoal to an almost impalpable powder for facing the molds. Back of these is a cinder mill, consisting of a plain revolving barrel cylinder, used for grinding cinders to powder, which falling down leaves all the particles of iron which have adhered to the coals in either the ladle or cupola. Next to these mills on the north, is the engine room, where a double cylinder engine, constructed at the slopes, furnishes the motive power. In the engine room is the blower, and also a patent hoist for the "breaker".

This consists of an upright cylinder receiving the steam at the bottom, forcing the piston up which bears at the end of its rod a double pully and chains, and which lead to the top of the frame of the "breaker" through a series of blocks and wheels, and elevate the hammer to the top of the frame work. In the engine room, also, is the machinery of an elevator which lifts a large cage from the ground to the level of the cupola platform, and by which all the iron is hoisted for the cupola-a decided improvement over the process of throwing up the iron, as at the old works. West of the engine room stands the pyramidal building over which hangs the "breaker". Old wheels are placed in this building, the chains attached to the hammer, the hoist turned on and the hammer elevated to the top of the frame, when it is disengaged, much as in a pile driver. It descends some 30 feet, with great force, and crushes the old wheels into pieces suitable for the cupola.

In front of the cupola, which has a capacity of 12 tons of metal, is swung on pivots a ladle, with a capacity of six tons of metal. This ladle is tilted by a worm gear that is, an endless screw, working on a semicircle of cogs. In front and on either side are two massive cranes, and in a circle within the radius their arms describe are thirty moulds, each of which consists of three circular cast iron rims, within which, as they stand bolted and keyed together, are the car wheel moulds. Ten other moulds stand in the east corner of the foundry. In the west corner are forty pits, sunk in the earth, their tops rising to a terrace level, one foot above the foundry floor.

These forty pits are each forty inches in diameter, and have cast iron curb rims. The pits are built of brick, laid in a species of mud found near the city, and readily prepared. This mud bakes as hard as the bricks, and is thought by the workmen to be as good, if not better than fire clay. It has for a long time been used for cupola and ladle lining. There are 78,000 bricks in the pits, and each pit will hold ten wheels, making a holding capacity of 400 wheels. In the center of the circle of pits is another crane for lifting wheels into and out of the pits, and next [to] the line of molds in the east corner is a fourth crane for reaching these molds.

When the casting is made it is allowed to stand until it sets, when the mold is stripped, and the wheels placed in the pits and well buried, and a cast iron cover fastened down. The casting remains in the pit ten days that it may be properly annealed by slow cooling. The capacity of the foundry is forty wheels per day, filling four pits. Thus, when the last four pits are filled, the first four of the forty, filled ten days before, are ready to be opened.

Yesterday, all the machinery was found to work well, and to the satisfaction of Mr. Allen, the foreman of the department, who is one of the most experienced found-rymen in the country. The cupola was charged with two-third old wheels and one-third new pig-iron. Heat was put on early in the day; at 2:55 P.M. the blower was turned on, and all the men stood at their posts waiting for full heat. A number of railroad men and heads of departments were present to do honor to the occasion among whom we noticed Master Mechanic Stevens, and Messrs. Wilkinson, Watson, Welch and others, besides numbers of citizens who came over to witness the opening of the works. It was the remark of all present that a more conveniently arranged foundry had not been seen nor a more perfectly constructed cupola. It was almost noiseless in its operation, and with a blower running at full speed there was no hum, thunder, or buzz. In sixteen minutes after the blower was put in, the probe found liquid, metal, and from that time out it ran freely. The six ton ladle was one hour and five minutes filling, but when a little practice is had in charging the furnace this time, it is expected, will be greatly reduced. On the platform, and looking into the huge cupola, was a sight worth the trouble many times over – a steadier burning furnace was never built, nor a more perfect operating blower attached. Looking down into the foundry one could see, through the dashing sparks, shooting and scintillating in clouds, the brawny workmen standing around the circles of the molds, or by crank, or ladle, or puddler, or skimmer, each one ready for the word. Presently it came. The huge cranes swung round, bearing the kettle ladles, the receiving ladle was tilted and ran out the molten iron. Carts ran forward and received the filled pots, and bore them to the most distant molds. Here another crane picked them up and whirled them into place. Skilled workmen tilted them and the heated mass rolled into the funnel of one mold after another until ten were filled. Then, while other workmen went on with the thirty nearest the cupola, a gang of men began to uptrip the steaming molds, just filled. One after another was overturned, and out fell the red-hot car wheel. The first one showed a crack, and was rejected; the other nine were sound and with celerity were transferred to the pit, packed in and covered over. By the time this was done, ten others were ready for a like process, and so on until the four pits were filled. It was a first day's work, with new machinery, new sand, untested furnaces and all, but it developed the fact that, if necessary, the capacity of the works can be extended without change of machinery to fifty wheels a day, so admirably is the foundry arranged for working a large body of men. This new foundry will give employment to from twenty-five to thirty-five men additional to the old foundry force. No car wheels will be cast in the old foundry, and the force there will be just as busy as ever on other work. It will thus be seen that the new foundry is a new addition to the industries of the city. As the foundry opens on all sides, it is easy to witness the operations of casting, stripping and pitting any afternoon, and they will well repay for the trouble.

The *Sacramento Daily Record*
August 2, 1879

THE NEW DEPOT
A Very Large Building-Its Location-What The Erection of the Building Suggests

The work upon the new depot of the Central Pacific Railroad is now rapidly progressing. The floor joists and sills are nearly all laid, and before next Saturday the skeleton form of the building will be a prominent object. The depot is being built upon the margin of China Slough, just south of the railroad shops, and will face almost directly to the north. The western end of the main pile will be upon the projected east line of Second Street, and the structure will extend easterly for 414 feet and over. It will be one of the largest depots in the West, and will be nearly 15 feet longer than the great machine shop at the railroad company's works. The style of the building will be Gothic, and without being elaborately ornamental, will be architecturally attractive, avoiding the extremes of severe plainness or excessive ornamentation. It will consist of a central pile of buildings- a portion being two stories in height-faced by a depot arcade or sheltered avenue 70 feet in width and 414 feet 4 inches in length, through which three tracks will be carried, the main one being almost in a line east and west with the entrance to the new bridge, while one will lead around to and along Front street. South of the track arcade and joined to it, the pile of buildings will be located and will be in length 164 feet and 4 inches. Beginning on the extreme east on the main floor, the plans show first a sleeping-car store-room 20 by 21 feet, with water closets attached; next west is a four foot hall running across the building, and opening upon a front landing, from which a broad passage way extends along the front of the buildings on the west of the hall, and opening upon the passage way or promenade is the bar-room 21 feet in length by about 15 in width; next to it is the gentlemen's waiting-room, 35 feet 4 inches by 35 feet 7 ½ inches in size. Attached to it is a series of water

closets and a wash-room. Next on the west is the ladies waiting, which is 36 feet 11 ¾ inches by 35 feet 4 inches in size. Attached to it are a retiring-room 12 by 14 feet, washroom and series of closets. Between the gentlemen's and ladies' waiting-rooms and sitting half its size in each room, is to be built the ticket office, which is to be 14 feet front by 20 feet in depth. Back of the ticket office and fronting in both waiting-rooms are to be two lunch counters, these forming the rear division line between the two waiting-rooms, the ticket office separating them in front. Back of the two waiting-rooms, with broad doors opening into either, is to be the dining-room, 40 by 65 feet, with a kitchen on the east of it 20 by 20 feet, and a store-room with pantry, closets, etc. The lunch counter between the two waiting-rooms makes an oval sweep through the north wall of the dining-room, and thus has a frontage in the latter apartment also. On the extreme west and connecting with the central building by the passage or promenade named is to be the baggage-room, 20 by 30 feet in size and fitted with closets and racks and with a patent lift to carry baggage to be stored to an upper chamber.

The second floor is over the gentlemens's and ladies' waiting-rooms, the baggage-room and the kitchen. Over the baggage-room on the front is to be a baggage store-room 15 feet 3 inches by 17 feet 4 inches in size and into which the patent lift mentioned is to open. Back of this room, and separated from it by a 4-foot hall, is to be a conductor's room, 16 feet 6 inches by 20 feet in size. Over the ladies' waiting room is to be found the Superintendent's office, opening upon the balcony which extends over the longitudinal passage way or promenade on the first floor level. This room is 14 feet 4[3/8] inches by 23 feet 2 1/8 inches in size and is fitted with two closets 6 feet 8 by 8 feet each, with washrooms, etc. Back of and connecting with it is to be the Train Dispatcher's room, 14 feet 4 3/8 inches by 31 feet 9 5/8 inches in size. Next a hall 5 feet wide crosses the building, and on the opposite side going east we come on the front to the stationery room, the dimensions of which will be 14 feet 4 3/8 inches by 15 feet 7 1/16 inches. Next to it on the east is the office of the Superintendent of the Sacramento Valley Railroad Company, and the room is to be the same size as the stationery room, and back of it is a second office of like size, with closets, etc. In the rear of the stationery room will be the "time" room, and of the same size. The stairways lead up to this story from the promenade at either end.

Over the kitchen is to be a second floor, containing four rooms for servants, with a small hall and closets, the approach being by a stairway leading out of the storeroom below.

At each end of the structure will be built ornamental towers, rising somewhat above the roof. From the ground to the eaves of the building will be 27 feet 1 7/8 inches, and thence to the apex of the gable roof 22 feet [3/4] inch, making the total height of the structure 49 feet 2 5/8 inches. The building will be constructed entirely of wood, and will rest upon heavy brick foundation walls, which have frequent deep sunken piers of masonry.

While it is a subject of general congratulation that the company is to abandon the old time sheds on Front Street and erect a depot building of a character becoming to the city, it is a common regret that the exigencies of ground space compel it to place the structure on the very margin of China Slough. In a day or two the frame work will go up, and be erected within 16 feet of the northern sloping bank of the slough, where the water seeps into the porous soil, and passes off in vapor, carrying with it, according to medical testimony, the most noxious exhalations. The depot extends from the east line (extended) of Second Street in an easterly direction with its front to the north, nearly direct. Therefore from the waiting-room windows, the dining-room windows, and from the entire south side, the visitor will have a splendid view of those architectural beauties which line the south bank of the slough while his nostrils will receive odoriferous whiffs of the volatile

perfumes which rise to delight the sense of smell from the graceful banks of the Lake "of" Como, where the accumulated filth of Chinatown and a good part of the rubbish of the rest of the city is daily dumped. When the visitor to Sacramento leaves the depot to come into the city proper he will pass down Second Street directly along the westerly margin of the slough, and for several minutes have a fine view in close perspective of the rear end of the elegant cottage residences of the Chinese, whose summer villas line the banks of the beautiful pond. If this does not give him a forcible first impression of Sacramento, he will be hard to please. One of the railroad engineers has suggested that if the city still neglects to clear out what three Boards of Health have formally condemned as a public nuisance, that it is likely the railroad company may be induced to erect high board fences along the north and west banks of the slough to lead visitors to think there is a "deer park" within. If none of the more forcible and sanitary reasons for abating the nuisance existed, the fact of the new depot location should alone incite active effort to wipe out that standing reproach to the neatness and cleanliness of the city-the pestilential rookeries and the foul deposits on the south bank of China slough.

❧ ❧ ❧

The *Sacramento Daily Record*
November 28, 1887

THE RAILROAD SHOPS - THE LARGEST MANUFACTURING ESTABLISHMENT WEST OF THE MISSOURI RIVER

The largest machine shop and construction works west of the Missouri River is the railroad works in this city, which now gives employment to 2,000 skilled workmen and laborers, and when the various extensions are completed, which will be inside of a year, this force will increased to 3,000 men. An additional 1,000 workmen means the addition of 1,000 families to those already in our city, means an increase of the population [of] at least 5,000, and necessitates the erection of 1,000 cottages. The increase of 1,000 workmen means additional stores and tradesmen of all kinds, means prosperity, internal improvements, and advance in real estate, means a solid, permanent growth, free from all boom agencies. The railroad shops in this city are constructing and will construct and repair all the rolling stock of the Southern Pacific Company and their leased lines. The constant addition of new roads and the rapidly increasing traffic on all the lines will continue to add to the extent and importance of this great establishment. It has been demonstrated that better engines can be built here for less money than in the East, and to-day the work turned out of the Sacramento shops has a national reputation. Orders have been received for the construction of twenty-eight new locomotives and there still remains an unfilled order for sixteen, making thirty-four to build.

❦ ❦ ❦

BIBLIOGRAPHY

American Railway Engineering Association. *Proceedings of the Thirteenth Annual Convention of the American Railway Engineering Association.* Vol. 13, Chicago, 1912.

Bain, David Haward. *Empire Express.* New York: Penguin Books, 1999.

Best, Gerald M. *Snowplow.* Berkeley, CA: Howell-North Books, 1966.

California Department of Parks and Recreation. "The Folsom Powerhouse." http://www.parks.ca.gov/?page_id=22909.

Gaddis, Malcom R. Interview by John Howard, November 6, 1996, California State Railroad Museum Oral History Collection, Sacramento, CA.

Howes, Edward H. "Three Weeks That Shook the Nation and California's Capital." Conference of the California Historical Societies. University of the Pacific, Stockton, CA, 2003. http://www.californiahistorian.com/articles/pullman-strike.html.

Horner, Jody and Rick Horner. *The Golden Hub Sacramento.* Sacramento, CA: Century Books, 2008.

Jennings, Sherry E. *Images of America - Truckee.* Charleston, SC: Arcadia Publishing, 2011.

Joslyn, D.L. *Sacramento General Shops Southern Pacific Company—Pacific Lines,* 1948.

Loveland, Jim A. *Dinner Is Served-Fine Dining on the Southern Pacific.* San Marino, CA: Golden West Books, 1996.

McGowan, Joseph A. *History of Sacramento Valley.* Vol 2, Lewis Historical Publishing Company, 1961.

Morse, John Frederick. *The First History of California,* Sacramento Book Collectors Club Number 3, 1945.

Northwest Railroad Museum Weblog. 2009. "Barney and Smith Car Company." http://trainmuseum.blogspot.com/2009/05/barney-and-smith-car-company.html.

Sacramento History Online. "Sacramento History Timeline." http://www.sachistoryonline.org/resources.timeline.html.

San Diego Health Archive. "Third Biennial Report (1874-1875)." http://sandiegohealth.org/state/biennial/1874-1875_3rd_Biennial Report.pdf.

Secrest, C. M. "P.F.E. Co. Takes Great Stride in 14 Years." *Southern Pacific Bulletin,* 1921.

Signor, John R. *Donner Pass. Southern Pacific's Sierra Crossing.* San Marino, CA: Golden West Books, 1985.

Singer, Kent. "Just What is a Railroad Watch?" National Association of Watch and Clock Collectors. http://ph.nawcc.org/Railroad/Railroad.htm, 2000.

The Basics of Railroad Wheels and Wheel Inspection 3rd Edition. Omaha, Nebraska: The Railway Education Bureau, 2008.

The Daily Journal of the United States Government. "Railroad Safety Appliance Standards. Miscellaneous Revisions." https://www.federalregister.gov/articles/2010/07/02/2010-16153/railroad-safety-appliance-standards-miscellaneous-revisions#h-9.

Truckee Donner Historical Society. "The Town of Boca Played a Significant Role in Truckee's Past." http.//truckeehistory.org/historyArticles/history13.htm.

Ueberall, E. and K. Singer. "Railroad Time Service." *National Association of Watch and Clock Collectors Bulletin,* October 2001.

Ueberall, E. and K. Singer. "So What Is a Railroad Watch?" *National Association of Watch and Clock Collectors Bulletin*, December 2001.

United States Department of Health and Human Services Centers for Disease Control and Prevention. "The History of Malaria and Ancient Diseases." http://www.cdc.gov/malaria/about/history/ross.html.

White, John H. *The American Railroad Freight Car*. Baltimore: The Johns Hopkins University Press, 1993.

White, John H. *The American Railroad Passenger Car Part 1*. Baltimore: The Johns Hopkins University Press, 1985.

White, John H. *The American Railroad Passenger Car Part 2*. Baltimore: The Johns Hopkins University Press, 1985.

Whyte, H. T. "The Story of a Half-Million Cakes of Ice." *Southern Pacific Bulletin*, February 1927.

ACKNOWLEDGMENTS

Any compilation of data, descriptions, accounts, and quotations requires that someone feel the subject is important enough to gather information and write about in the first place. The journalists who wrote the articles used in this book were people who lived at the time or had access to records that allowed an accurate description of the railroad's operations, and, as evidenced by the large number of articles available, must have sensed the value of writing on this subject.

I was aided by the staff at the California State Railroad Museum Library early in the process of collecting information, and am thankful for this assistance. Cara Randall and Chris Rockwell provided help with copies of photos.

Another source of old newspapers and articles from Southern Pacific's *Bulletin* was the California State Library.

Peace of mind regarding use of the *The Bulletin* articles came from Patricia Labounty of the Union Pacific Museum in Omaha. Her clear and quick correspondence clarified that the articles are in the public domain. Steven Davis assisted with this effort.

Learning the details of using photo-editing software to make *The Bulletin* articles as clear as possible was a long process. This part of the project could not have been completed without the suggestions from my long-time friend, Tully Simoni.

As always, review and proofreading is important. I was helped with this task by my young friend, Rosalina Fry.

My first introduction to the publishing process came from Gerald Ward of the Sacramento Library's I Street location. I knew nothing at the start, and he explained details and guidelines surrounding the publication process that helped move this book along.

I was naïve enough to think I could write a book with little help and did not need an editor. Then, my friend Cindy Sample, who is a wonderful fiction writer, said she has used an editor for her work. So, I pursued one for myself and was fortunate to find Robin Martin of Two Songbirds Press in Fair Oaks, California. Thanks to her input and guidance, I have confidence that this book has clarity, organization, and professionalism.

When I became discouraged during the course of this long project, I required support to build my confidence and persist. My advocate and supporter for the book – and for life's issues, my therapist, assistant, guide, listener, and, above all, best friend – was my dear mom.

This book is dedicated to my mother, Eunice.

Index